Mastering PC Tools™ Deluxe

Paul Dlug

TRADEMARKS

dBASE®	Ashton-Tate, Inc.
Epson ™	Epson America, Inc.
IBM ™	International Business Machines, Inc.
Lotus 1-2-3®	Lotus Development Corp.
PC Tools ™	Central Point Software, Inc.

FIRST EDITION
THIRD PRINTING

Library of Congress Cataloging-in-Publication Data

Dlug, Paul.
Mastering PC Tools Deluxe / by Paul Dlug.
p. cm.
ISBN 0-8306-3578-5
1. Utilities (Computer programs) 2. PC Tools deluxe (Computer program) I. Title.
QA76.76.U84D59 1990
005.369—dc20 90-31459
CIP

TAB BOOKS offers software for sale. For information and a catalog, please contact TAB Software Department, Blue Ridge Summit, PA 17294-0850.

Questions regarding the content of this book should be addressed to:

Reader Inquiry Branch
Windcrest Books
Blue Ridge Summit, PA 17294-0850

Acquisitions Editor: Stephen Moore
Book Editor: Kellie Hagan
Production: Katherine G. Brown
Book Design: Jaclyn J. Boone
Cover Design: Lori E. Schlosser
Cover Illustration: Greg Schooley, Mars, PA

To my wife Rita Marie, and our children, Paul, Laura, and Debra

Contents

Introduction

This book is a tutorial manual consisting of business and home applications that use the utilities in the *PC Tools Deluxe 6.0* software program by Central Point Software. The program can be run on the IBM PC or compatible microcomputers. Many business and utility applications can be solved with PC Tools' powerful command structure.

You can take full advantage of the program's powerful features without any special training or prior computer knowledge, because this book is designed as a tutorial and not a reference guide. You could easily use it as a reference guide, however, by referring to the index at the back of the book.

This book covers the three major parts of PC Tools Deluxe: the desktop manager, the data recovery and DOS utilities, and the hard-disk backup system. It is not necessary to read the chapters in sequence because each chapter describes utilities that you can use without reference to the rest of the book.

This book is designed for both the complete novice and the experienced programmer who needs to solve a particular problem or use a certain utility. It emphasizes a "hands-on" approach, or "learn by doing." Each problem starts with a business application, or utility, and uses the commands that pertain to that utility. Each chapter contains a series of figures that show what the screen should look like as you proceed through the chapter.

PC Tools Deluxe 6.0 is a utility program that contains a DOS shell, file manager, desktop manager, backup capability, and text searcher. The program supports Token Ring and Novell networks, as well as IBM's systems application architecture (SAA) interface. The desktop manager has a notepad for writing and saving text, a database that can create dBASE-compatible files, an appointment scheduler to locate free time or find appointments, a calculator that does both scientific and financial calculations, a telecommunications module that automates log-on and dialing sequences, a clipboard to "cut" and "paste" documents, an outliner to create report outlines for presentations, and a keyboard macro that can record keystrokes.

The PC shell provides access to file managment and text retrieval functions. It lets you view a file and show multiple directories in tree form. It lets you sort directories, recover lost data, and move files from one directory to another. It lets you undelete files, save your hard disk partition table, rebuild a damaged hard drive, compress data on a disk, and use other DOS commands.

The desktop manager includes utilities for a mini word processor called a notepad, an appointment scheduler, a database, several different calculators, telecommunications, a clipboard, macros, an outliner, and fax board support to send and receive faxes with your computer. It has full mouse support and has moveable and resizable pull-down windows. Its user interface and functions are, indeed, user-friendly.

The desktop workspace includes dBASE IV and Lotus 1-2-3 file compatibility, a constant time display, short-cut function keys, and an autodialer for its communications module.

The data recovery system and DOS utilities include a DOS shell, a viewer and launcher for dBASE and Lotus files, an undelete command, a mirror and rebuild command to construct a copy of a damaged hard drive, a compress command to unfragment files on a hard disk, caching capability to increase the speed of commonly used commands, an encryption command for sensitive data, and format command that is more useful than DOS formatting, and a DeskFix utility that diagnosis and repairs both hard and floppy disks in order to restore lost directories and files.

The hard-disk backup utility quickly backs up a hard drive, compresses the files as it backs up, and estimates how long the backup will take, including the number of disks for the backup. It can verify the writing of disks while backing up, and create backup reports. You can back up either a whole disk, selected directories, or selected files on both disk and tape drives.

PC Tools supports many basic functions, such as a formatter, calculator, notepad, and database. Once you've mastered these, you can go on to more specialized tools, such as text retrieval, DOS management, and file backup.

PC Tools can be used by homeowners, students, educators, business people, salespersons, engineers, or anyone who wants to learn how to use a simple but powerful utility package for desktop management, data recovery, DOS utilities, and hard-disk backup.

1
CHAPTER

Setting up PC Tools Deluxe

This chapter explains the hardware requirements of PC Tools Deluxe and the method of installing and setting up the program.

Hardware requirements

To run *PC Tools Deluxe 6.0*, you must have a PC, XT, AT, PS/2, or 100% compatible, with DOS 3.0 or higher and a minimum of 512K of RAM. You need a minimum of one disk drive, either $3^1/_2$-inch or $5^1/_4$-inch, but the program works better with a hard disk. The floppy disk drives can be either low- or high-density format. The computer can also back up to another hard drive, a Bernoulli box, or a DOS-compatible tape drive.

PC Tools Deluxe is designed to work with a mouse. It supports Microsoft mouse drivers, version 6.14 or higher, or Logitech/Dexxa mouse drivers, version 3.4 or higher. The mouse driver must be installed in the AUTOEXEC.BAT or CONFIG.SYS file. PC Tools Deluxe can operate on a floppy drive system, but was designed to operate with a hard drive and a mouse.

Before using PC Tools Deluxe

Before you begin using PC Tools, you need to know the type of video controller card in your computer, and the make and model of your printer. You will also need several formatted floppy disks. If you are using a floppy disk system, you will use one formatted floppy disk to store your notepad documents, database files, and outlines that you create. If you are working with a hard disk, you will have to determine the name of the directory in which you want to store PC Tools, such as C:/PCTOOLS.

The setup program

The PC Tools program is available on both $5^1/_4$-inch and $3^1/_2$-inch disks. You will use the set of disks appropriate for your computer. The $5^1/_4$-inch disk set contains six disks, while the $3^1/_2$-inch disk set contains three disks.

Before you can run PC Tools, you must run the setup program. The setup program installs the program by copying PC Tools, along with the printer and monitor information, onto the working disk or the hard drive. It also copies the spelling checker, help information, and other utilities. If you decide to change the printer and/or monitor settings at a later date, the setup program enables you to do so.

If you are using a hard-disk system, all the files will be copied onto the hard disk. You will not exchange floppies when the program is run. The program assumes that DOS is already on the hard disk and that your hard drive is drive C.

Use the following steps to install PC tools on either a floppy disk system or a hard drive system:

1. Boot up your computer system with DOS.
2. Remove DOS (floppy systems) and insert Disk#1 in drive A.
3. At the A> prompt, type PCSETUP and press the Enter key. The PC Tools Installation menu appears, as shown in Fig. 1-1. This Installation menu lets you install PC Tools files on either a hard disk or network server, or change the configuration of a previous installation. When you press 1, Fig. 1-2 appears.
4. You can now selectively install one, two, or all three of the major parts of PC Tools to your hard disk. When you select one, a checkmark will appear to the left of that particular item. You are also given the required disk space for the item. Press C to continue with the installation, or Q to quit.
5. If you press C to continue, Fig. 1-3 appears. PC Tools accepts PCTOOLS as the name of the default directory, but if you want to type in the name of a new directory, type it in. Press the Enter key when you are finished.

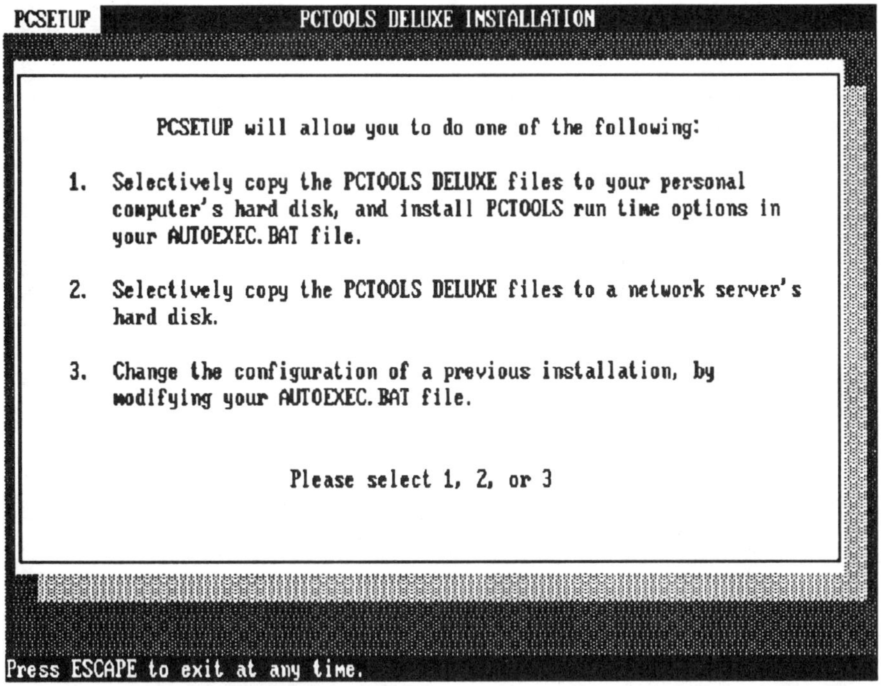

PCSETUP PCTOOLS DELUXE INSTALLATION

PCSETUP will allow you to do one of the following:

1. Selectively copy the PCTOOLS DELUXE files to your personal computer's hard disk, and install PCTOOLS run time options in your AUTOEXEC.BAT file.

2. Selectively copy the PCTOOLS DELUXE files to a network server's hard disk.

3. Change the configuration of a previous installation, by modifying your AUTOEXEC.BAT file.

Please select 1, 2, or 3

Press ESCAPE to exit at any time.

1-1 PC Tools Deluxe Installation menu

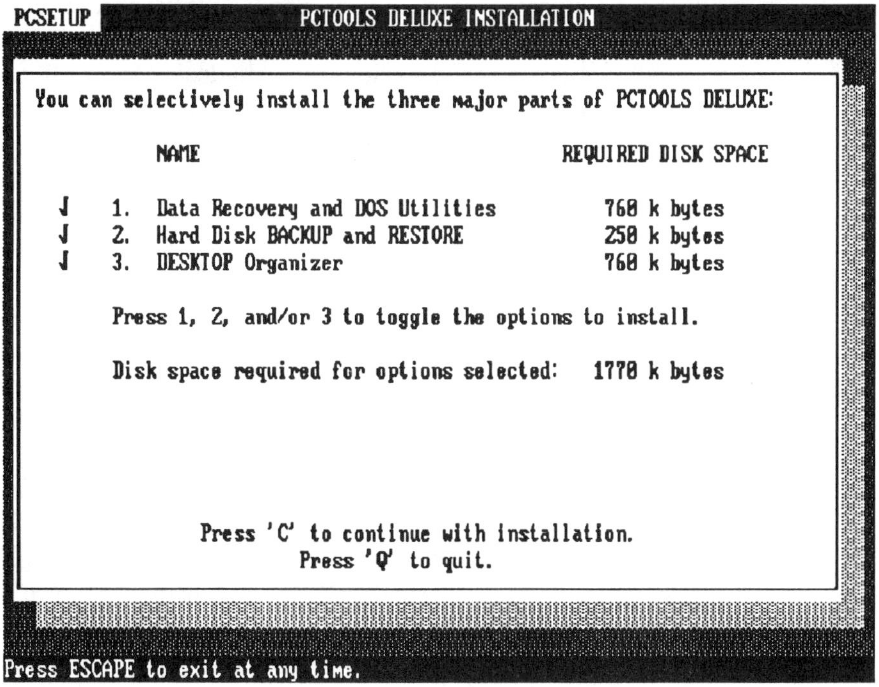

PCSETUP PCTOOLS DELUXE INSTALLATION

You can selectively install the three major parts of PCTOOLS DELUXE:

	NAME	REQUIRED DISK SPACE
√	1. Data Recovery and DOS Utilities	760 k bytes
√	2. Hard Disk BACKUP and RESTORE	250 k bytes
√	3. DESKTOP Organizer	760 k bytes

Press 1, 2, and/or 3 to toggle the options to install.

Disk space required for options selected: 1770 k bytes

Press 'C' to continue with installation.
Press 'Q' to quit.

Press ESCAPE to exit at any time.

1-2 Three major parts of PC Tools

```
PCSETUP                    PCTOOLS DELUXE INSTALLATION

Please type below the full pathname of the directory
where you would like PCSETUP to put the PCTOOLS files.

To accept C:\PCTOOLS as the directory for the new files:
      - press the <return> key.

To specify a different directory:
      - Use the backspace key to delete characters
      - Type the new directory name ( for example: C:\PCTOOLS )
      - press the <return> key.

C:\PCTOOLS

NOTE:   Your original AUTOEXEC.BAT file will be renamed AUTOEXEC.SAV.
        If you have an old version of PCTOOLS in a subdirectory called
        "PCTOOLS", the old directory will be renamed to "PCTOOLS4".

Press ESCAPE to exit at any time.
```

1-3 Directory where PC Tools is installed

6. You will now be prompted to install PC Shell. You have the option of running PC Shell as a memory-resident program, as a DOS shell, or not having it at all.Notice, in Fig. 1-4, that if you do decide to install it, PC Tools places the word "installed" next to the option letter. If you install it as memory-resident, it will always be there to be activated, but will take up some memory. You will always have access to it, even while running other programs. The DOS format command will also be renamed in this installation. Press I to install PC Shell as resident, press S to install it as a DOS shell on the disk, press R to remove PC Shell from memory, and C to continue without change. When PC Shell is loaded, you will be asked for the amount of memory you want to allow for it, as shown in Fig. 1-5. The larger the memory, the faster it will load. Press T for tiny (9K), S for small (70K), M for medium (90K), L for large (170K), or C to continue without change.
7. You will then be asked whether you want to have the program Mirror installed on your hard disk, as shown in Fig. 1-6. Mirror saves your hard disk's partition table, root directory, and file allocation table (FAT) to protect against any accidental format. It is run automatically as you turn on your computer. I highly recommend that you select this item, as it makes the reconstruction of a damaged disk easier. Press I to install Mirror as resident, R to remove Mirror from memory, or C to continue without change.

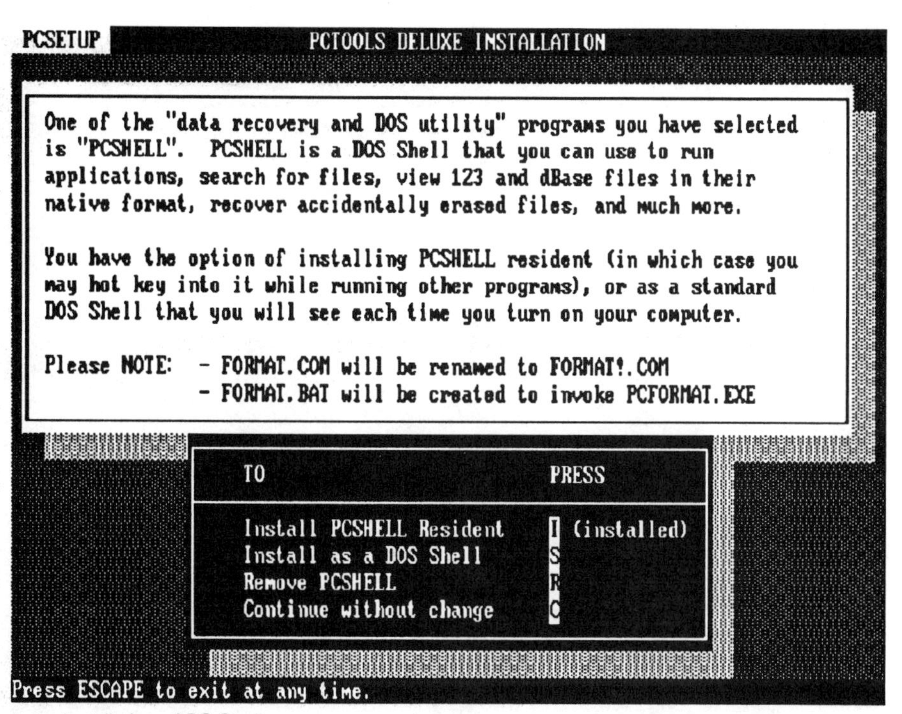

1-4 Installation of PC Shell

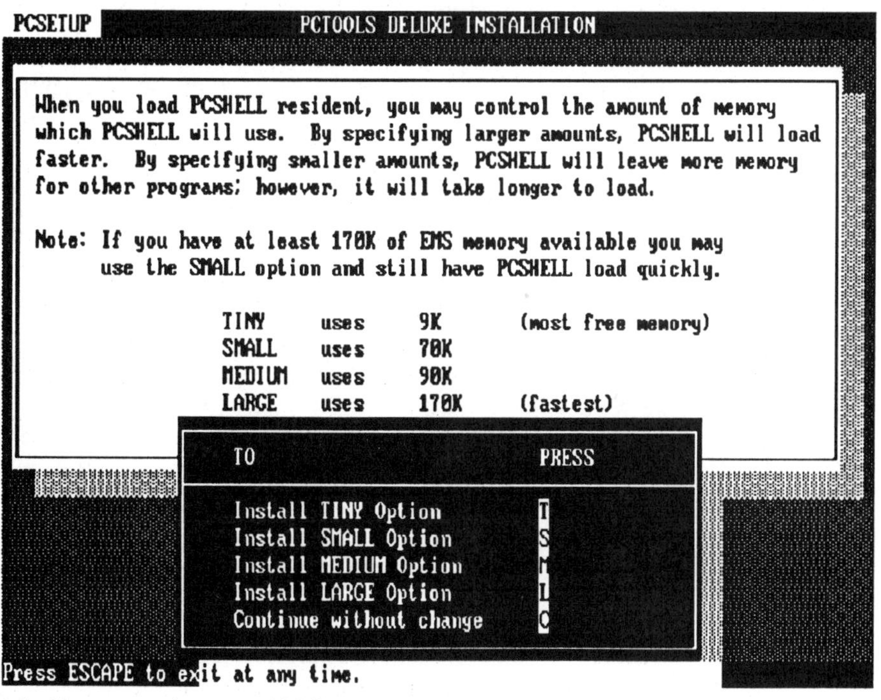

1-5 Memory requirements of PC Shell

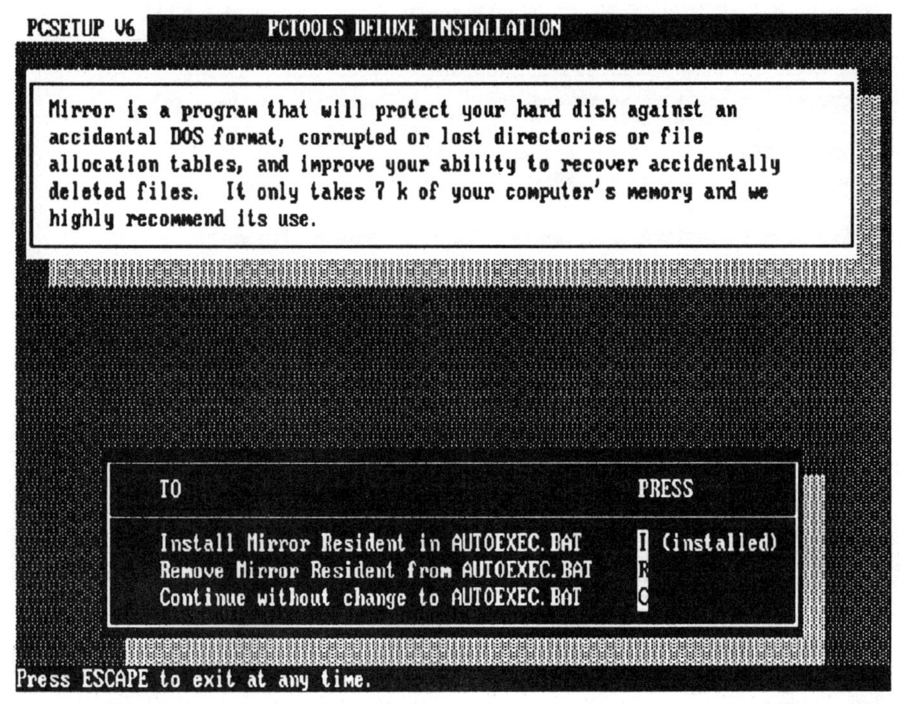

1-6 Installing Mirror

8. Figure 1-7 appears asking you if you want the PC Cache program installed. PC Cache stores frequently used information in memory, so your computer doesn't have to access the disk drives so frequently. It supports expanded and extended memory. Press I to install PC Cache, R to remove PC Cache, or C to continue without change.
9. You are asked if you want the Desktop program to be added, as shown in Fig. 1-8. The Desktop program contains a notepad, database, outliner, appointment scheduler, telecommunications program, macro editor, clipboard, and calculator. If you install it as a memory-resident program, you can always call it by pressing the Control and Space "hotkeys." Press I to install the Desktop as resident, press R to remove the Desktop, or C to continue without change.
10. Figure 1-9 appears, asking you if you want to install Backtalk. Backtalk is the communications program that lets you send and receive files while in other programs. You can specify the communications port to which you want it installed by typing in a number, 1 through 4. You can also uninstall it by typing R to remove Backtalk from memory, or typing C to continue the installation without change. A list of the files that are transferred is displayed in Fig. 1-10. You are then prompted to insert the remaining disks.

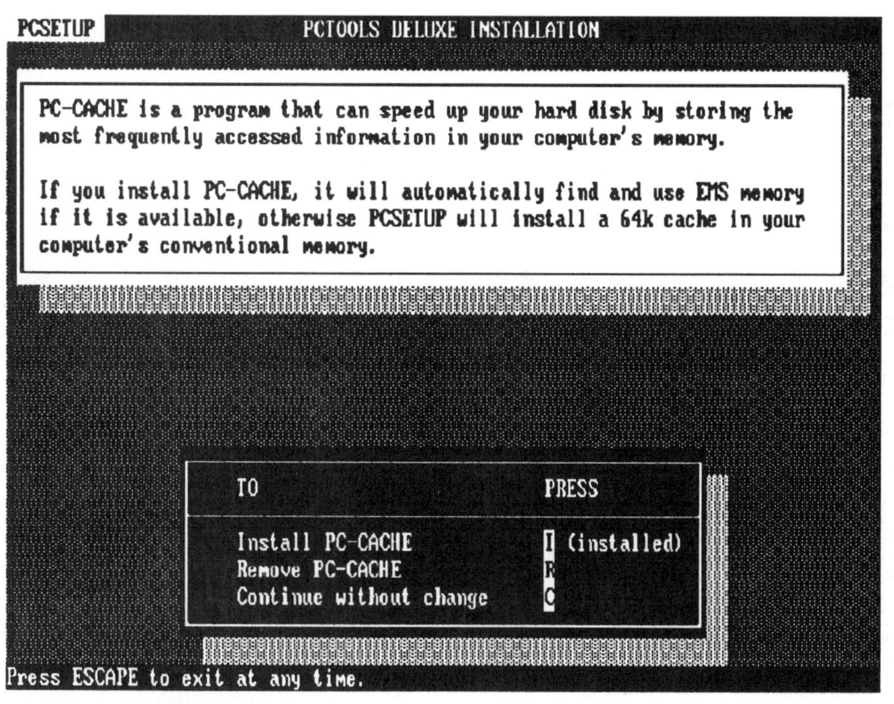

1-7 Installing PC Cache

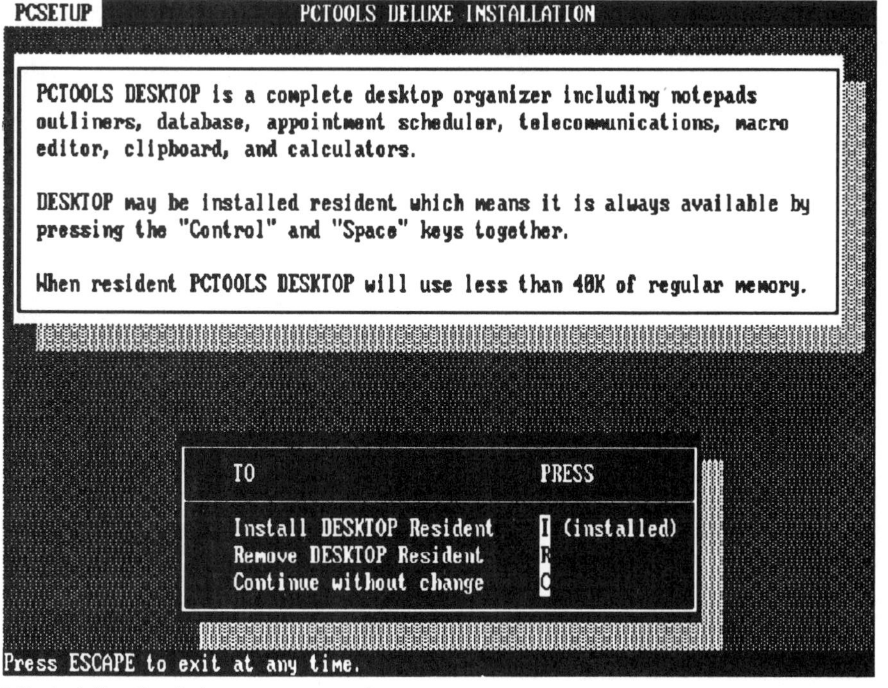

1-8 Installing the desktop

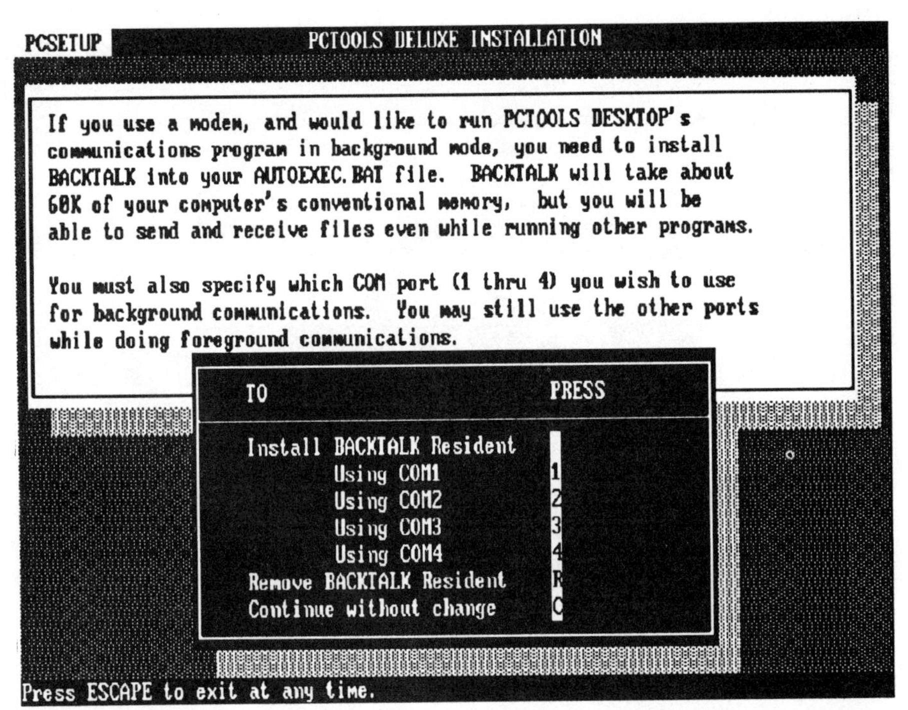

1-9 Installing Backtalk

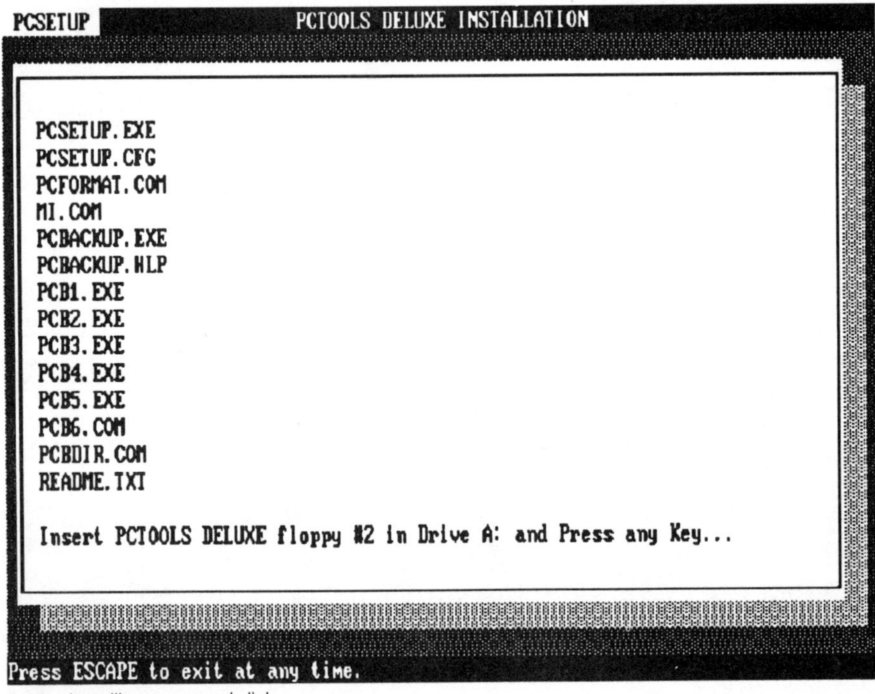

1-10 Installing a second disk

11. When the program is successfully installed, Fig. 1-11 appears, telling you to press any key to decompress the spelling checker on the disk. If you are updating PC Tools, it will tell you that it is done and ask you to press any key to continue.

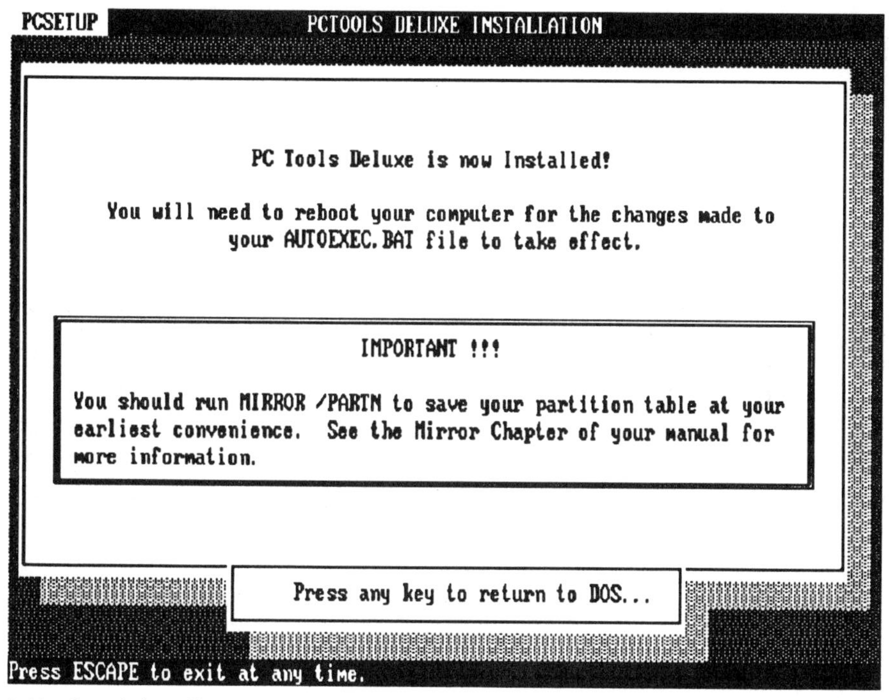

1-11 Completion of installation

Removing PC Tools from memory

If you load an application and it gives you an error reading because of insufficient memory, you have to remove PC Tools from memory. Type the command KILL at the DOS prompt and it will remove PC Tools Desktop, PC Shell, and Backtalk, if they were installed as memory-resident programs. The program is still on the hard disk, but is erased from memory.

Installing PC Tools on a network

PC Tools can be installed on a Novell NetWare network or an IBM Token Ring network server by using a system login script such as:

SET PCTOOLS = h:\courses\$user_name

The directory COURSES is available to all users, but is write protected.

2
CHAPTER

Notepads

In this chapter, you will learn how to use the notepad utility of PC Tools—how to create and edit a small document. In the edit mode, you can insert and delete characters, words, lines, and paragraphs. The notepad utility contains both a spelling checker and a find-and-replace function. With the notepad, you can create your own document, save it to disk, print it out, and load it from disk. You can also change window colors and active windows.

Notepads allow you to create your own letters, memos, resumes, and notes, and perform other word processing functions. Word processing can be thought of as typing with a computer. The way text is entered into a computer in order to obtain the finished product is nearly the same process as using a typewriter, but you will find the job easier with a computer. Essentially, the computer becomes a typewriter that is capable of doing many things. The words are contained in the computer's memory, rather than on paper, so you can easily correct and adjust your text. You can correct keying errors; add or delete words, sentences, paragraphs, or pages; set margins; define page lengths; and perform many other functions that involve the manipulation of the written word. Word processing makes it easy to produce quality printed output.

Word processing, however, is not just typing and printing. You can compose a customized letter made up of standardized paragraphs, each of which is stored separately in the computer's memory. You can insert data into form letters and generate mailing labels or envelopes. You can store copies of text to have ready for instant retrieval and printing months later.

Any individual who routinely writes any kind of document benefits most from word processing, or the PC Tools notepad. Business people, especially, save time by producing their own letters on a word processor. There is no need for dictation, no wait for someone to type and then proofread the document. All adjustments and improvements can be made immediately.

The PC Desktop notepad is truly a mini word processing program that enables you to create simple letters, notes, memos, etc.

Loading the desktop

PC Desktop can be loaded as a stand-alone application by typing the word DESKTOP at the DOS prompt. In this mode, it does not tie up any memory when you use it. PC Tools can also be loaded as a memory-resident application by typing DESKTOP/R. The added feature of this mode is that the program will always be in memory, even when you use other applications or programs.

If you have a mouse and want to disable it, load the desktop by using the command DESKTOP/IM. If you are left-handed and want the left and right mouse buttons exchanged, use the command DESKTOP/LE.

After PC Desktop is loaded into memory, you can activate it by pressing the Ctrl and Space keys simultaneously. They are called the "hotkeys", and enable you to get into the desktop from other applications.

Removing the desktop

You can remove the PC Desktop program from memory at any time by typing the command KILL at the DOS prompt. This will free up memory so it can be used in other applications.

Desktop menu

After you press the hotkeys, Ctrl and Space, the Desktop menu appears, as shown in Fig. 2-1. It lists all of the utilities in the desktop, including notepads. You will be using notepads as your mini word processor.

If you have installed PC Shell before you load the desktop, PC Shell will appear as one of the choices in the Desktop menu.

Getting help

If you need help at any time, just press the F1 (help) key. The screen shown in Fig. 2-2 appears, and explains how to access the help information. The rectangular box containing the menu is called a dialog box. As you can see, options at the bottom of the dialog box enable you to access the help index. PC Tools provides

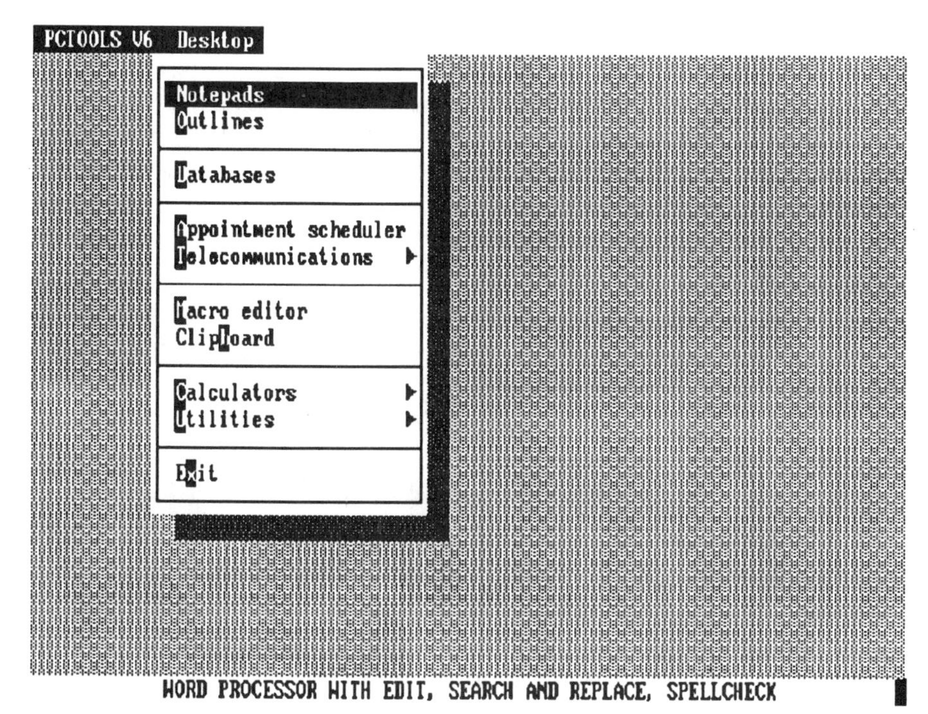

2-1 Desktop menu

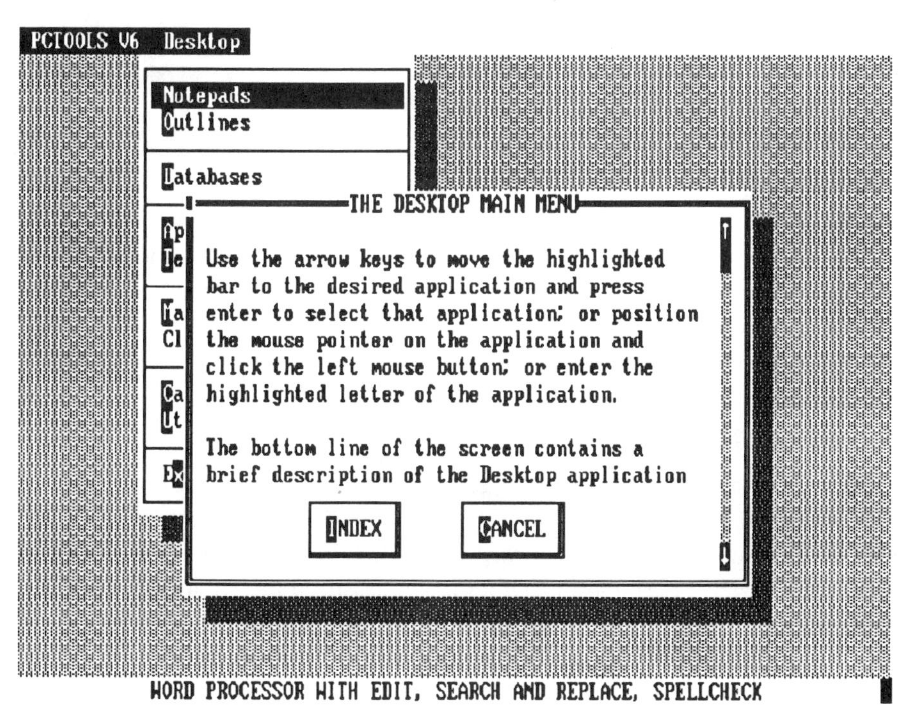

2-2 Help screen

help screens pertaining to what you were doing just before you pressed the Help key. For example, if you were trying to save a document and pressed the help key, PC Tools would present help screens on how to save a document.

The Help key lets you access the help index, shown in Fig. 2-3. With this index, you can access help in other topics. Highlight the topic in which you want instruction and press the Enter key.

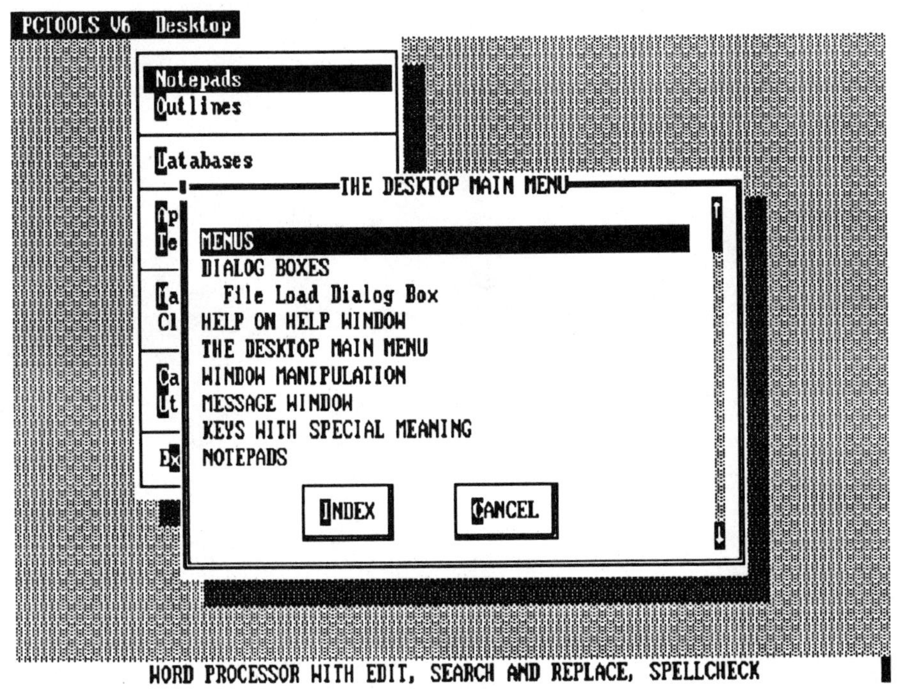

2-3 Help index

To exit the help screen, press C for Cancel or, if you have a mouse, click on the Cancel box. You will be returned to the screen you were on when you pressed the help key. Use the information you obtained from the help screen to respond to your question or solve your problem.

Starting the notepad

When you select notepads from the Desktop menu, Fig. 2-4 appears. It gives you three options: load an old program, cancel back to desktop, or create a new document. If you are using the keyboard, pressing the Tab key will advance you from one box to the next. If you are using a mouse, position the cursor on the box that you want to execute and click. Notice that this information is given to you at the bottom of the screen in a line called the message bar. It contains messages to help you through an application.

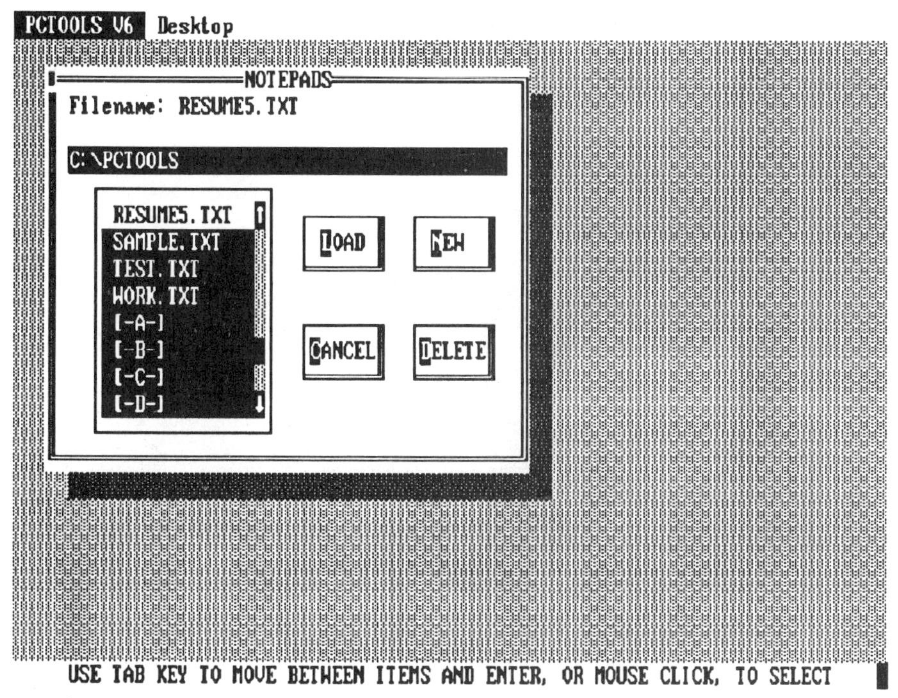

2-4 Starting notepads

To create a new item in the notepads, type in a filename, up to eight letters, highlight New, and press the Enter key. All files in notepads are given the extension .TXT. This makes them easy to identify if you are getting a directory of all files created by PC Tools.

A notepad file or document can hold up to 60,000 characters.

Entering text

After the dialog box appears on the screen, you can type information into the notepad, as shown in Fig. 2-5. The notepad contains a ruler at the top of the dialog box or window to help you in typing. The ruler line can be eliminated by going into the Controls menu and selecting Tab ruler display. To bring the ruler line back, go into the Control menu again and select Tab Ruler Display. It is a very good idea to leave the ruler line on the screen, as it will come in handy when you want to set tabs.

The horizontal menu bar at the top of the screen contains the names of the pull-down menus and online help. The message bar at the bottom of the screen tells you what keys to press and provides other helpful hints.

The dialog box displays the status line, which gives the current position of the cursor by giving the line and column position. It gives the name of the file in the

2-5 Typing in information

top right corner. When you are starting a document that has no name, the default name is WORK.TXT. Otherwise, the name you give the document is in the top right corner. "INS" indicates that you are in the insert mode of typing, rather than the overstrike mode. When you begin PC Tools, you are in the insert mode—ready to insert text at the cursor. Every character you type will be inserted at the cursor position. At the bottom left-hand corner of the dialog box, PC Desktop will display the time, assuming you have a built-in clock.

Several of the items on the screen pertain to mouse operations. The scroll bars on the right side of the window let you scroll through the file. The resize box in the bottom right corner of the window is used to resize the window. The close box in the upper left-hand corner of the window closes or removes the window. The selection bar is on the left-hand side of the window. The blank space in the first column selects text.

The screen shown in Fig. 2-5 represents part of a resume that you will enter. Before you start entering text, however, you need to look at the cursor key movements and the other menus necessary for text editing.

Cursor keys

On the right side of the keyboard are the four arrow, or cursor, keys. Move them around to get a feel for how they work. The right and left arrows move the cursor one character to the right and one character to the left, respectively. The up and down arrows move the cursor up or down one line.

The Backspace key erases the character to the left of the cursor. The PgUp and PgDn keys move your text up or down one window at a time. Pressing the Home key returns the cursor to the beginning of the line, and the End key moves the cursor to the end of the line. By simultaneously pressing the Ctrl and Home keys, you move the cursor to the beginning of the document. By simultaneously pressing the Ctrl and End keys, you move the cursor to the end of the document.

The Tab and Shift–Tab keys move the cursor to the right and left. The first tab stop is half an inch from the left margin.

The cursor can also be advanced to any line by using the Goto command in the Edit menu. A dialog box will be displayed asking you to type the number of the line you want to advance to. Type in the line number and select O for OK.

Saving a document

You can save a document at any time by accessing the File menu and pressing S for Save, as shown in Fig. 2-6. It prompts you for a name, if you have not already given one. All files in the notepad are automatically saved when you press the Esc key or click the mouse button in the dialog box.

2-6 Saving a file

The Save command saves the current document, replacing the old one. It will keep the same filename. In other words, the document RESUME will always contain your latest changes and updates. You might name each successive version RESUME1, RESUME2, and RESUME3. RESUME3 would be the most current update, and the previous versions are left on the disk for later reference.

A file or document can be saved in two different formats. The first is the PCTOOLS Desktop format, which is the default. This format preserves all tabs, headers, and footers that have been made in the creation of the document. The second format is the ASCII format. Most word processing documents can translate ASCII files. To send a PC Notepad file to another word processor, you will have to send it as an ASCII file. When converting a file to ASCII, however, all the formatting features are lost. The character and paragraph formats, margins, tab settings, and so forth are stripped away when the file is saved to ASCII.

The Make backup file option makes a backup file for the document, with the file extension .BAK. This is useful if you want to make sure there is always an original copy of the document available.

The name of a document can be no more than eight characters long. For clarity, the name of the document should represent what is in the document. For example, you could name a document A or any combination of letters or numbers, but you would probably forget what the document is at a later date. PC Desktop automatically adds the extension .TXT to the end of your filename if you don't specify a three-letter extension.

To save a document, press the S for Save, or highlight the Save box with the mouse button and click.

Desktop File menu

The Desktop File menu is displayed in Fig. 2-7. It can be accessed at any time from the desktop by pressing Alt−F. If you cannot highlight a menu item, you can always press the Alt key. A single letter of each of the menu items is highlighted so you can access that particular menu or command.

It is through this menu that you can load, save, and print programs. The Print menu is shown in Fig. 2-8. It prompts you for how many copies of the document you want to print and the printer port or device you want to use. LPT ports are the parallel ports that can be used, and COM ports are the serial ports to print the file to. The disk file writes the text to disk for later printing.

The Autosave feature is shown in Fig. 2-9. It automatically saves files at a predetermined interval. It is a good idea to leave this feature on because it will periodically save your files to disk. The default time is five minutes, but you can set it to any time you want. The on/off option serves as a toggle, so you can turn the autosave on or off at your leisure.

2-7 File menu

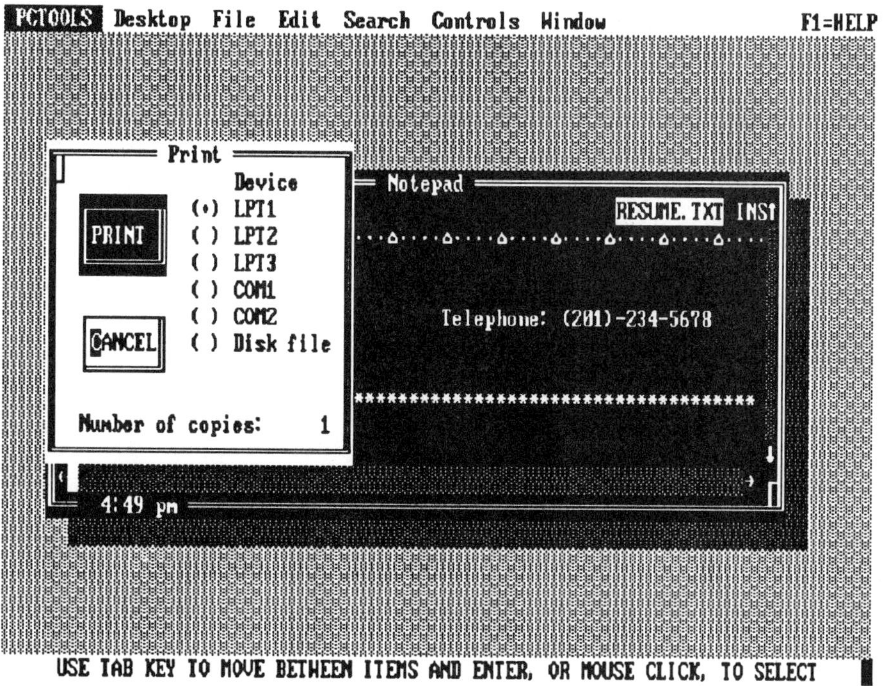

2-8 Print menu

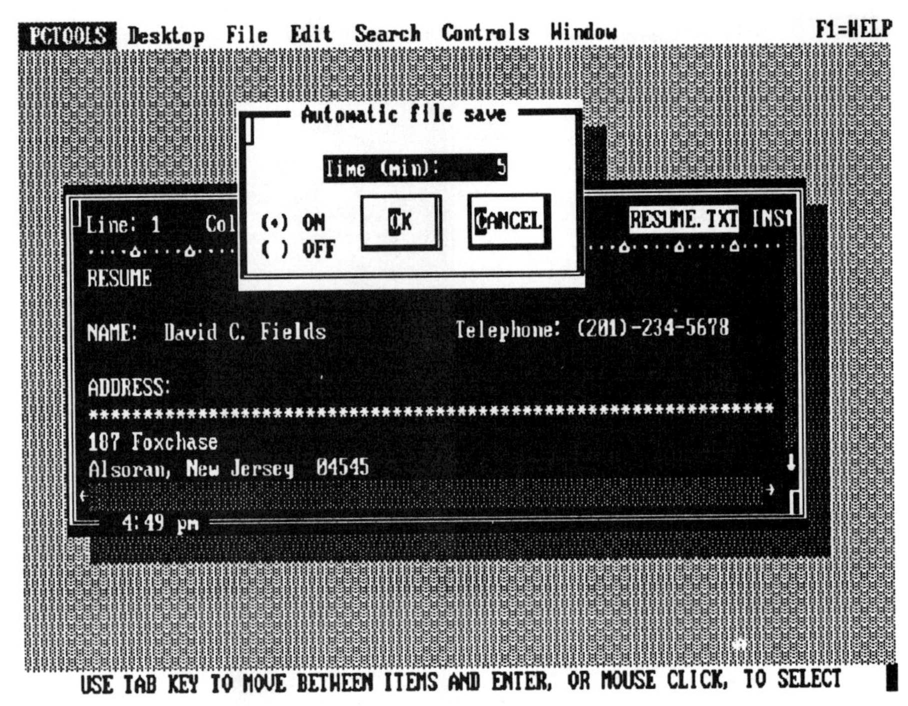

2-9 Automatic file save

Edit menu

The Edit menu, shown in Fig. 2-10, lets you cut, copy, and paste text to a clipboard. The clipboard is a temporary storage area that holds text until it is ready to be copied or moved. The clipboard is limited in memory to about 45 lines of text. If you try to move more than it can hold, it will give you a warning message.

The Edit menu also lets you delete text, insert one document into another, or go to a specific page in a document. The spelling checker also resides here, and enables you to check the spelling of a word, the text on the screen, or the whole file.

Search menu

The Search menu, shown in Fig. 2-11, lets you find selected text in a document. It also allows you to replace one piece of text with another.

Controls menu

The Controls menu, shown in Fig. 2-12, lets you adjust the page layout for printing or display. It lets you create headers and footers, set and delete tabs, hide the ruler display, change type mode, set wordwrap, and auto indent.

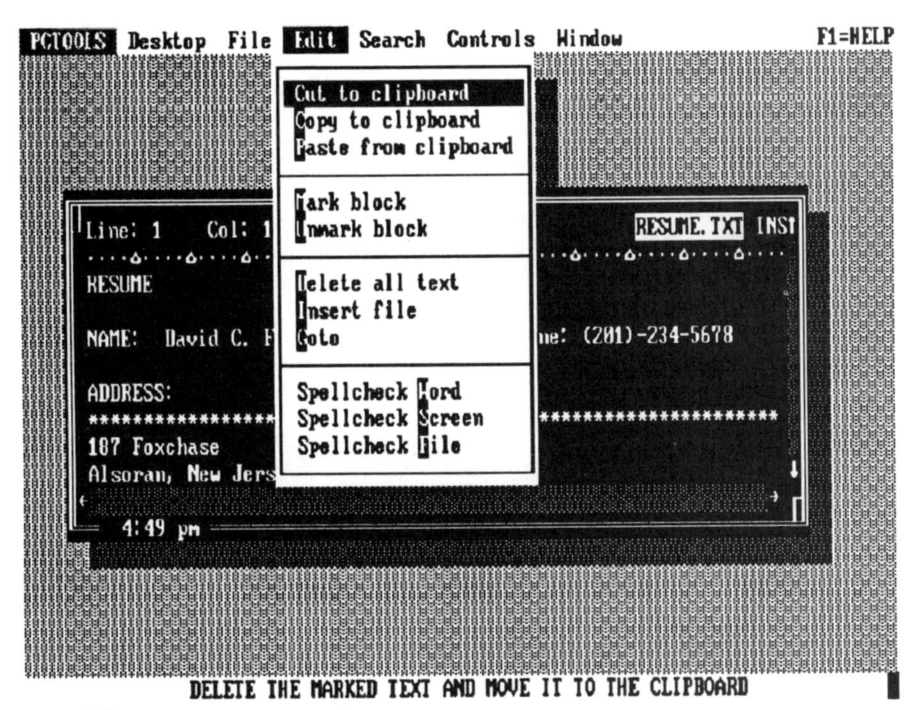

2-10 Edit menu

2-11 Search menu

2-12 Controls menu

Window menu

The Window menu, shown in Fig. 2-13, lets you change the colors of the windows, and move and resize a window. The zoom command lets you expand a window to fill an entire screen.

Completing text

You can now start entering a word processing document in the work area. I will provide some intentional errors and misspellings; you will correct these later. Type the complete document (with intentional errors) shown in Fig. 2-14. As you type the resume, the top part will scroll off the screen. If any document is long, it is impossible to see all the text on the screen at one time.

When you have finished typing the document, use the PgUp or PgDn keys to move the cursor up and down the document, one window at a time. Use Ctrl–Home to move the cursor to the beginning of the document, and Ctrl–End to move the cursor to the end of the document. Experiment with all the cursor movements listed in Table 2-1.

When you begin to use PC Notepad, you are in the insertion mode—ready to insert text at the cursor. Every character you type is inserted at the cursor location.

2-13 Window menu

Look at the resume: the word Army has two As. In order to delete one, move the cursor so it is over one A and press the Del key. This key deletes the character under the cursor. You can also use the Backspace key, but it deletes the character just to the left of the cursor.

The word "classes" is also misspelled, with only one s instead of two. Move the cursor to the s and type another s. The letter is automatically inserted and all the other letters are adjusted around it.

You can insert the word "Drive" after the name of the street by placing the cursor in that position and typing the word.

The word "Real" appears twice. There are several ways to eliminate one of them. You can place the cursor on the letter R and press the Del key four times, or you can place the cursor on the space to the right of the letter d and press the Backspace key four times.

Copying text

It is easy to insert and delete characters and words, but how do you select large portions of text so they can be copied or moved? The row of asterisks in line 6 can be copied and inserted below the other headings, such as Personal Data, Education, etc.

```
RESUME

NAME:  David C. Fields          Telephone: (201)-234-5678

ADDRESS:
***************************************************************
187 Foxchase
Alsoran, New Jersey  04545

PERSONAL DATA
Date of Birth: 6-11-48

EDUCATION:
1966 - 1972         B.A. New jErsey University Real Estate

1962 - 1966         High School Academic Diploma
                    Blue Jay High School

EXPERIENCE:
1980 - Present      Real Real Estate Broker
                    Take-A-Home Realty
                    Strawberry Foelds, New Jersey

1974 - 1980         Credit Manager
                    Red Star Oil Company Cutty Shark, New York

MILITARY SERVICE:
1972 - 1974         U.S. AArmy - Corporal; Quartermaster Corp;
Taught Business  clases to to servicemen during term of service;
Served one year in Spain as a Spanish interpreter; Received Ary
Commendation Ribbon

LICENSE:
1972 - New Jersey Brokers License

AWARDS:
Outstanding Take-A-Home Real Estate Broker for 1988; Awarded to
the broker who is both civic minded and has the most sales
```

2-14 The complete resume

First select the text to be copied by accessing the Edit menu and selecting Mark block. Highlight the row of asterisks using the arrow keys. If you are using a mouse, hold down the left mouse button and move the mouse to the end of the block. Release the mouse button. Select Copy to clipboard at the Edit menu. If you make a mistake marking text, select Unmark block from the Edit menu and start again, or press the Esc key.

Move the cursor to where you want the row of asterisks inserted (below "Personal Data") and select Paste from clipboard at the Edit menu, or use the speed key Shift-Ins. The row of asterisks will immediately be copied. The row of asterisks will stay in the clipboard until you put more text in it.

In order to place the row of asterisks after the headings EDUCATION, EXPERIENCE, MILITARY SERVICE, LICENSE, and AWARDS, you don't

Table 2-1 Cursor movement

Key	Result
Left arrow	One Character to left
Right arrow	One Character to right
Up arrow	One line up
Down arrow	One line down
Ctrl – left arrow	Word left
Ctrl – right arrow	Word right
PgUp	Window up
PgDn	Window down
Home	Beginning of line
End	End of line
Ctrl – Home	Beginning of file
Ctrl – End	End of file
Home (twice)	Beginning of window
End (twice)	End of window

need to reselect the text and copy it again. Move the cursor to the location where you want the row of asterisks copied and select Paste from clipboard. Insert the row of asterisks after all the headings listed above. The final document should look like that shown in Fig. 2-15. There are still a few errors in the document, but you can fix them later.

Moving text

You saw in the last section that copying text creates a duplicate copy of the text in another area of the notepad. But what if you want to *move* the Military Service section to the end of the resume. Select the text by choosing Mark block from the Edit menu. Highlight the text to be moved. Select Cut to clipboard or use the speed key Shift – Del. The highlighted text will now be moved to the clipboard.

Move the cursor to the end of the resume and select Paste from clipboard in the Edit menu. The new resume should look like that shown in Fig. 2-16.

Deleting text

Look at the resume. The Personal Data section is discriminatory and should be removed. Select the Personal Data paragraph by choosing Mark block at the Edit menu, highlighting the marked text, and then selecting Cut to clipboard. The paragraph should now be deleted, as shown in Fig. 2-17.

Be careful! If you select Delete all text from the Edit menu, it will delete all text in the notepad. The notepad will be open, but there will be no text in it.

```
RESUME

NAME:  David C. Fields            Telephone: (201)-234-5678

ADDRESS:
***************************************************************
187 Foxchase  Drive
Alsoran, New Jersey  04545

PERSONAL DATA
***************************************************************
Date of Birth: 6-11-48

EDUCATION:
***************************************************************
1966 - 1972       B.A. New jErsey University Real Estate

1962 - 1966       High School Academic Diploma
                  Blue Jay High School

EXPERIENCE:
***************************************************************
1980 - Present    Real Estate Broker
                  Take-A-Home Realty
                  Strawberry Foelds, New Jersey

1974 - 1980       Credit Manager
                  Red Star Oil Company Cutty Shark, New York

MILITARY SERVICE:
***************************************************************
1972 - 1974       U.S. Army - Corporal; Quartermaster Corp;
Taught Business  classes to to servicemen during term of
service; Served one year in Spain as a Spanish interpreter;
Received Ary Commendation Ribbon

LICENSE:
***************************************************************
1972 - New Jersey Brokers License

AWARDS:
***************************************************************
Outstanding Take-A-Home Real Estate Broker for 1988; Awarded to
the broker who is both civic minded and has the most sales
```

2-15 Resume with inserted asterisks

Inserting one document into another

It is possible to insert one document into another. Figure 2-18 shows a document called Outside Interests that you can insert into the resume. To insert the file OUTSIDEI into your resume, position the cursor on the row and column in the resume where you want it inserted. Access the Edit menu and select Insert file. You will be given a list of files that can be inserted into the resume. Highlight OUTSIDEI and press the Enter key. The paragraph Outside Interests is inserted in the document at the cursor position, as shown in Fig. 2-19.

```
RESUME

NAME: David C. Fields               Telephone: (201)-234-5678

ADDRESS:
*****************************************************************
187 Foxchase  Drive
Alsoran, New Jersey  04545

PERSONAL DATA
*****************************************************************
Date of Birth: 6-11-48

EDUCATION:
*****************************************************************
1966 - 1972        B.A. New jErsey University Real Estate

1962 - 1966        High School Academic Diploma
                   Blue Jay High School

EXPERIENCE:
*****************************************************************
1980 - Present     Real Estate Broker
                   Take-A-Home Realty
                   Strawberry Foelds, New Jersey

1974 - 1980        Credit Manager
                   Red Star Oil Company Cutty Shark, New York

LICENSE:
*****************************************************************
1972 - New Jersey Brokers License

AWARDS:
*****************************************************************
Outstanding Take-A-Home Real Estate Broker for 1988; Awarded to
the broker who is both civic minded and has the most sales

MILITARY SERVICE:
*****************************************************************
1972 - 1974        U.S. Army - Corporal; Quartermaster Corp;
Taught Business classes to to servicemen during term of service;
Served one year in Spain as a Spanish interpreter; Received Ary
Commendation Ribbon
```

2-16 Resume after using Copy and Move

Spelling checker

There are still a few intentional spelling errors in your resume, and now is the time to correct them. Remember that the spelling checker of PC Desktop was developed for a utility program and not a full-fledged word processing program. It is quite limited in this respect.

To use the spelling checker, access the Edit menu. You can spellcheck a word, the screen, or the entire file. If you select Spellcheck file, it will highlight

```
RESUME

NAME: David C. Fields              Telephone: (201)-234-5678

ADDRESS:
****************************************************************
187 Foxchase  Drive
Alsoran, New Jersey  04545

EDUCATION:
****************************************************************
1966 - 1972         B.A. New Jersey University Real Estate

1962 - 1966         High School Academic Diploma
                    Blue Jay High School

EXPERIENCE:
****************************************************************
1980 - Present      Real Estate Broker
                    Take-A-Home Realty
                    Strawberry Fields, New Jersey

1974 - 1980         Credit Manager
                    Red Star Oil Company Cutty Shark, New York

LICENSE:
****************************************************************
1972 - New Jersey Brokers License

AWARDS:
****************************************************************
Outstanding Take-A-Home Real Estate Broker for 1988; Awarded to
the broker who is both civic minded and has the most sales

MILITARY SERVICE:
****************************************************************
1972 - 1974         U.S. Army - Corporal; Quartermaster Corp;
Taught Business classes to to servicemen during term of service;
Served one year in Spain as a Spanish interpreter; Received Army
Commendation Ribbon
```

2-17 Resume after deleting a paragraph

the words that are misspelled or that it doesn't recognize. Proper names and technical or specialized words are not in its dictionary. It will stop at words in the resume such as Foxchase, Alsoran, and diploma. It does not recognize names of cities and seldom-used words like diploma. Figure 2-20 shows the prompt for the misspelled word "Foelds."

A dialog box appears at the bottom of the screen with the misspelled word, asking you to make one of four choices. You can type I to ignore the word. Do this for words like Foxchase, Alsoran, and diploma. This means that the word is correct but is not in the dictionary. You can type C to correct the word. When you select this option, Fig. 2-21 appears with a list of suggested words, none of which are correct in this case. It doesn't realize that Fields is the correct spelling, so you

2-18 "Outside interest" paragraph

have to type it in. The third option is to type A for Add, to add the word to the dictionary. The last option is to type Q to Quit the spelling checker.

Notice that the spelling checker does not stop at incorrectly capitalized words such as "jErsey," and at repeated words like "to to." These will have to be manually edited.

Searching and replacing text

If you decide that the word "oil" should be replaced by the word "utilities," you could do this easily by just looking for every instance of the word. In a long file, however, this would be very time consuming. PC Desktop has an easier way.

You can replace certain letters, words, or groups of words with other words. For example, if you used a long word in a document, such as the name of the town Massapequa Park, you can abbreviate it as MP when you type it in . When the document is finished, you can perform a search, and replace each MP with Massapequa Park. You can save a lot of typing by abbreviating frequently-used long names in a document and replacing them later.

Access the Search menu and select Find and replace, as shown in Fig. 2-22. Type in the word Oil on the "Search for:" line and the word Utilities in the "Replace with:" line. You can make the search case sensitive, where it will

```
RESUME

NAME: David C. Fields                Telephone: (201)-234-5678

ADDRESS:
*****************************************************************
187 Foxchase  Drive
Alsoran, New Jersey  04545

EDUCATION:
*****************************************************************
1966 - 1972          B.A. New jErsey University Real Estate

1962 - 1966          High School Academic Diploma
                     Blue Jay High School

EXPERIENCE:
*****************************************************************
1980 - Present       Real Estate Broker
                     Take-A-Home Realty
                     Strawberry Foelds, New Jersey

1974 - 1980          Credit Manager
                     Red Star Oil Company
                     Cutty Shark, New York

LICENSE:
*****************************************************************
1972 - New Jersey Brokers License

OUTSIDE INTERESTS:
*****************************************************************
Vice-President of the local Strawberry Fields Real Estate
Association; Captain of the Gappha Bowling Team; Member of the
Alsoran Beer Can Collecting Team; Lifetime member of
Friends-Of-Animals Association; Member of the Love-To-Garden
Club; Member of local Dentist Advisory Group; Founder of Local
Kids-R-Great Association

AWARDS:
*****************************************************************
Outstanding Take-A-Home Real Estate Broker for 1988; Awarded to
the broker who is both civic minded and has the most sales

MILITARY SERVICE:
*****************************************************************
1972 - 1974          U.S. Army - Corporal; Quartermaster Corp;
Taught Business classes to to servicemen during term of service;
Served one year in Spain as a Spanish interpreter; Received Army
Commendation Ribbon
```

2-19 Expanded window

search for only upper- or lowercase. If you select Whole words only, it will search only for the exact word. If you don't select Whole words only, your word will be found even if it is embedded in other words. For example, Oil might be found in boiler, toil, spoil, etc.

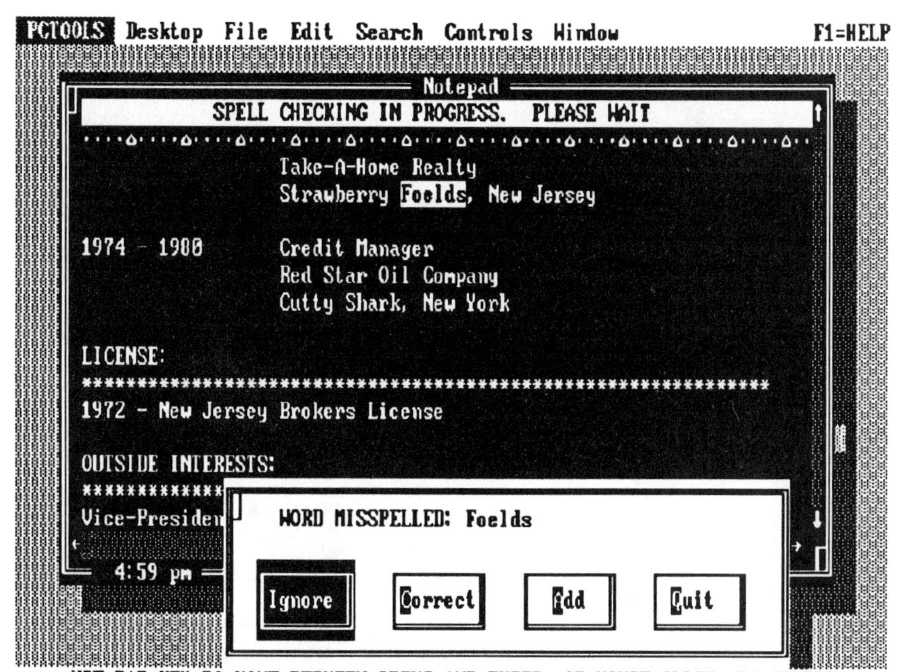

2-20 Spelling checker

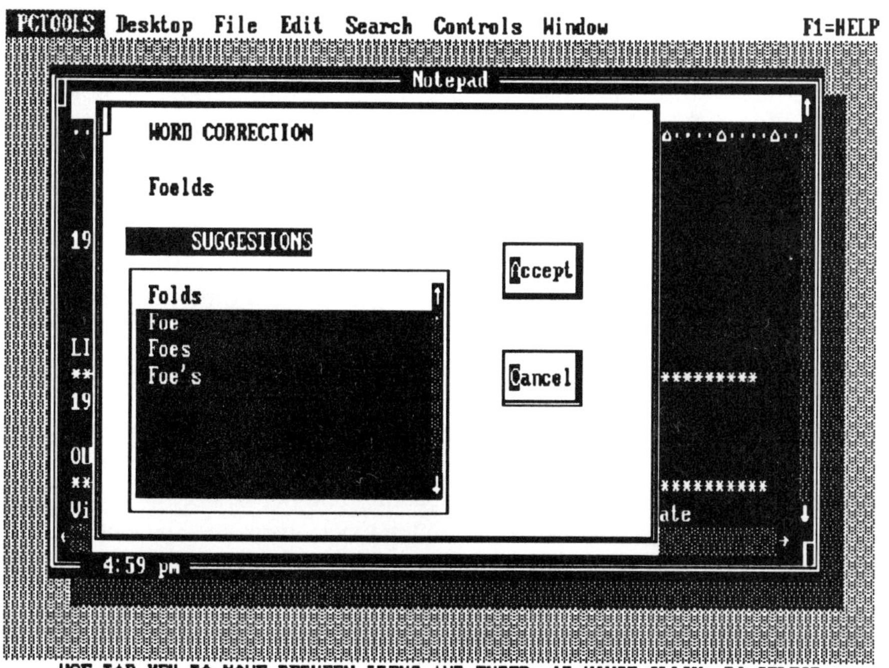

2-21 Some suggested spellings

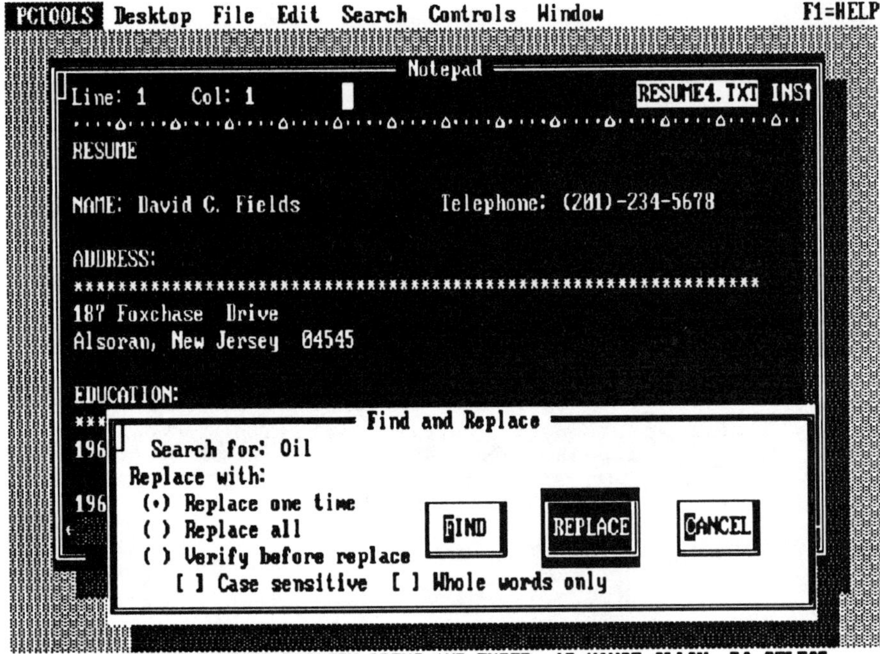

2-22 Find and Replace menu

On the left side of the dialog box you will be asked whether you want to replace one occurrence or all the occurrences of the word automatically. You can also have each word individually highlighted so you can see it before it is replaced. I strongly recommend that you choose the Verify before replace option so you do not replace unwanted words. The final resume is shown in Fig. 2-23.

The Find command is very similar to the Find and replace command, but it simply looks for the occurrence of a word in the file, as shown in Fig. 2-24. In this case, it will find the first occurrence of the word "School." If it finds it, it will ask you if you want to find the next occurrence of the word. Like the Find and replace, you can make the search either Case sensitive or Whole words only.

Page layout

If you do not like the way the resume is printed, you can adjust the left and right margins, top and bottom margins, line spacing, and paper size, as shown in Fig. 2-25. The Page layout option is located in the Controls menu. You can also adjust the line spacing to single or double spacing while in this option.

If you want to keep the page layout with the document, you can select Save setup, and the page settings will be saved with the document.

RESUME

NAME: David C. Fields   Telephone: (201)-234-5678

ADDRESS:
****
187 Foxchase Drive
Alsoran, New Jersey 04545

EDUCATION:
****
1966 - 1972   B.A. New Jersey University Real Estate

1962 - 1966   High School Academic Diploma
                     Blue Jay High School

EXPERIENCE:
****
1980 - Present   Real Estate Broker
                     Take-A-Home Realty
                     Strawberry Fields, New Jersey

1974 - 1980   Credit Manager
                     Red Star Oil Company
                     Cutty Shark, New York

LICENSE:
****
1972 - New Jersey Brokers License

OUTSIDE INTERESTS:
****
Vice-President of the local Strawberry Fields Real Estate Association; Captain of the Gappha Bowling Team; Member of the Alsoran Beer Can Collecting Team; Lifetime member of Friends-Of-Animals Association; Member of the Love-To-Garden Club; Member of local Dentist Advisory Group; Founder of Local Kids-R-Great Association

AWARDS:
****
Outstanding Take-A-Home Real Estate Broker for 1988; Awarded to the broker who is both civic minded and has the most sales

MILITARY SERVICE:
****
1972 - 1974   U.S. Army - Corporal; Quartermaster Corp; Taught Business classes to servicemen during term of service; Served one year in Spain as a Spanish interpreter; Received Army Commendation Ribbon

2-23 The final resume

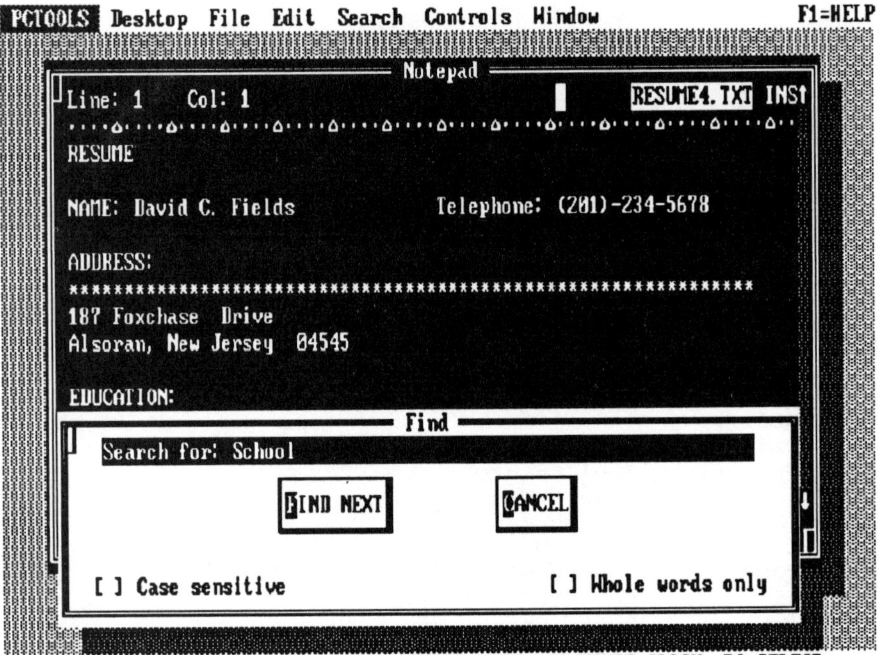

2-24 Find command

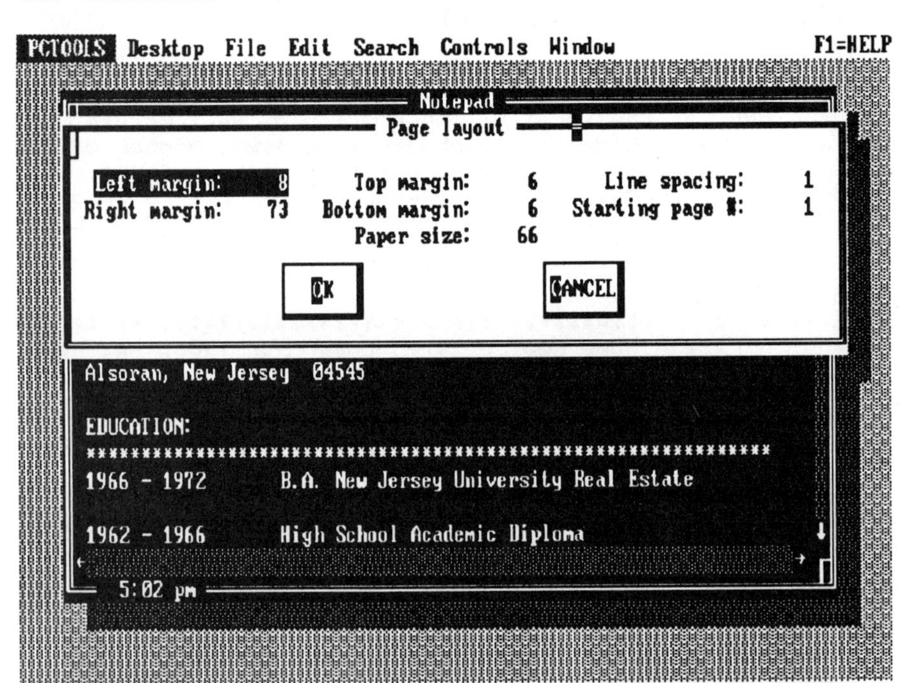

2-25 Page Layout menu

Page headers and footers

A header is one or more lines of text printed at the top of each page when the document is printed. It can contain chapter headings or any other pertinent information. A footer is one or more lines of text printed at the bottom of each page. It can contain text and, typically, contains page numbers. The pound symbol, #, in the footer is a page counter that automatically starts with the number 1. If you delete it, no page numbers will be printed in the footer. If you place the pound symbol in the header, page numbers will be printed in the header. In each file, you can have headers, footers, or both.

Headers and footers can be placed in the document with the Control menu by selecting the Page header & footer option. You can type the headers and footers into the space provided, as shown in Fig. 2-26.

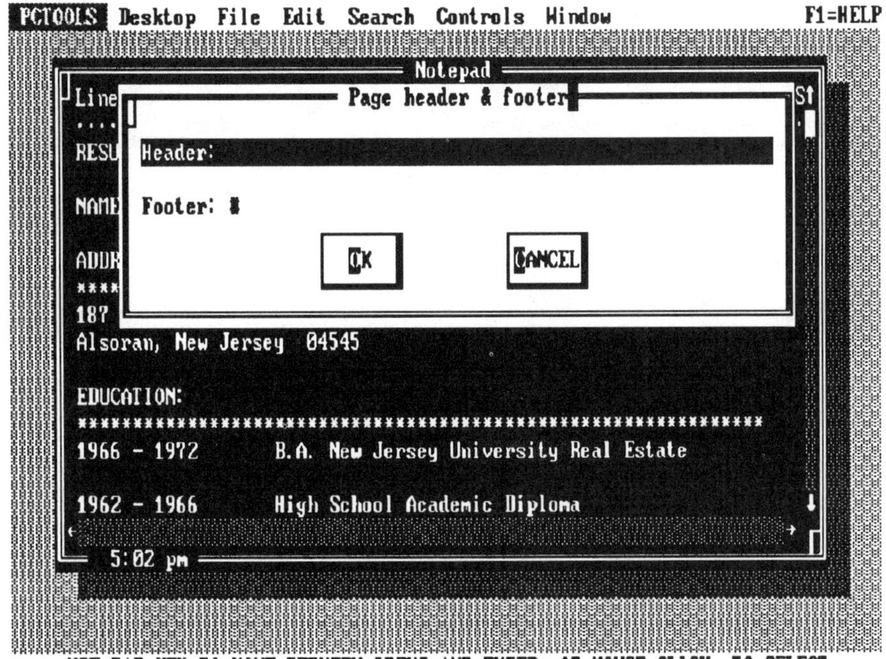

2-26 Page Header and Footer menu

Setting and erasing tabs

PC Desktop automatically inserts a tab every five columns. Tabs can be inserted or deleted with the Tab ruler edit option of the Control menu, as shown in Fig. 2-27. Notice the top line of the window says "Editing tab ruler." To insert a tab,

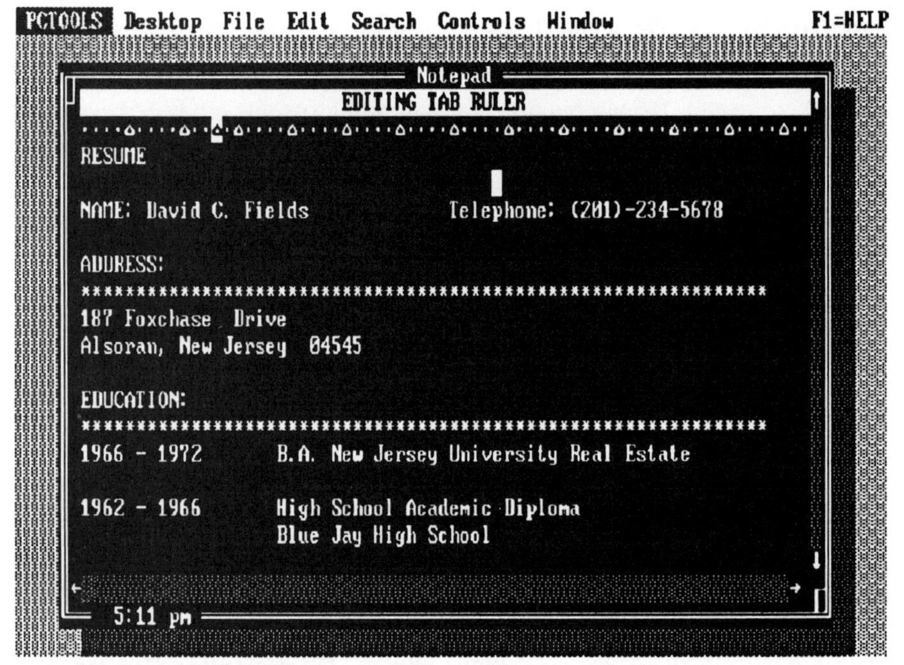

2-27 Setting and erasing tabs

move the cursor to where you want the tab inserted and press the Ins key. A triangle will be inserted on the ruler where the tab is specified. In Fig. 2-27, a tab is inserted at column 13.

To delete a tab, place the cursor over the tab that you want deleted on the ruler and press the Del key. The Tab will be deleted.

In the same menu, you can suppress the ruler by selecting Tab ruler display. This acts as a toggle to show or hide the ruler.

Selecting the overtype mode will place the cursor in a mode where the cursor types over characters instead of inserting them.

Control character display

When you select Control char display, all the carriage returns are shown as arrows and blank spaces as dots, as shown in Fig. 2-28.

Wordwrap and auto indent

Wordwrap lets you continue typing without pressing a Return at the end of each line. Words are automatically placed at the beginning of the next line. *Auto indent* automatically indents a line to align with the first character of the previous line. It is useful for indenting paragraphs.

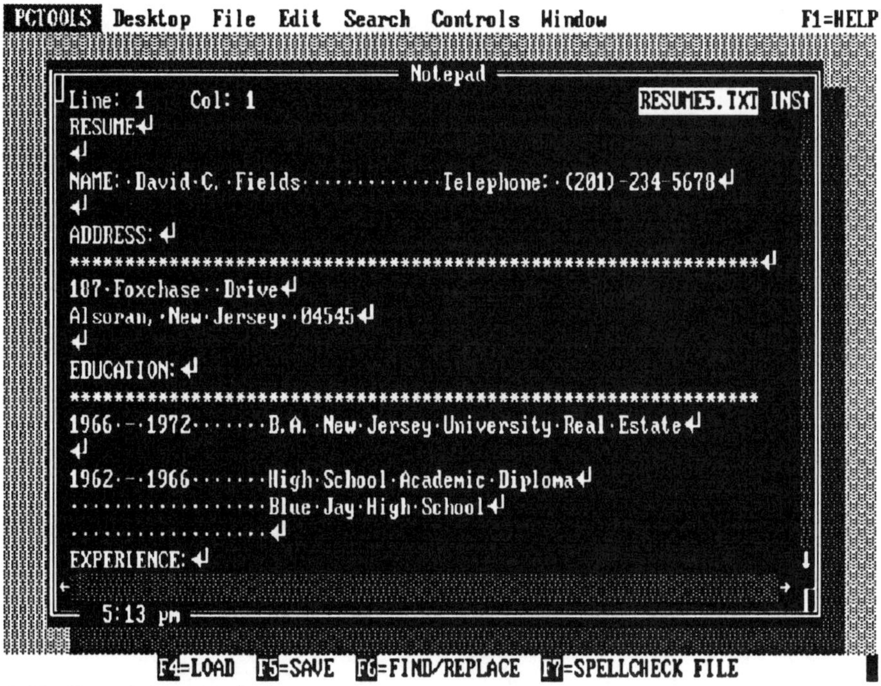

```
PCTOOLS Desktop File Edit Search Controls Window                F1=HELP
                              Notepad
Line: 1     Col: 1                                    RESUMES.TXT INS
RESUME

NAME: David C. Fields            Telephone: (201) 234 5678

ADDRESS:
*******************************************************************
107 Foxchase  Drive
Alsoran, New Jersey  04545

EDUCATION:
*******************************************************************
1966 - 1972       B.A. New Jersey University Real Estate

1962 - 1966       High School Academic Diploma
                  Blue Jay High School

EXPERIENCE:
    5:13 pm
        F4=LOAD  F5=SAVE  F6=FIND/REPLACE  F7=SPELLCHECK FILE
```

2-28 Control character display

Inserting IBM extended characters

Any members of the IBM extended character set can be inserted in a document if you press the Alt key and the corresponding number key located on the numeric keyboard. (Do not use the number keys located at the top of the keyboard.) For example, the square root symbol can be generated by pressing the Alt key and typing the number 251 on the numeric keypad. Figure 2-29 shows some of the characters that you can use in order to accentuate words, lines, or paragraphs.

Window menu

By selecting the Change colors option of the Window menu, you can change the colors of the background, the window border, the document text, and the ruler and status, as shown in Fig. 2-30.

The cursor keys can adjust the size of the window, as shown in Fig. 2-31. The changed window is shown in Fig. 2-32. The Move option lets you move the window around the screen. The Zoom option lets the window take up the whole screen.

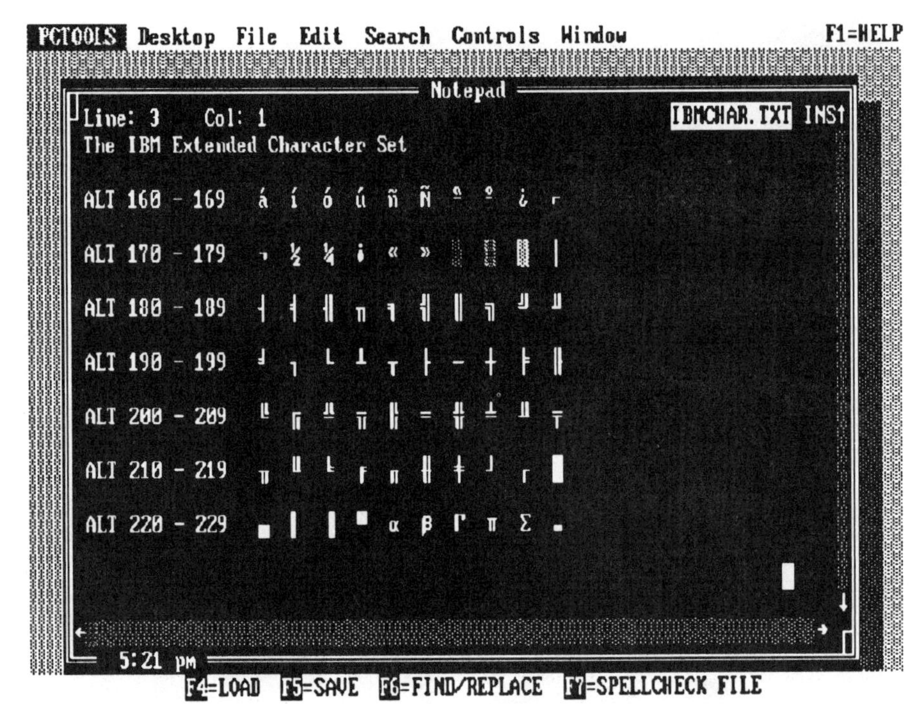

2-29 The IBM extended character set

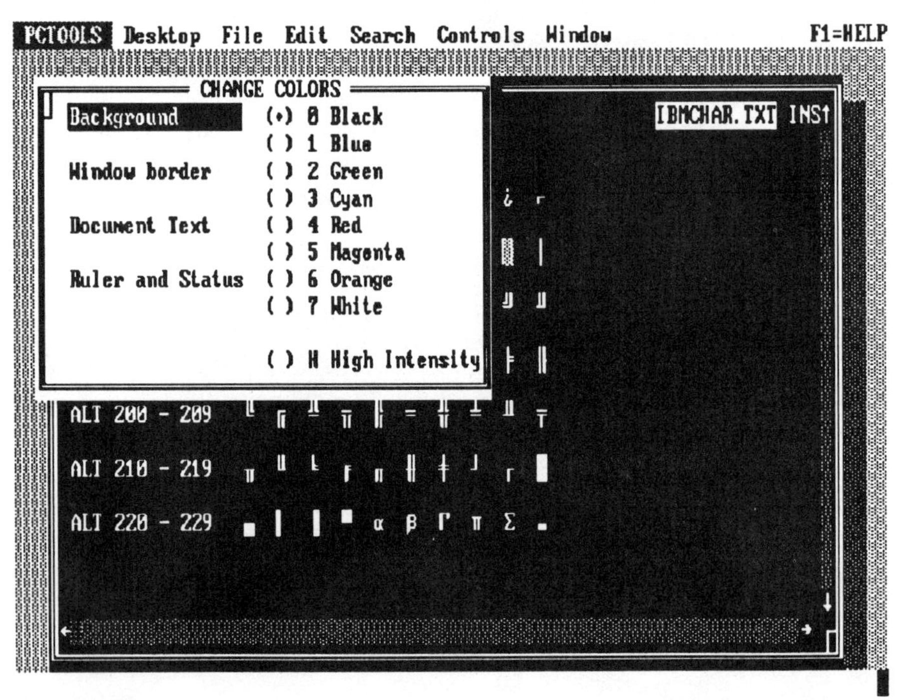

2-30 Change colors command

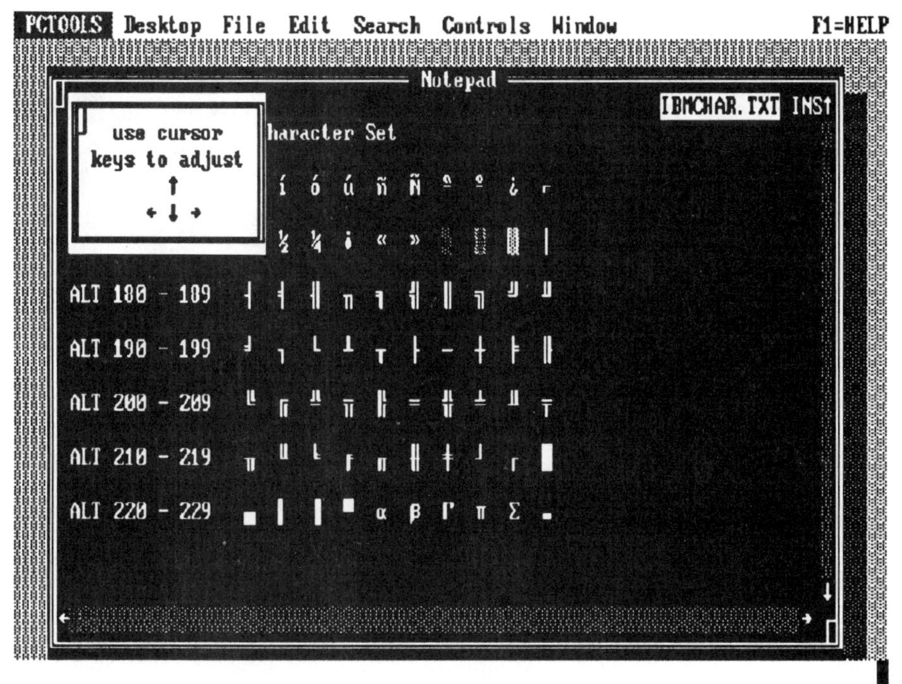

2-31 Adjusting the window size

2-32 Moved and resized window

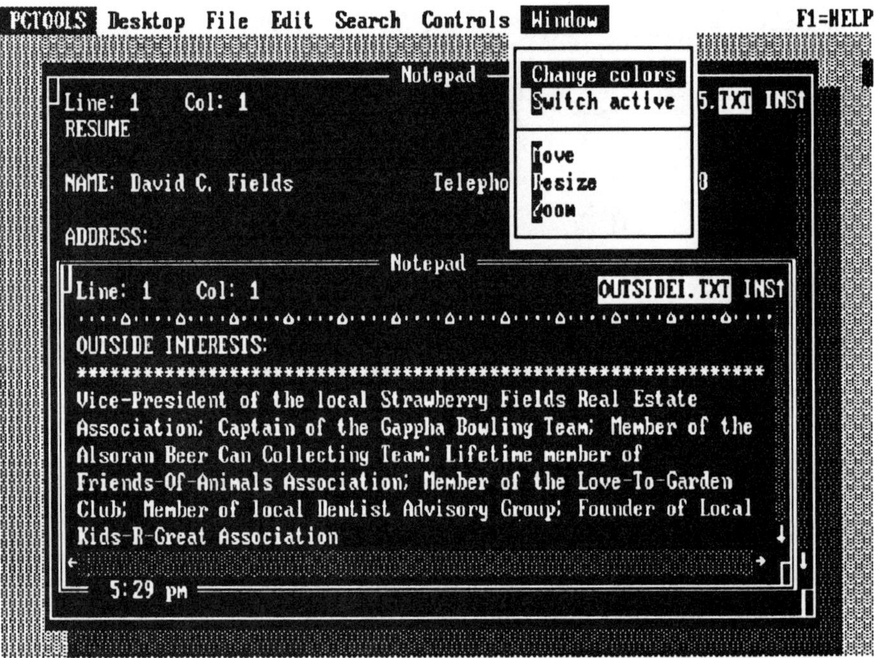

2-33 Notepad with two active windows

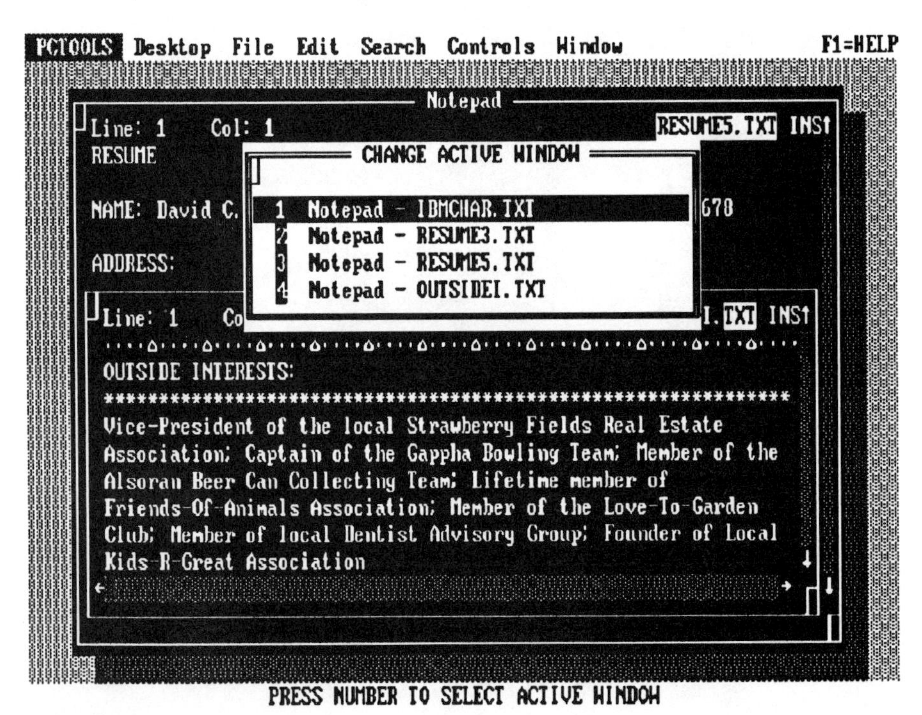

2-34 Change Active Window menu

Switching active windows

PC Desktop lets you work with up to 15 windows at once. You can bring several documents into the notepad by simply loading them in. Figure 2-33 shows two on-screen windows. The window with the double lined box around it is the active window. Compare Figs. 2-13 and 2-33. The Switch active option appears when you have more than one window active.

When you select the Switch active option, it gives you a list of the files you can look at as shown in Fig. 2-34. You can have up to 15 windows opened or active at once.

3
CHAPTER

Outlines

In this chapter you will learn about the outline facility of PC Tools—how to create, collapse, and expand outlines for structuring information in a document.

Why use an outline?

The notepad feature of PC Tools lets you write notes, letters and memos, while the outline utility of PC Tools makes it easy to structure information. As you gather information, you can use an outline to structure your document. It lets you focus on the overall organization by hiding less important details. It helps you identify critical topics and restructure sections. An outline lets you format your thoughts. It provides a structure within which you can organize and work with your ideas. When you identify what you want to accomplish, the goals of the project or proposal become the items of the outline. Each of these items can then be broken down into further detail.

Once the outline is created, you can reorganize it by cutting and pasting information from one section to another. The outline automatically manages the niceties of indentation—letting you arrange and organize your thoughts without worrying about format.

In the early stages of planning your document, an outline can serve as an electronic file for your ideas, expanding as you gather detailed information that you'll use later in the writing process.

If you use an outline to prepare a standard structure to organize the information you collect, you'll consolidate the information in your document and write more concisely.

Headlines

To access the outline utility from PC Desktop, highlight it at the Desktop menu and press the Enter key. The Outline menu appears, as shown in Fig. 3-1, along with the Outline window.

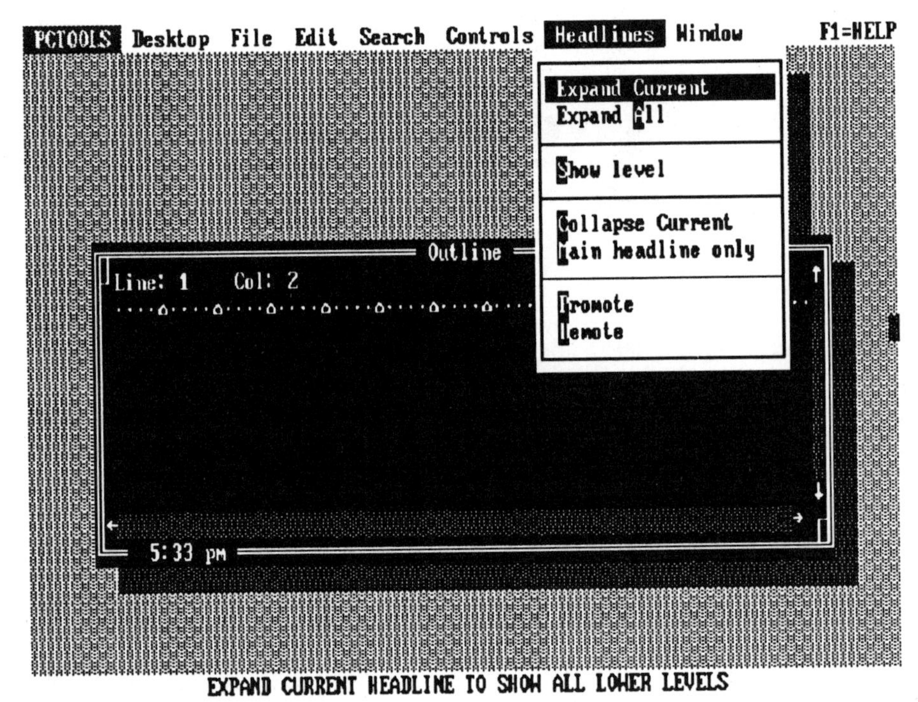

3-1 Outline menu

Outlines are composed of different levels. Level one appears at the left margin, level two appears next to the first tab, level three appears next to the second tab, etc. This structure is apparent when you look at the sample outline shown in Fig. 3-2.

Each of the levels is considered a *headline* level. When you finish typing in a line of text, press the Enter key and you will remain at the same level. If you want to change back to a previous level, press the Backspace key. Pressing the Tab key will advance you to the next level.

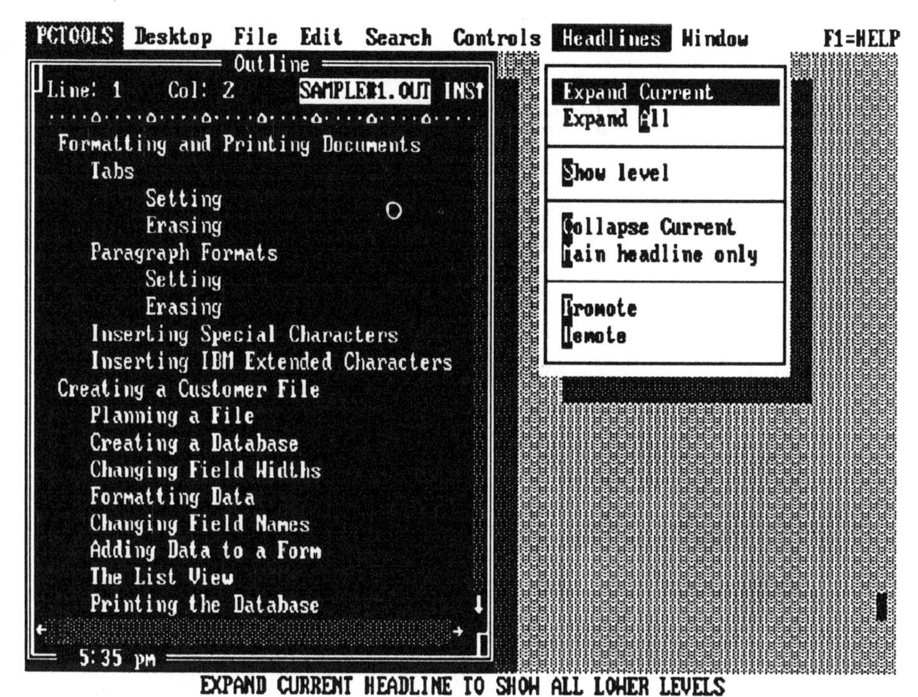

3-2 Sample outline

Showing levels and expanding all commands

Levels can be inserted, deleted, and edited using the same cursor keys in the notepad utility. If you get lost in an outline, select the Expand all option in order to see the entire outline with all its levels.

Levels can be collapsed and expanded. If you place the cursor on the line containing the word "Tabs" in Fig. 3-2 and select Show level, Fig. 3-3 appears. The Show level command, in this case, collapsed the outline to show the first two levels. The third level has been eliminated. This command is very useful if you want to create an abbreviated outline when the original outline might be too long for your purposes.

The Show level command displays the level the cursor is on while hiding any levels below it. In Fig. 3-3, all the level three lines have been hidden because the cursor was placed on a level two line (Tabs). The symbol to the left of the first two lines in level two indicates that text has been hidden. If you want to see the complete outline again, select Expand all.

Showing main headline only

The Main headline only command shows only the first levels of an outline, as shown in Fig. 3-4. The symbols to the left of each level indicate that all subsequent levels are hidden.

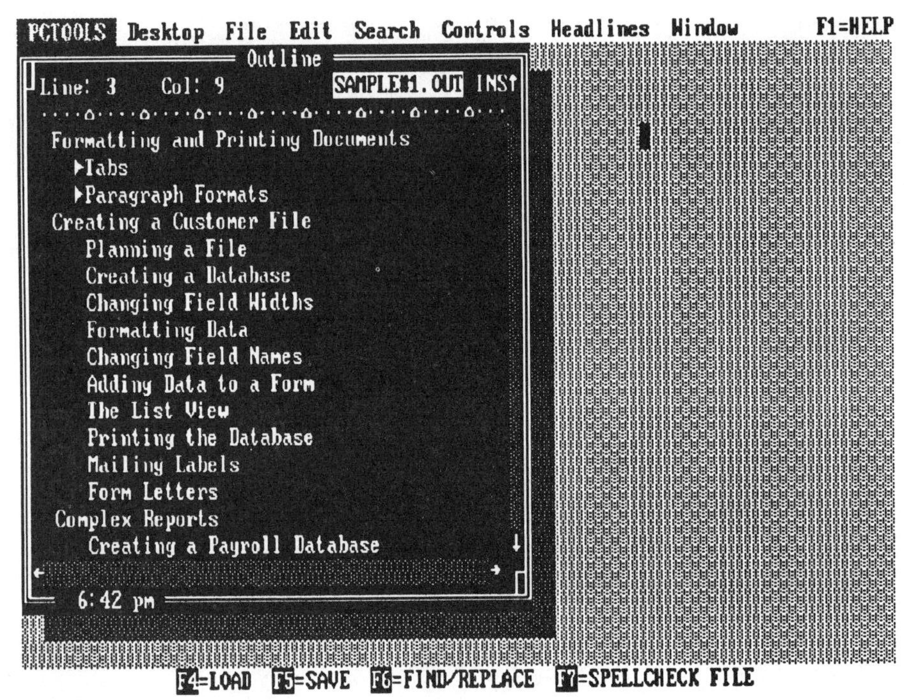

3-3 Show level command

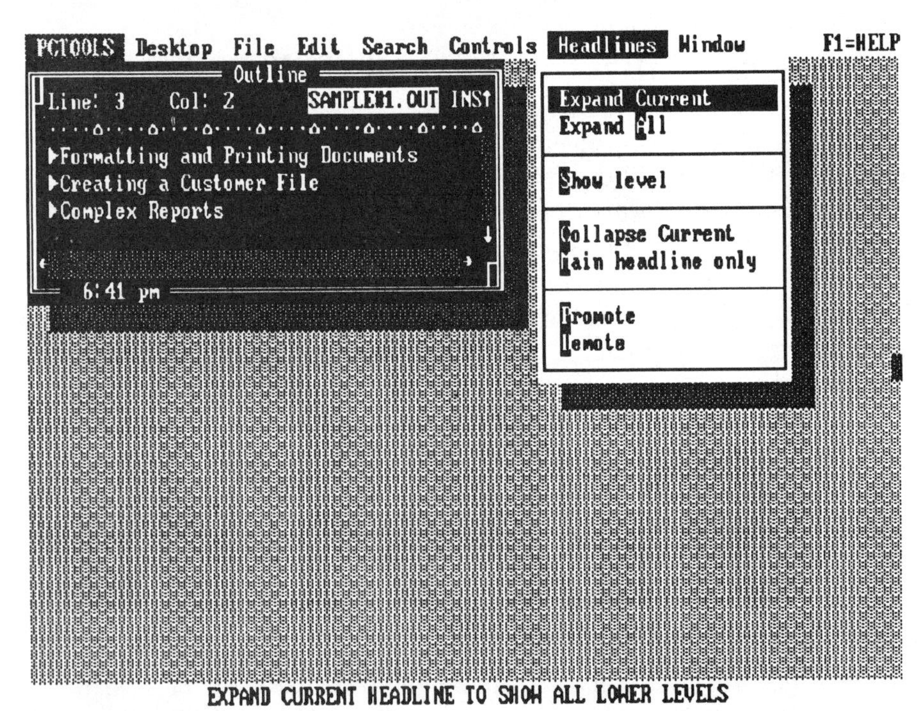

3-4 Main headlines only command

Collapsing to the current level

The real test of an outline's power is its ability to help you keep track of a large outline by quickly and easily switching between an abbreviated view of its structure and a closely focused look into its detail. PC Outline lets you hide or reveal text and headings separately, so you can focus your view of the outline. The Collapse current command, shown in Fig. 3-5, hides any levels that are below the cursor and lets you condense all the subitems of a level. In this case, the cursor is on a first level item. All subheadings below that level are hidden, as indicated by the symbol to the left of "Creating a Customer File."

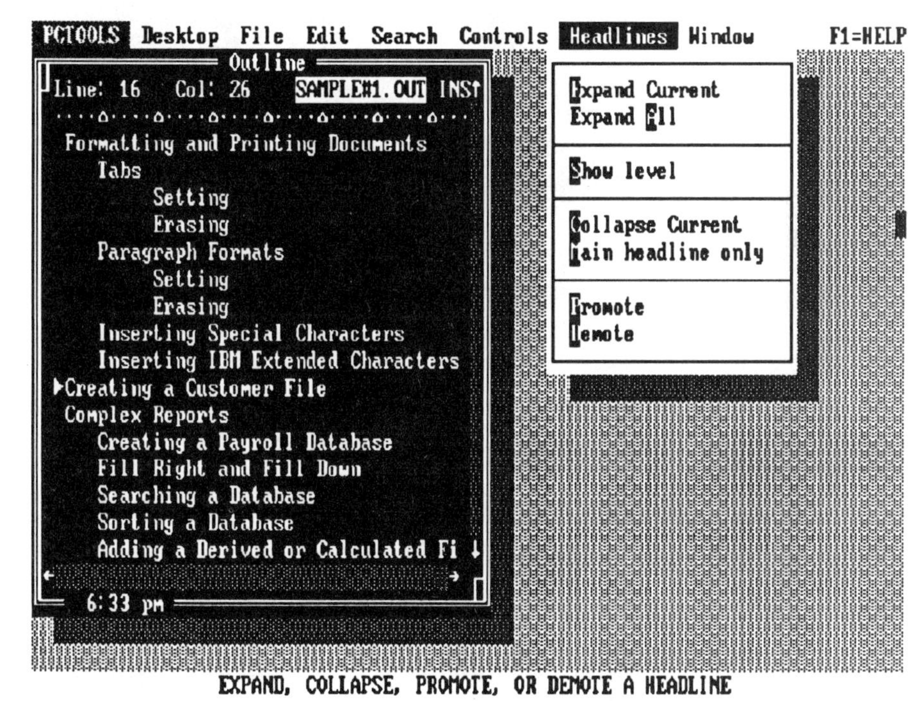

3-5 Collapse current command

Expanding the current level

The Expand current command shows all the subheadings or levels below the cursor. In Fig. 3-6, the cursor is on the first line of text in the window when the Expand current command is executed. All the subheadings for that level are immediately shown. This lets you expand a single headline to all subheadings.

Promoting and demoting

You can either upgrade a level to a higher level or downgrade it to a lower level. Figure 3-7 gives an example of the Promote command. With the cursor on level

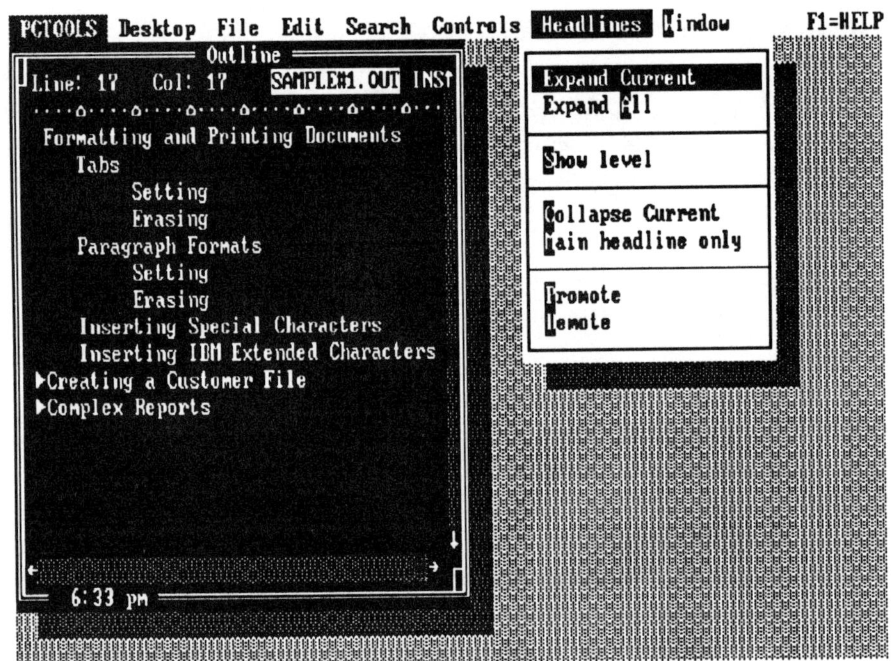

3-6 Expand current command

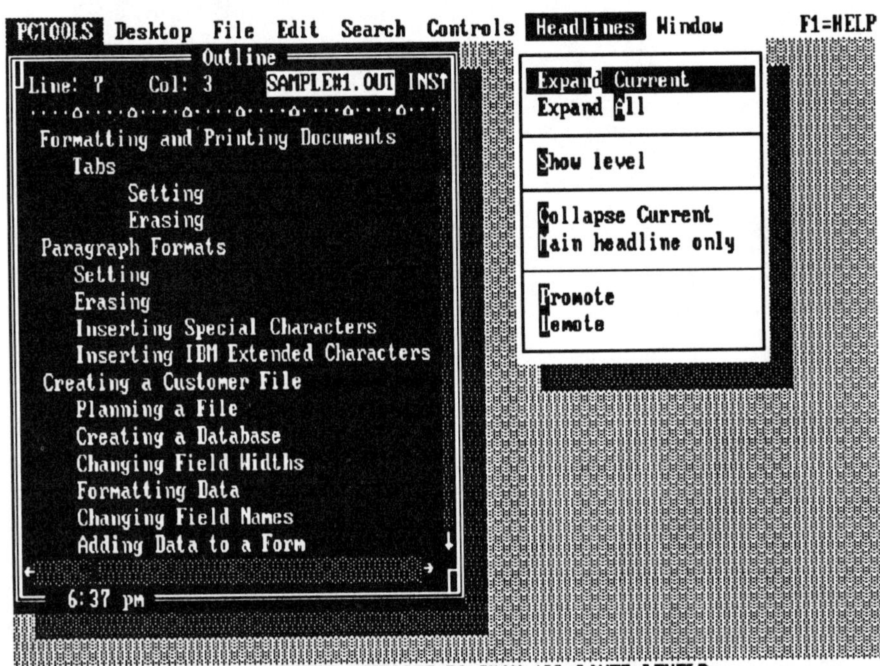

3-7 Promote command

one, Paragraph Formats, execute the Promote command. The two lines that were on level three in that headline (Setting and Erasing) have automatically jumped to level two. The Promote command brings a heading to a higher level. To bring the two lines back to where they were, select Demote. The demote command brings headings to a lower level. The cursor must always be on the line that you want to Promote or demote.

Printing an outline

Outlines can be printed by selecting the File menu and selecting the Print option.

4

CHAPTER

Databases, Mailing labels, and form letters

In this chapter, you will create, add data to, and print a customer database. You will learn how to delete, undelete, hide, and pack records. Once the file is created, you can then create mailing labels and form letters from information contained in it.

Background

The primary role of the database application utility in PC Tools is to provide an easy way to create and maintain files, and to retrieve data from those files.

A *database* is a collection of information organized for easy retrieval and addition of data. A telephone directory is an example of a database. Another example is a list of customers who are patrons of a local VCR rental store. Figure 4-1 shows the structure of the database you will create in the next section.

Each of the items listed in Fig. 4-1 is called a *field.* Each field is capable of

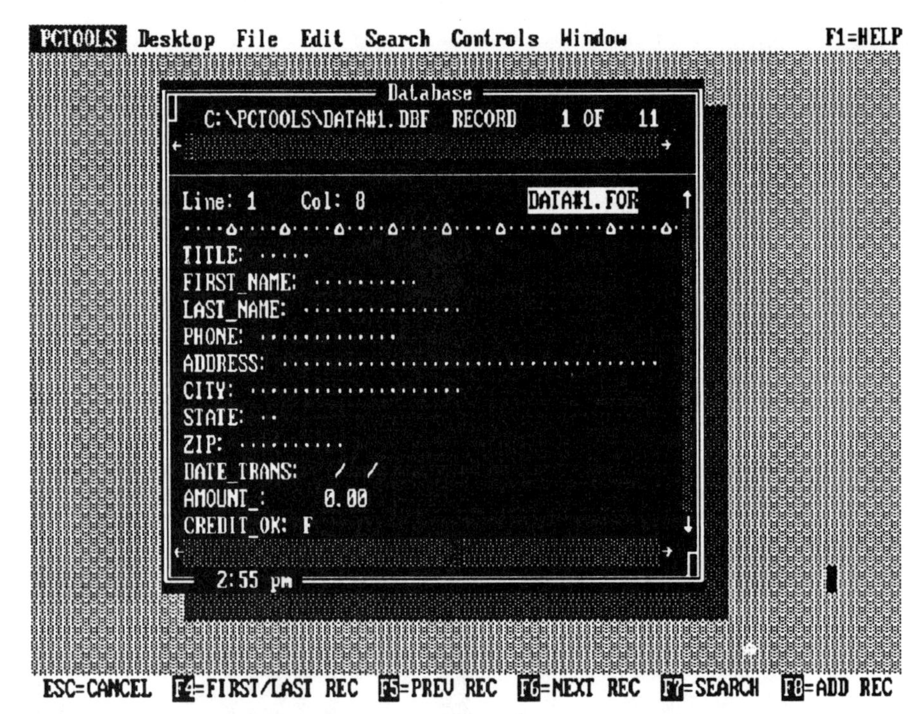

4-1 Structure of a VCR database

holding information from many sources about one specific part of a transaction or business event. A field is a piece of information. Some common fields in your example are Last Name, First Name, Address, City, State, and Zip. Notice that the fields are empty. The information will be entered later in the chapter.

The set of all fields that relate to each customer is called a *record.* A record is a set of information about one object or person. PC Tools can hold 3,500 records per database. A group of records forms the database. A database is a collection of organized information.

What you have, then, is a database of customers. Information about one customer is called a record, and a record can be broken down into fields. PC Tools can manage 128 fields per record, or up to 4000 characters per record. These definitions will become a little clearer when you start entering information into your database. Database information can also be easily updated, searched, sorted, retrieved, and printed out. A mailing list, home or business inventory, and a stock market portfolio would all make excellent databases.

Planning a database

The first step in planning a database is to determine how it will be used. Changes can always be made in a database, but it is much better to put as much planning as you can into the initial creation. Changes take a lot of time and can create many avoidable errors.

The first thing to determine is the kind of information you need to store. In other words, "What fields do you need to create?" In the VCR store, for example, you need to determine what forms are filled out, what records are kept, what data is collected, and what data is missing that you wish you had. The names of the fields, or pieces of data you want to collect, is of extreme importance. Planning now will make life much easier later on.

The next item you need to address is how the information is going to be used. Will you create mailing labels that show the customers first name first? Will the labels be sorted by ZIP code to help save money in a bulk mailing? Will notices be sent out to all customers who haven't paid their bills within the last thirty days? Which fields will be required for every record, and which fields will be optional? (Try to group required information at the beginning of the record and optional at the end of the form.)

You might find it helpful to design your database "form" on paper before you type it into the machine. Although this step might seem unnecessary, it can save time and trouble later.

Creating a database

Suppose you're an owner of a video rental store and want a file of all the customers who order tapes so you can create a mailing list and send out form letters announcing new arrivals. From the Desktop menu, access the Database option. Figure 4-2 appears asking you if you want to load an old file, create a new file, or cancel the request. If you want to create a new file, type in a filename of not more than eight letters and press N for new file. All filenames are automatically given the extension .DBF, which stands for database file. It is not necessary to type in the extension; PC Tools will add it for you.

After you press N, Fig. 4-3 appears, asking you for a field name. Field names can be up to ten characters long, but must start with a letter. No spaces are permitted in a field name. You can type the field name in upper- or lowercase letters, but they are automatically converted into uppercase.

Field types

PC Tools assumes that all the data you type into the form will be character data. A *character* field can be composed of letters, special symbols, or numbers. Telephone numbers, ZIP codes, and social security numbers are all character fields. Although they can contain numbers, they are treated as characters, because they sometimes contain special symbols. You cannot do any type of arithmetic operations in character fields. There can be a maximum of 70 characters in this type of field.

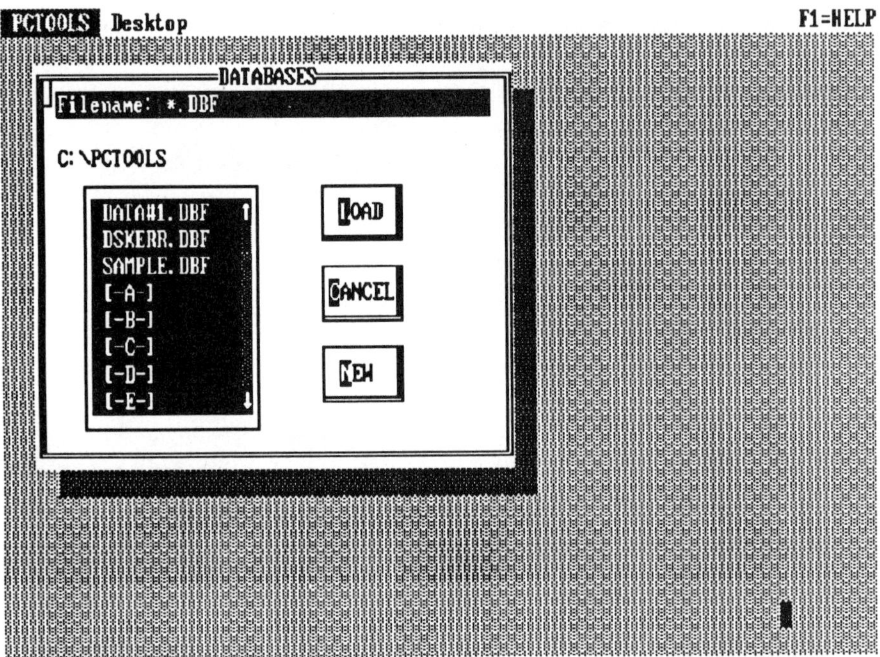

4-2 Database menu

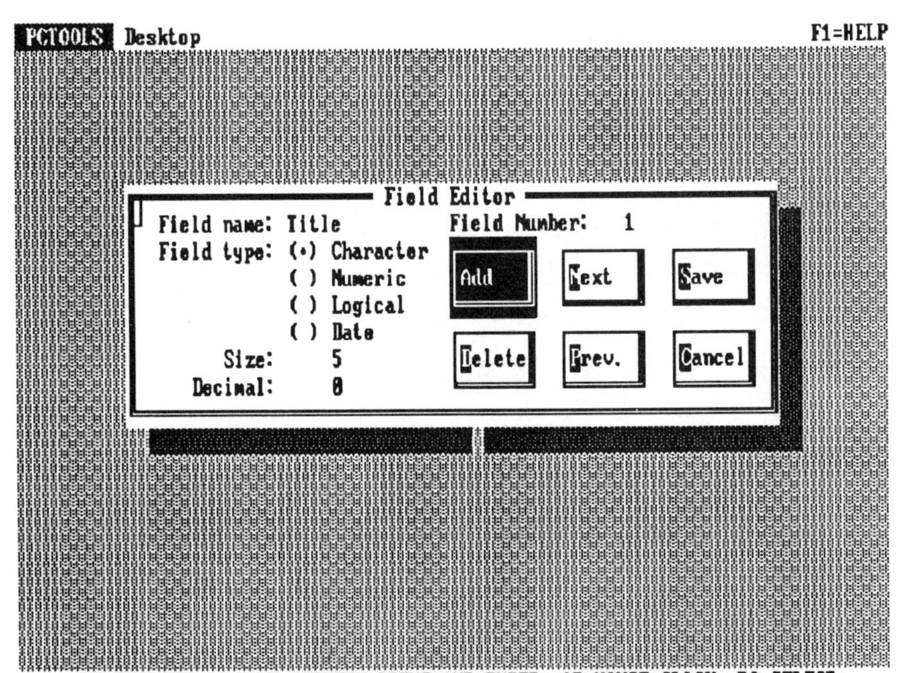

4-3 Field editor

Besides character, PC Tools has three other types of fields: numeric, logical, and date. Each one is stored in the memory of the computer differently, and they all take up less space than text fields. Setting up the proper field types helps PC Tools in its sorting process.

Numeric fields contain numbers that can be added, subtracted, multiplied, or divided. Salaries ($55,000), sales prices (5.75), and baseball batting averages (0.310) are all examples of numeric fields. PC Tools is a limited functioning database because no arithmetic operations can be performed on any field. A numeric field, which can be up to 19 characters long, can also contain a plus sign, minus sign, and a decimal point.

Logical fields are useful to store information that requires only a yes or no answer. For example, "Are you a member of a union?" would be a logical field. Either a member is a member of the union or he is not. Acceptable answer formats are Y or T for yes or true, and N or F for no or false.

Date fields simply contain the date in the form MM/DD/YY.

Size, decimal, and field numbers

The size of the field is determined by how many columns the largest piece of data takes. For example, if the longest town name is Massapequa Park, you would want the size of the City field to be at least 15 columns long. It is always a good idea to add a few extra columns to the size to avoid making a mistake. A character text field can be up to 70 columns in length.

The maximum size of a numeric field is 19 columns, including the decimal point. For example, the number 123456.78 takes up nine columns and has a size of 9 with 2 decimal places. No dollar symbols ($) or commas (,) are permitted in a numeric field.

A logical field has a size of one and a decimal value of zero. This is because a logical field can only have a value of T or F.

A date field has a size of eight with zero decimals because it has the form MM/DD/YY, which is eight columns long.

It is important that you design a database with the proper field sizes and decimals. Once a field is created, you can change only its name. The field type, size, and decimal place cannot be changed.

The field number is the number of the current field. In Fig. 4-3, the Title field is the first field to be entered. You have the following options once the field is created: You can add the field to the database, delete it, save it, go on to the next field, go to the previous field (impossible, in this case, since this is the only field that had been created), or cancel the database creation.

The field names should be straightforward, so that data can be entered with no confusion. The title (Mr., Mrs., Ms., Dr.) of the person is entered as a separate field, so you can use it in the creation of mailing labels and/or form letters. Continue adding the fields to the database, making the size and decimals large enough to accommodate your sample data. When you are finished entering all the fields, select the Save option. Once the database definition is saved, you can add data to it.

Adding data to a record or form

To add data to a form in order to create records, press the F8 key (Add Record) or select Add new record from the Edit menu, as shown in Fig. 4-4. The Edit menu lets you add, delete, undelete, hide, pack, edit, sort, and search a database. Figure 4-5 shows data entered into a record without the Edit menu. Notice that in the order of fields, the Phone field seems out of place, sandwiched in between Last_Name and Address. PC Tools interprets the first three consecutive numbers in a database as a phone number. In Fig. 4-5, if you positioned the Phone field later down in the list of fields, PC Tools would assume that the street number, 302, is a phone number and dial it using the autodialer. Therefore, all fields containing phone numbers should be placed as high as possible in the list of fields.

Add about ten records to a form so that you can investigate some of the options in the Edit menu.

Printing forms

The data in Fig. 4-5 is the default form of a database. A database can have many forms. A form can contain all or several fields of a database in any order. Later on in this chapter, you will create several forms for mailing labels and form letters, but before you do that, print out the form. The Print Selection dialog box is contained in the File menu, as shown in Fig. 4-6. If you create several forms for a database, you can choose the form by selecting Load form. A list of the forms you created is shown in a window (see Fig. 4-7). You simply highlight the form to be loaded and select the Load key.

If you select the Print option, Fig. 4-8 will appear. You can print selected records, the current record, or field names. Print current record will print the record on the screen. Print field names will print the field names, along with their types, sizes, and decimals. If you select any of the Print field options, Fig. 4-9 appears. It asks you for the printing device to send the file to. The default device is LPT1, but you can save it to either a disk file or a COM1 or COM2 port for transmission over telephone lines. You are also prompted for the number of copies to print.

PC Tools has the print driver specifications for the Epson FX-80 printer, the IBM Proprinters, all Panasonic printers, and the Hewlett-Packard Laserjets. Other kinds of print drivers will be discussed in the chapter on macros.

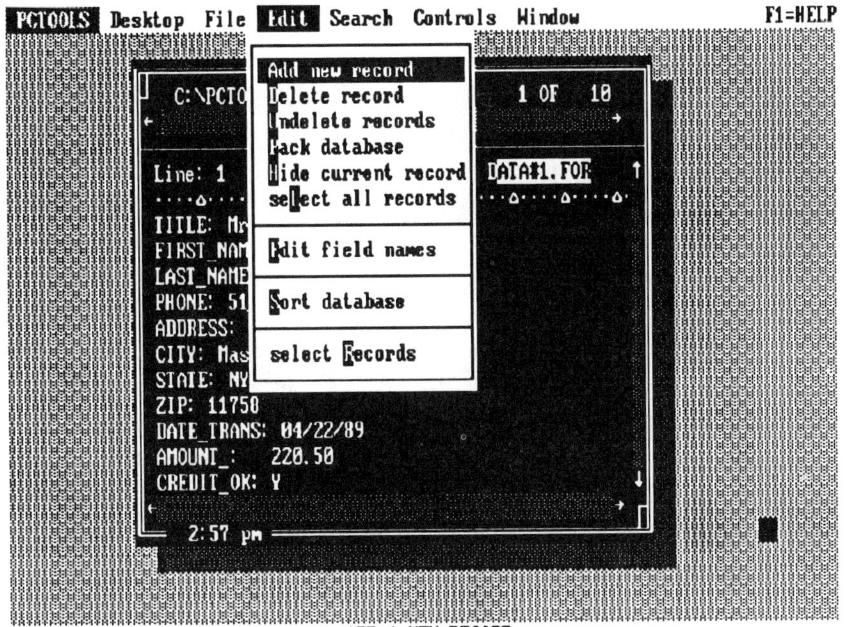

4-4 Database Edit menu

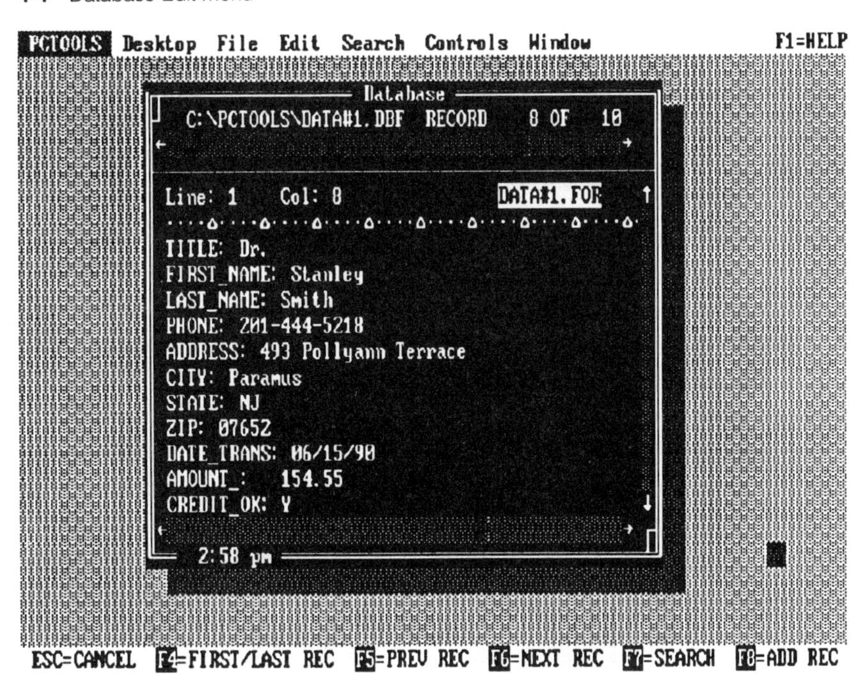

4-5 Data entered into a record

4-6 Database Print menu

4-7 Load Form menu

4-8 Database Print Form menu

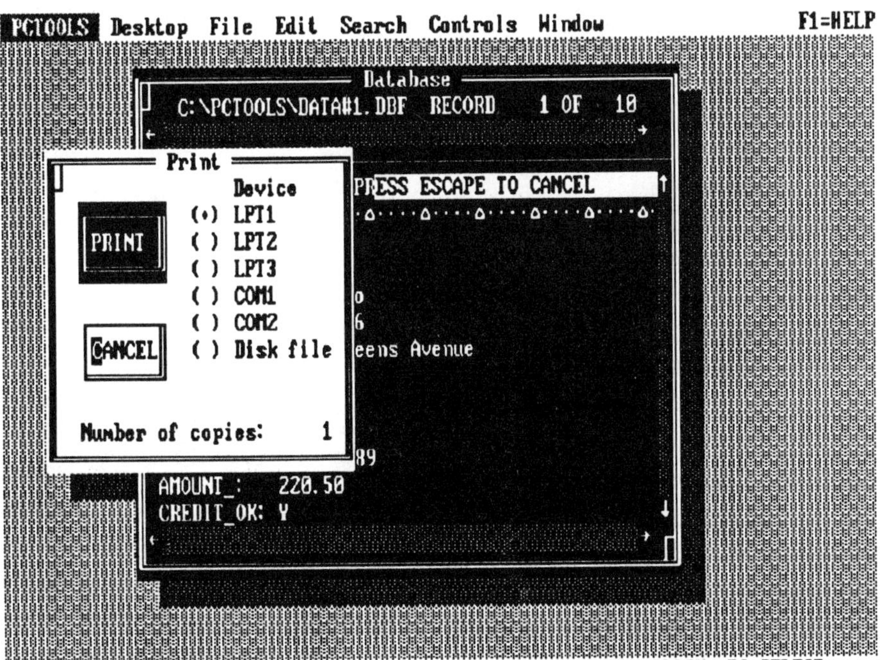

4-9 Database Print menu

The Page layout option of the Controls menu lets you control the left and right margin, top and bottom margin, paper size, line spacing, and starting page number. These options are exactly the same as the Control menu notepad, discussed in Chapter 2. Figure 4-10 shows the printout when the Print field names option is selected. It is a good idea to print out this database structure because it shows the type, size, and decimal.

FIELD NAME	TYPE	SIZE	DEC.
TITLE	C	5	0
FIRST_NAME	C	10	0
LAST_NAME	C	15	0
PHONE	C	13	0
ADDRESS	C	35	0
CITY	C	20	0
STATE	C	2	0
ZIP	C	10	0
DATE_TRANS	D	8	0
AMOUNT_	N	8	2
CREDIT_OK	L	1	0

4-10 Print field names option printout

Editing field names

The Edit field names option of the Edit menu lets you change the name of any field, as shown in Fig. 4-11. This option lets you scroll to the field name you want to change and change it. In this manner, you could change the field name City to Town or the field name Last_Name to Surname. It is important to realize that you can change only the name of a field and *not* its type, size, or decimal. The whole database would have to be recreated to accomplish this. No easy task! As mentioned before, planning the database is of the utmost importance. Notice the new option Modify, in Fig. 4-11. In this mode, it will modify the field name only.

Sorting records

Records do not need to be entered into a database in any special order. When reports are printed, however, they are normally sorted according to certain criteria. For example, with bulk mailings, the mailing labels should be sorted by ZIP code. Once a database is sorted, all future commands, such as printing, are performed on the sorted database.

Figure 4-12 shows the result of selecting the Sort database option from the Edit menu. The Sort Field Select window appears. It lets you scroll through the fields to select the field to sort with. In this case, you are sorting with the Last_name field. Once the Sort option is selected, the records are presented in the new sort order.

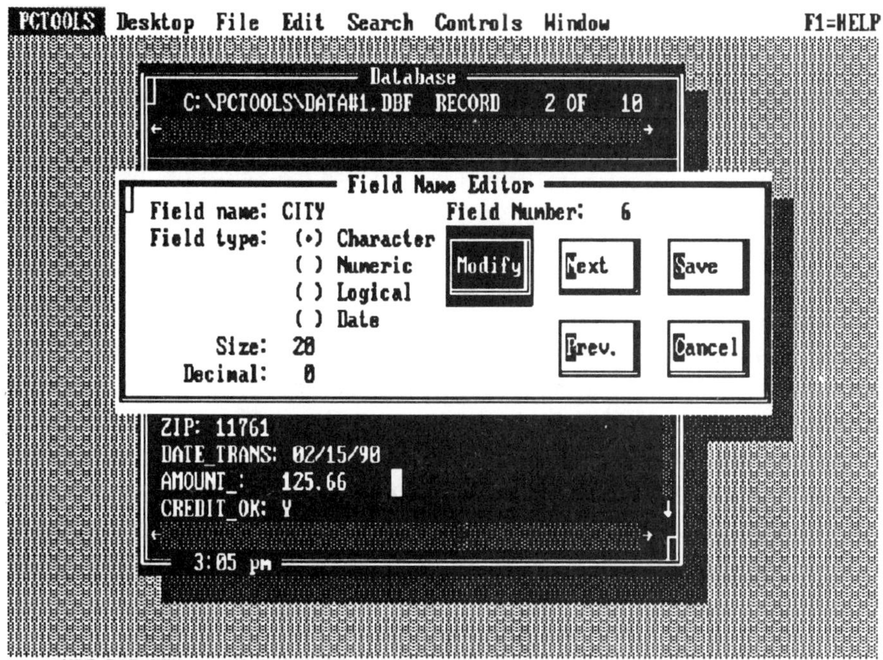

4-11 Field name editor

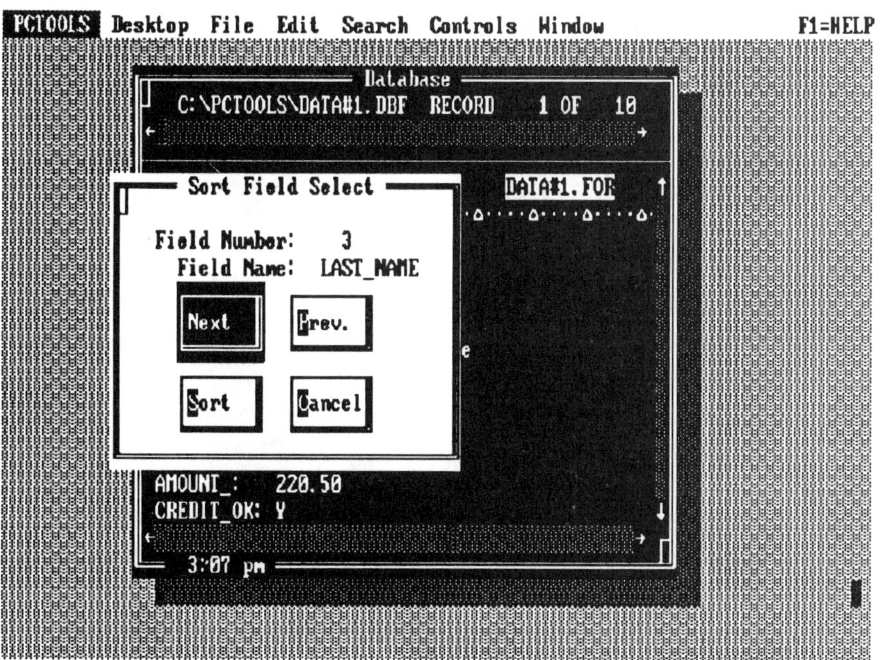

4-12 Sort database option

Selecting records

After you enter the data, you can then search it to find particular records by selecting the Select records option in the Edit menu, and being presented with the screen in Fig. 4-13. This command lets you find all people who live in the state of New Jersey from your database. When you choose the Select option, it will display them one at a time, if any satisfy that criteria.

You can search for any one field or any combination of up to eight fields. Figure 4-14 shows specifications for another search. The search specification W???Y in the Last_name field represents a wildcard search. Each question mark is a wildcard character. This particular search would retrieve Waszy, Worry, Wacky, Willy, or any other words in the file that have four letters, start with W, and end with a Y. The results of this search are shown in Fig. 4-15.

You can also search for any combination of fields, for example, all people who live in Aquebogue, New York and whose credit is OK. You can specify ranges for fields like L..Z. Ranges can be typed in either upper- or lowercase. Assuming you type it in the Last_name field, you would find any people whose last names start with the letter L through the letter Z. If you type 100..500 in the Amount field, you would find all amounts greater than or equal to 100 and less than or equal to 500. If you type ..500 in the Amount field, you would find all amounts less than or equal to 500.

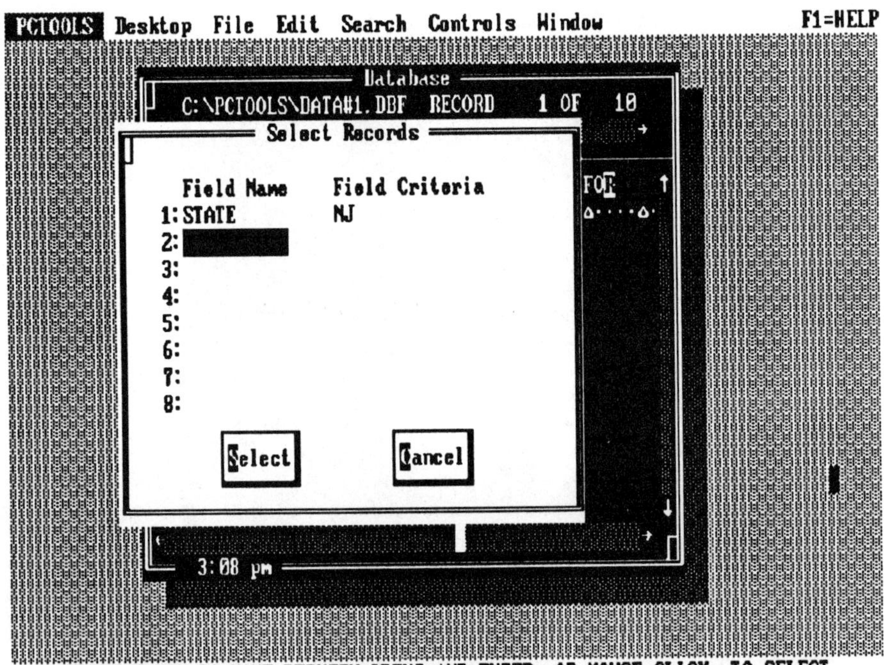

4-13 Select records option

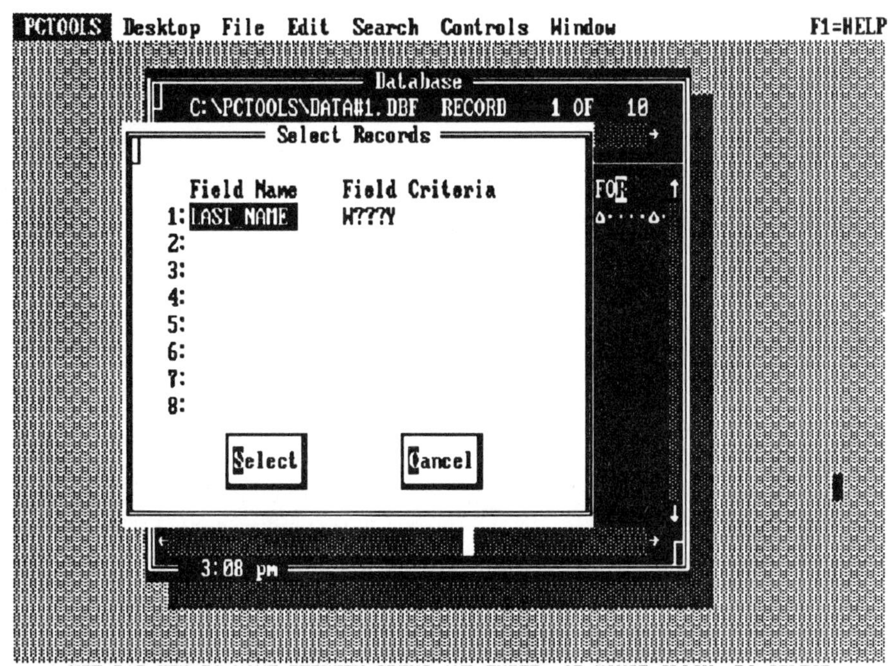

4-14 A wildcard search

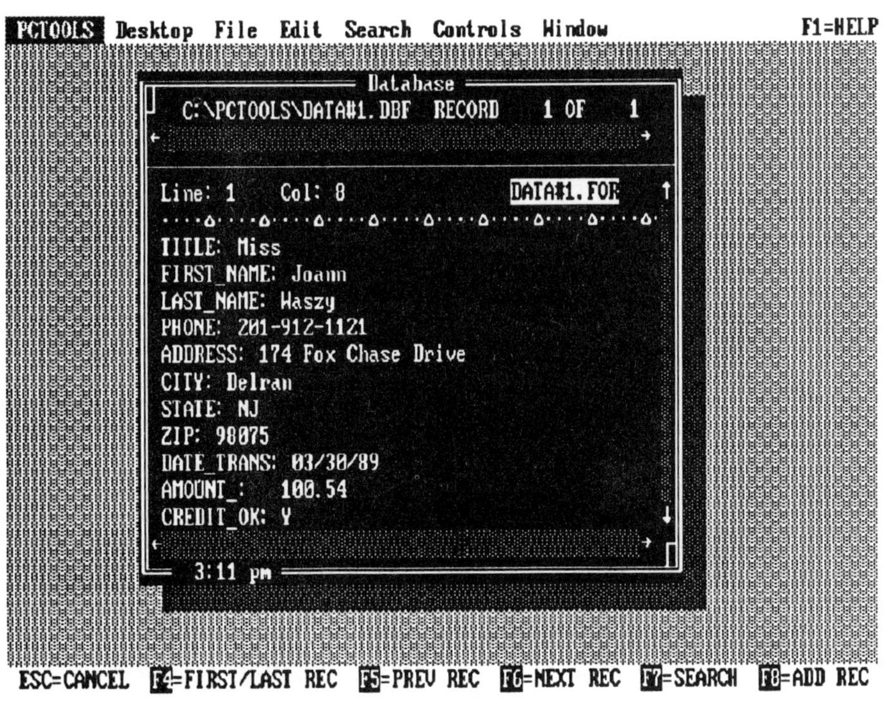

4-15 Result of a wildcard search

Searching for information

The Select records option is useful for finding information in certain fields. The Search command can find data in any field of the database. The Search menu gives you three options. The first, Find text in all fields, generates a Search All Fields window, shown in Fig. 4-16. This command will search for any string of characters in all fields of the database. The Search all records command searches for the string of characters, DIANE, in all of the fields of the database, whether it be a name, or part of an address or city.

The Search selected records command searches for DIANE in previously selected records. Suppose, for example, you have selected all the people living in Aquebogue, New York. You can now look for the word DIANE in those selected records.

The Search from current record command searches the database from the record the cursor is positioned on to the end of the database. Select the Search command to execute the search, or the Cancel command to cancel it.

The Find text in sort field command is the second option in the Search menu. It is very similar to the Find text in all fields command. Selecting this command gives you the Search Sort Field window, shown in Fig. 4-17. This search is much faster than searching unsorted fields. It is very useful if you sorted by Last_name and wanted to find the person's address, phone, transaction, and amount. The three options are the same as those for the Find text in all fields option.

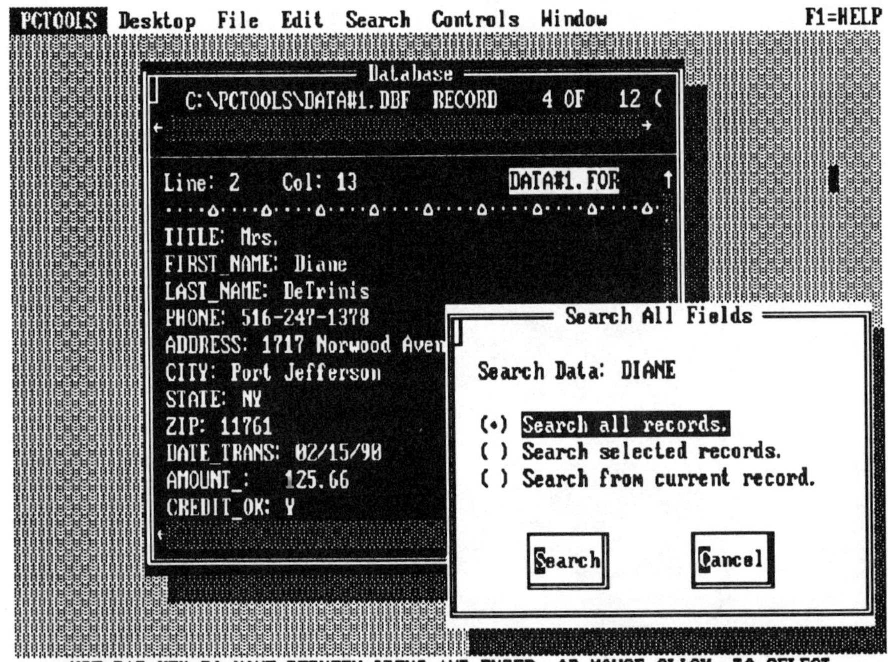

4-16 Search all Fields dialog box

4-17 Search Sort Field dialog box

The Goto record command is the third option in the Search menu. It lets you go to any record number, as shown in Fig. 4-18. This is a very useful command if the database is sorted and you know that the record you want is near the end of the database.

Deleting, undeleting, hiding, and packing records

The Edit menu contains other useful commands to manipulate records. The Delete records command lets you mark a record so that it will not be displayed on the screen. It is not, however, physically removed from the disk. You can view it again by selecting the Undelete records command. These commands are very useful because, if you accidentally mark the wrong record for deletion, you can undelete it to bring it back. The undelete command brings back *all* deleted records and makes them visible on the screen.

You must have at least one viewable record in a database. If you try deleting or hiding this last record, you will see a message saying there are no viewable records. It will ask you to add a record, select all records, or undelete records. The Select all records command lets you view all hidden or deleted records.

When business accounts are normally closed, associated records are deleted from the database, but not physically removed until all records are printed. This will help leave an audit trail so errors can easily be corrected.

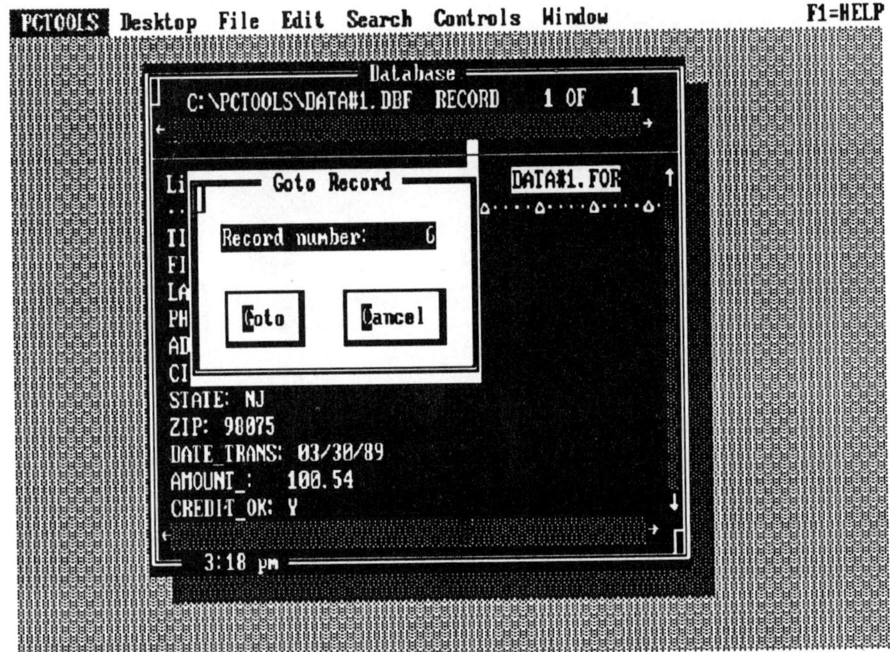

4-18 Goto record command

When the time comes to physically delete records from the database, you can use the Pack database command from the Edit menu. This permanently removes records from the disk. The maximum number of records in a database is 3,500. Packing a database will consolidate space by physically removing all deleted records.

Hiding records and selecting all records

The Hide current record option will hide a record from being either viewed on the screen or printed. You might want to use this option to prevent people from looking at certain records. The Select all records command will display all previously hidden records. It acts as a toggle with the Hide current record option. The Pack command does not delete hidden records from the database.

Mailing labels

You now have a file of the names and addresses of all your customers, along with associated information. If you have an upcoming sale, you can generate mailing labels and form letters. The PC Tools Desktop database utility can generate a file in any form you want, which can be used to display and print the database. The default form was shown in Fig. 4-1, and has the same filename as the database but with the extension .FOR.

To create mailing labels, create a word processing document in the notepad utility that shows the form of the mailing labels, as shown in Fig. 4-19. The field names can be placed anywhere on the screen, but they should be in the order you want them printed. The brackets ([ ]) around the field names indicate that the information is obtained from another file. This data file must be opened as one of the 15 windows, so that data can be extracted from it.

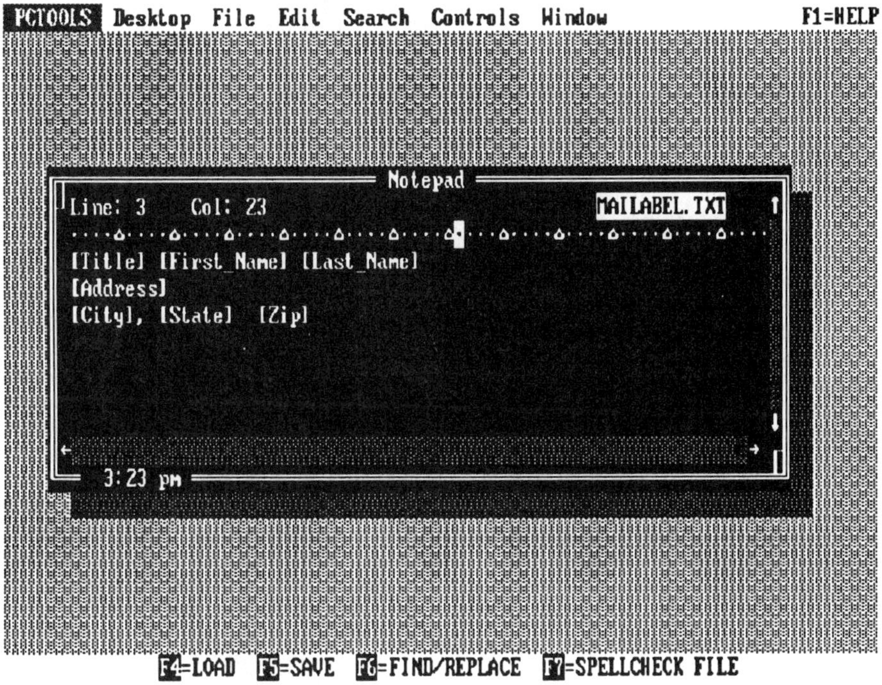

```
PCTOOLS Desktop File Edit Search Controls Window                    F1=HELP

                              Notepad
 Line: 3    Col: 23                                      MAILABEL.TXT
 [Title] [First_Name] [Last_Name]
 [Address]
 [City], [State]  [Zip]

   3:23 pm

        F4=LOAD  F5=SAVE  F6=FIND/REPLACE  F7=SPELLCHECK FILE
```

4-19 The mailing label form

Before you can print the labels, you must change the Page layout option of the Controls menu, shown in Fig. 4-20. The Controls menu of the database is the same as the Control menu of the notepad. Change the paper size to six. You might have to adjust the paper size, depending on the type of labels you are using. Select OK when finished. Figure 4-21 shows how one label appears on disk after the list is generated. The data is extracted from the database and inserted in the proper fields for the mailing labels.

Form letters

Just as you created mailing labels, you can also create a form letter that inserts the name, address, city, state, and ZIP code of individual records into a letter, as shown in Fig. 4-22. The form letter is created in the notepads utility. You can also include any other fields in the body of the letter.

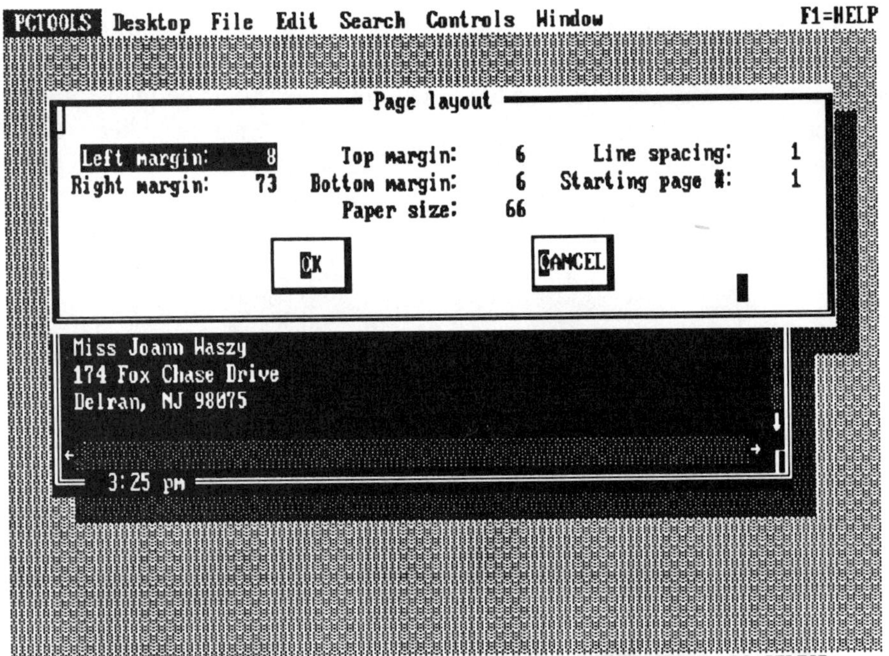

4-20 Page layout for mailing labels

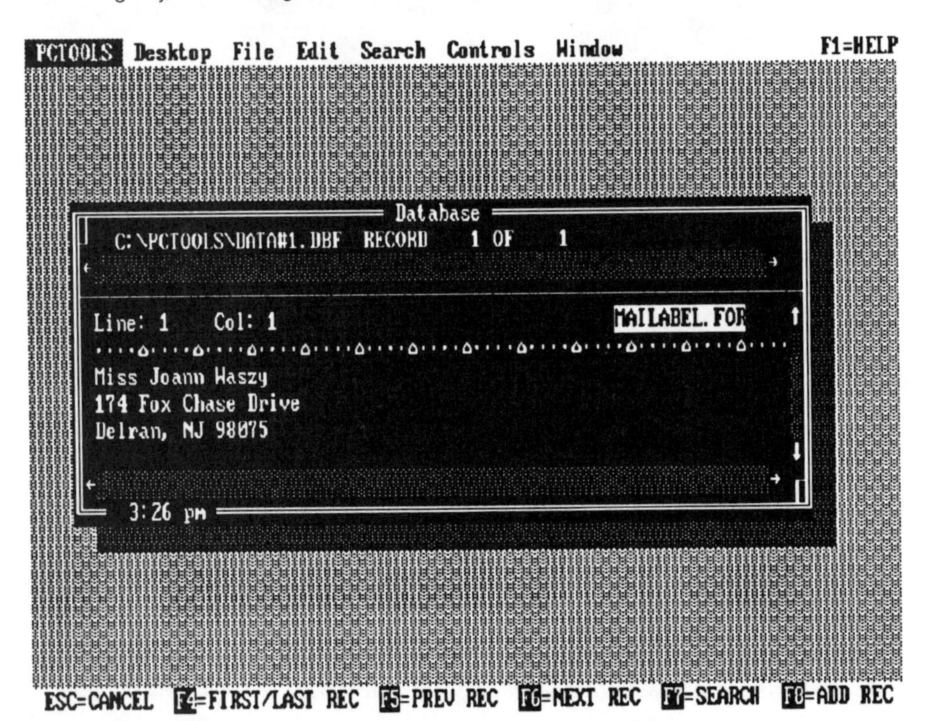

4-21 The first mailing label

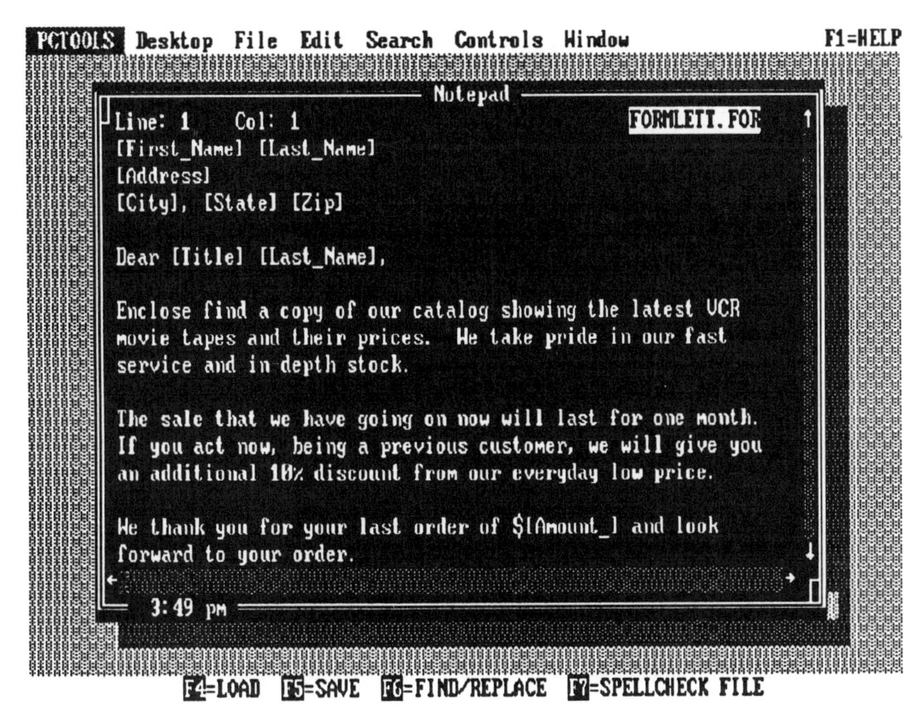

4-22 A form letter

The text enclosed in brackets are the fields to be transferred from the database into the letter. Fields can be repeated twice in the letter. The Title field gives a professional touch to the letter. The Amount of Transaction field is also embedded in the body of the letter.

The first form letter is shown in Fig. 4-23. All the fields have been integrated into the letter, and all the spacing adjusted. The form letter automatically adjusts the spacing to allow for long or short names and addresses.

Autodialer

Having the phone number as a field in a database is very useful, because the autodialer located in the Control menu can dial the phone number from a record. As mentioned before, the autodialer will dial the number of the first field that has three consecutive numbers in it. Therefore, the phone number field should be placed before any other numbered fields. For example, if a field containing social security numbers were placed before the phone number field, the autodialer would try dialing the social security number.

The phone number can contain many characters, including parentheses, dashes, spaces, and x's for extension numbers. You can insert the @ symbol or the letter W before the phone number to ensure that the computer waits for a dial tone before proceeding.

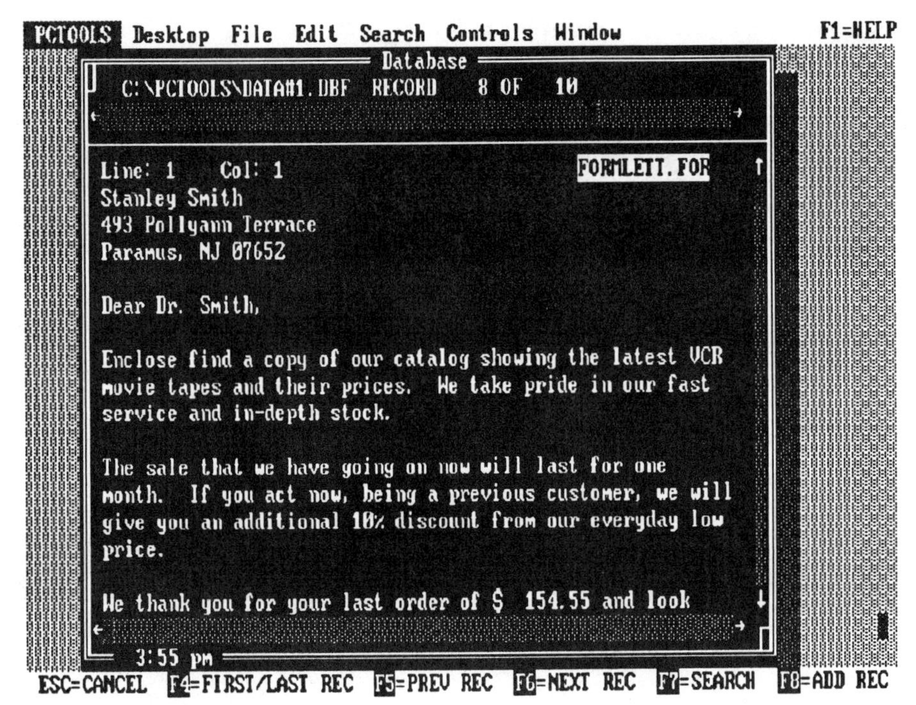

4-23 A form letter with data

Your phone setup must be configured before you use the autodialer, as shown in Fig. 4-24. The Configure autodialer command is located in the Controls menu. You must select whether you have a touch-tone or rotary (pulse) dial, what communications port your modem is connected to, and the correct baud rate (the speed of computer transmission). The baud rate is the speed of a modem, the maximum rate at which it can transmit or receive data. This rate is measured in bits per second (bps), or baud. In most cases the two terms are synonymous. A 300-baud rate is considered low speed for a data transfer rate. Medium speed is from 1200 to 9600 bps. When you are finished with the settings, select OK.

Once you choose the settings for your modem, you are ready to dial. The modem must be a Hayes-compatible modem. To have the computer dial the number, select Autodial from the Controls menu. PC Tools will look for the phone number in the first field of the record with three consecutive numbers and dial it. A message will be displayed on the screen telling you what to do. The modem can be disconnected only after it has rung the number by pressing the Esc or Enter key, or selecting Disconnect modem from the dialog box. If you try cancelling before the connection is made, the system will freeze and you will have to reboot the system to start again.

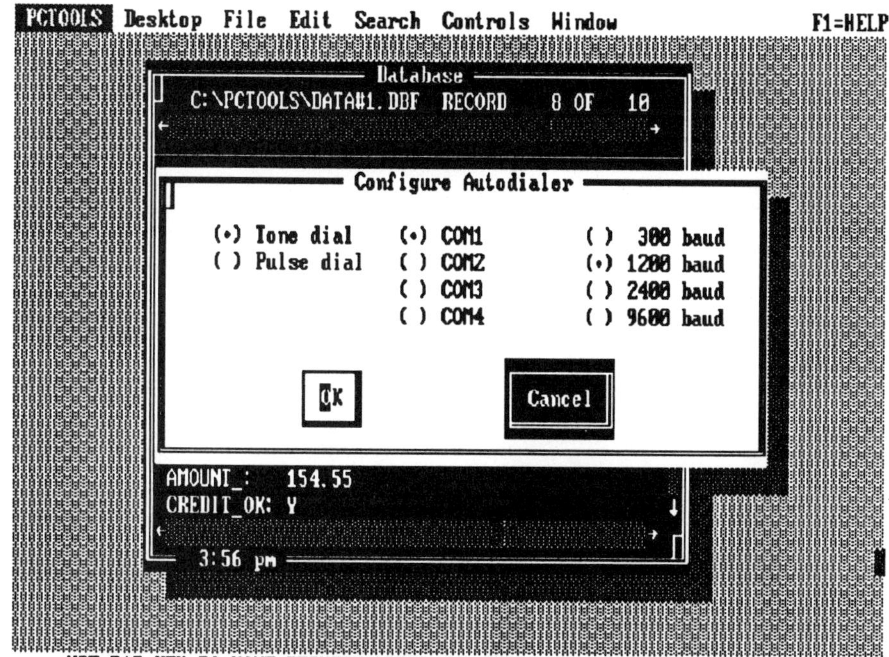

4-24 Configure autodialer dialog box

Importing from dBASE

The databases of PC Tools Desktop are compatible with dBASE files. Several limitations, however, do apply. If you try to load or import a dBASE file with more than 3,500 records, only the first 3,500 records are loaded into the PC Tools database. PC Tools will ignore all dBASE memo fields because they do not exist in the PC Tools database. PC Tools databases allow 70 columns per character field, while dBASE allows 254. PC Tools, therefore, can import only the first 70 characters of any text or character field from dBASE.

5
CHAPTER

Appointment scheduler

The appointment scheduler consists of three parts: a monthly calendar that lets you check dates, days of week, and future or past months; a to-do list that lets you enter reminders to yourself; and a daily time scheduler that lets you schedule future appointments. Timers can be included in the appointment scheduler, so you can have a reminder pop up on the screen to remind you of an imminent project or appointment.

When you select Appointment scheduler from the Desktop menu, the program asks you to create a new file or load an old file. If you want to create a new file, type in a name and select New. Figure 5-1 appears. As you can see, it contains the three parts mentioned above. The names of all files in the appointment scheduler have the extension .TM. Each person using the computer can set up their own file or set of schedules.

The horizontal bar at the top of the screen is the menu bar, and contains all the pull-down menu selections. Notice the similarities and differences between the other desktop utilities. The Appointment menu and To-do menu are entirely different, and are specific only to the appointment scheduler. The other pull-down menus (Desktop, File, Controls, and Windows) are the same as the other utilities.

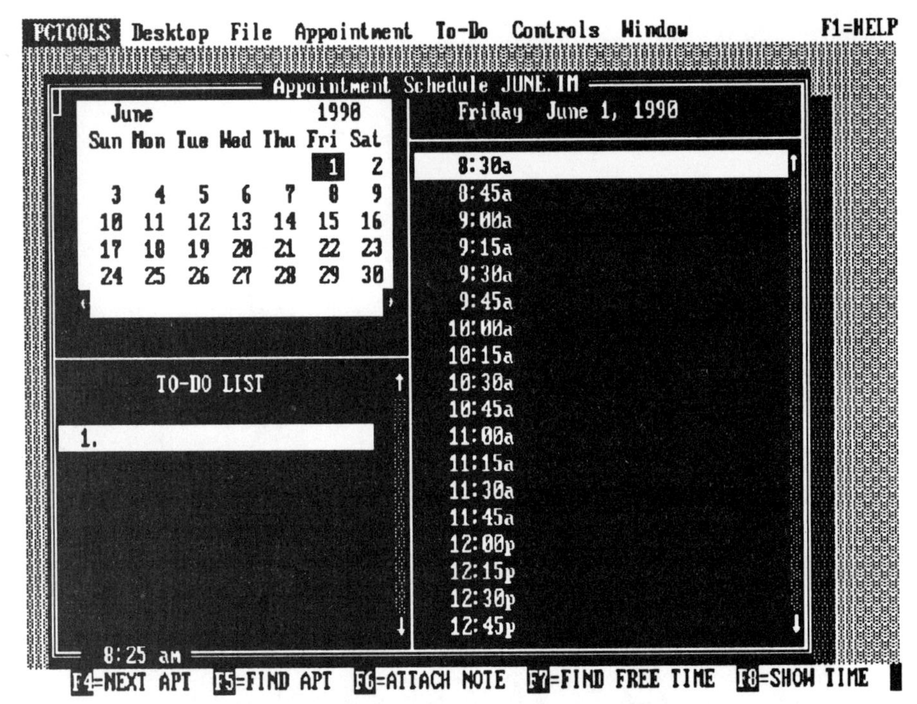

5-1 Appointment scheduler

The Help key, F1, shown in Fig. 5-2, gives the same context-sensitive help as the other applications. It also gives a help index so you can browse through other information.

A mouse can be used to move or close the windows. A mouse can scroll through the months, forward and backward, and to scroll through the daily schedule and to-do list. You can move between the three scheduler sections with the Tab key, and change the perpetual monthly calendar by pressing the PgUp key to advance one month and the PgDn key to precede one month. The Home key will always give you the present month and day, assuming you have a clock in your system. The present calendar day is always highlighted on the monthly calendar. Years can be changed by pressing the Ctrl and PgUp or PgDn keys.

The daily scheduler gives you the times of the day to schedule appointments. The beginning and ending times can be changed to suit your work day, with the Appointment settings command of the Controls menu. You can use the up and down arrows to scroll through the daily scheduler.

The to-do list lets you enter projects or things to do. You can enter up to eight items per screen, and up to 80 items in the to-do list. You can also scroll through the to-do list with the up and down arrow keys.

The message bar on the bottom of the screen displays some of the keys you can use to display next appointment, find appointments, attach notes, find free

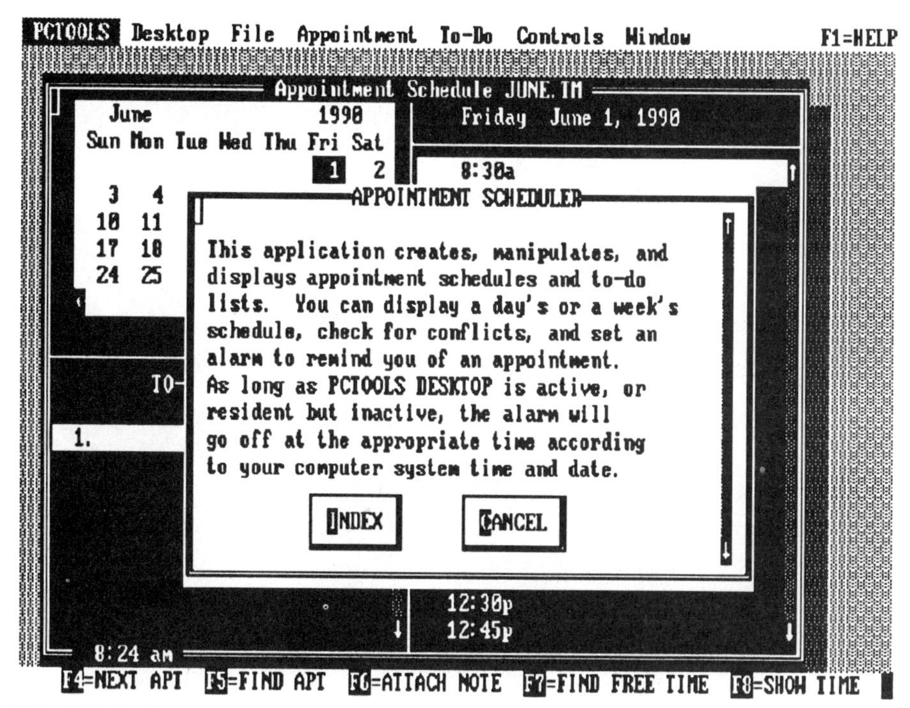

5-2 Help for the appointment scheduler

time, and show time. These functions are also in the menus, but are here to save time. The time is always displayed above the message bar.

Appointment menu

You can make appointments by accessing the Appointment menu, as shown in Fig. 5-3, to make, edit, and delete appointments. When you select the Make new appointment command, Fig. 5-4 appears. The Make Appointment screen lets you choose your starting and ending date. The starting date is useful only if you want to have the same appointment every day for several days in a row, a week, a month, etc. An appointment can be selected for one day, every day, only work days, weekly, monthly on a fixed day, or monthly on a fixed weekday. Monthly on a fixed day is the best choice if you want to schedule a meeting on a certain day of the month. Monthly on a fixed weekday is best if you want to schedule an appointment on, for example, every Monday of the month.

The Make Appointment screen lets you schedule both the time of an appointment and a note of up to 24 characters, which describes the appointment. The note you enter will be displayed on the daily schedule. The type of appointment, a single character field, can also be entered. You can create your own code for each type of appointment, like L for lunch, C for conference, and B for board meeting.

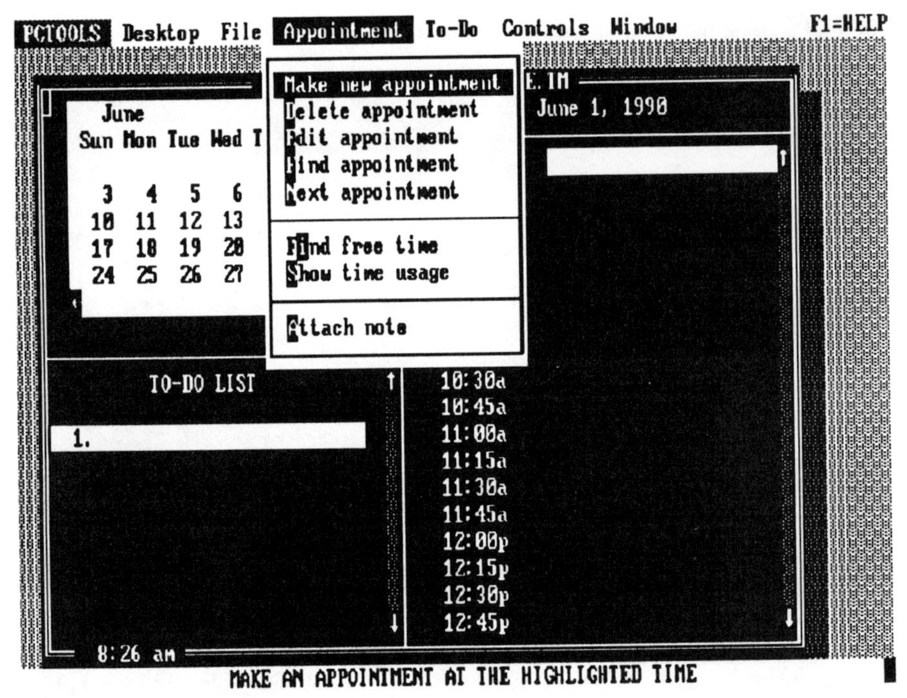

5-3 Appointment menu

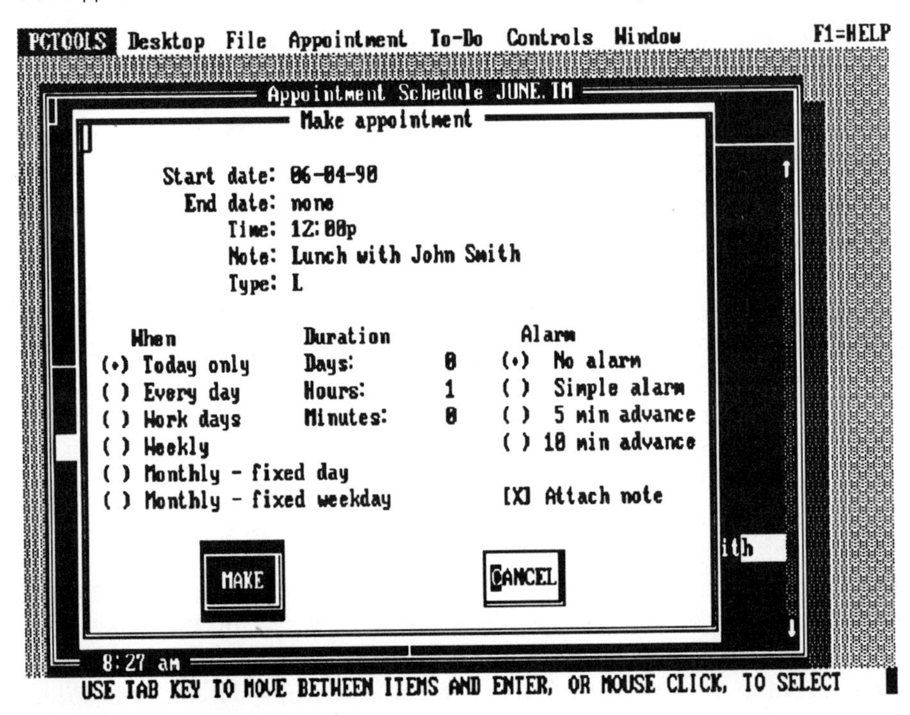

5-4 Make appointment command

The Duration facility lets you choose how long the appointment will be. You can type in the number of days, hours, or minutes you want the appointment to last. It is best to round out the minutes to the nearest 15 or 30.

The alarm is an interesting feature of this utility. You can arrange it so that, when you are working on another program, an alarm will go off reminding you of an approaching appointment. This alarm is composed of a beep and a dialog box that appears on the screen reminding you of an appointment. Since the appointment scheduler is resident in the machine, it compares the times you set in the Make Appointment selection to the system clock. When they match, an alarm will be sounded if you select that option.

The alarm has four choices. You can select No alarm, which specifies you will never be interrupted by the resident program. The Simple alarm means you will be reminded of an appointment at the exact time of the appointment. The 5 and 10 minute advance will sound the alarm early to give you time to prepare for the appointment. As mentioned before, the alarm is a beep through your computer's speakers and a dialog box on the screen reminding you of an appointment. Whatever note you type into the Make Appointment screen will appear on the screen. If you select OK when the dialog box appears on the screen, it will be erased from the screen. If you select Snooze from the dialog box, the alarm will clear from the screen, but redisplay in five minutes.

The Attach note command lets you attach a Notepad file to the appointment, as shown in Fig. 5-5. This is very useful to not only schedule appointments, but also attach outlines and other notes you might need at the meeting. To close the note, press the Esc key or use the mouse to click on the box in the upper left-hand corner of the window. Notice that this window has a double line around it, which means it is the active window.

The daily scheduler in the bottom right of the screen of Fig. 5-5 contains the time of the appointment and a statement taken from the note on the appointment. Notice that there is a little musical note to the left of the appointment. This signifies that there is an attached alarm. The letter N on the side of the note signifies that a note is attached to that particular appointment. A double musical note signifies that there is an alarm attached to a recurring appointment.

If you make and save an appointment that occurs at the same time as another appointment, you will receive a warning (see Fig. 5-6), but still be allowed to save the appointment. You will also get a warning if you try to make an appointment prior to today's date. As before, you will be warned but still allowed to save the appointment.

The scheduler can automatically be set so that the day's appointments will automatically appear on the screen. If you install the desktop with DESKTOP/RA, the appointments, if any, will automatically be displayed on the screen. If there are no appointments for that day, the desktop will be displayed.

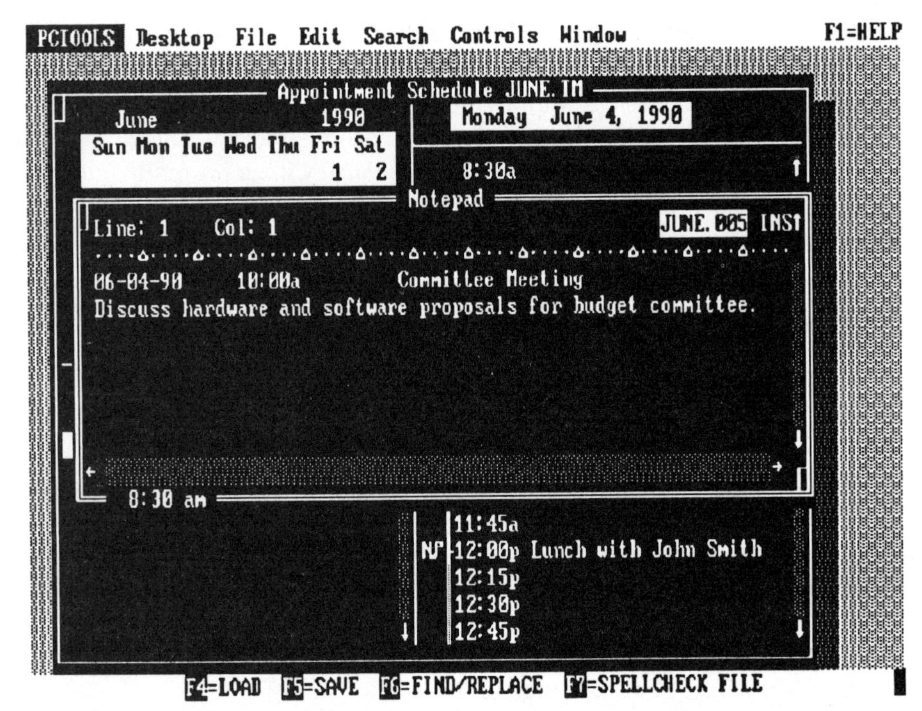

5-5 Attaching a note to an appointment

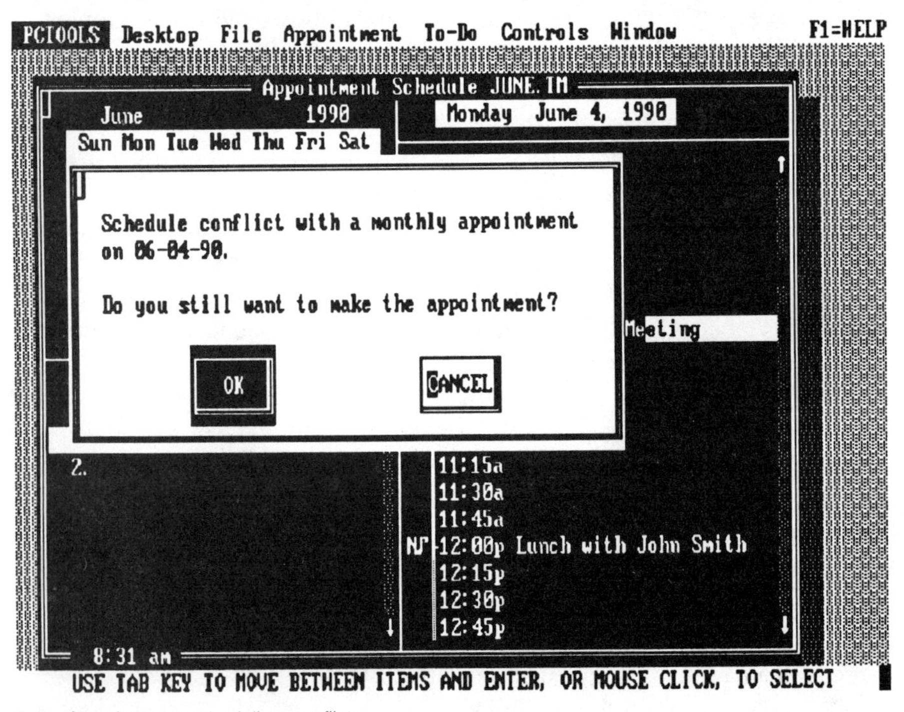

5-6 Appointment scheduling conflicts

Appointments can be deleted by selecting the Delete appointment command from the Appointment menu (Fig. 5-3). You must be positioned on the appointment you want deleted before choosing this option. Select OK from the dialog box to remove the appointment.

Editing an appointment is done the same way. You must be on an appointment to edit it. Select Edit appointment from the Appointment menu, and Fig. 5-7 appears. Select Delete, Edit, Alter Note, or Cancel. The appointment appears if you select Edit, and you can change any information you want.

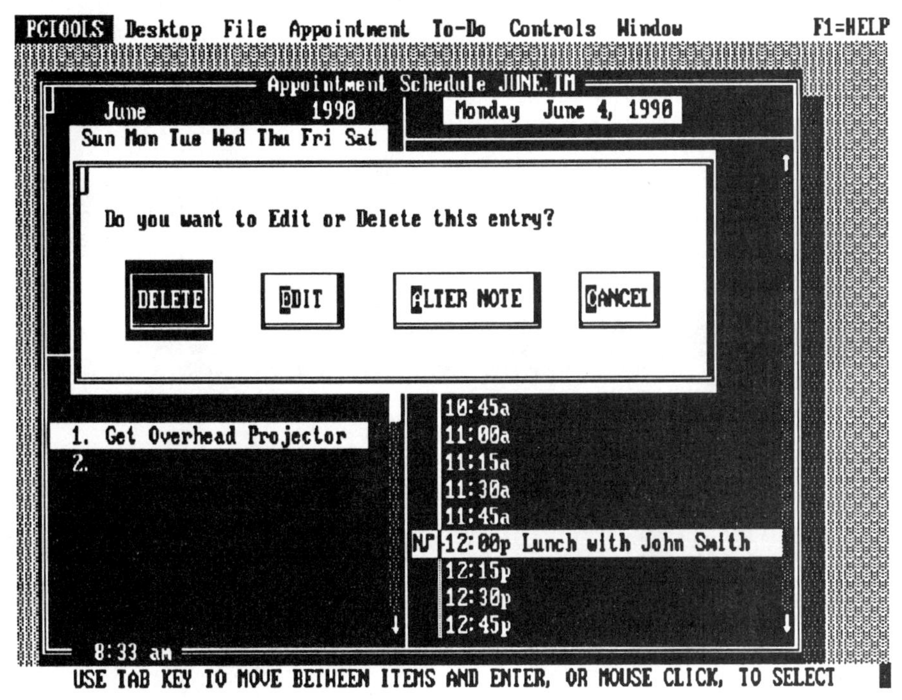

5-7 Editing an appointment

The Find appointment command displays a dialog box, shown in Fig. 5-8, that lets you enter the text of an appointment, as well as the type, date, and time of the appointment you are trying to find. You can enter any or all pieces of information in the various fields you want to search for. It will find all appointments that meet these conditions on any date after today's date. The information can be typed in upper-or lowercase letters because the search is not case-sensitive.

If you don't enter any information in the Find Appointment dialog box, all appointments will be displayed in sequential order. When conducting a search, pressing the F4 key will display the next appointment.

The Find free time command, selected from the Appointment menu and shown in Fig. 5-9, lets you find the next available block of free time that meets the specifications you set. You can request the starting time, stop time, any day or

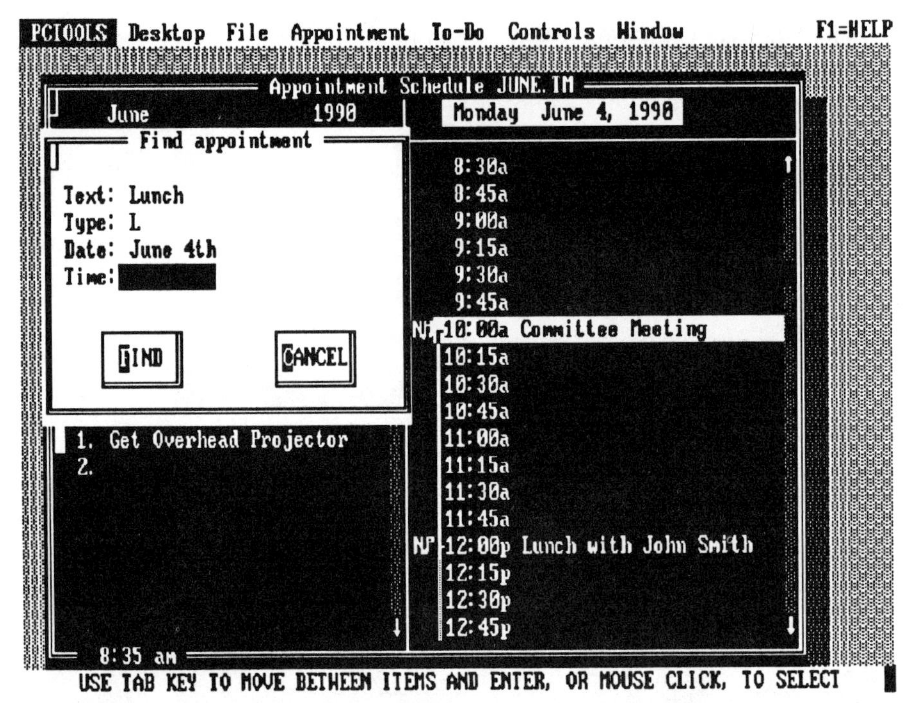

5-8 Finding an appointment

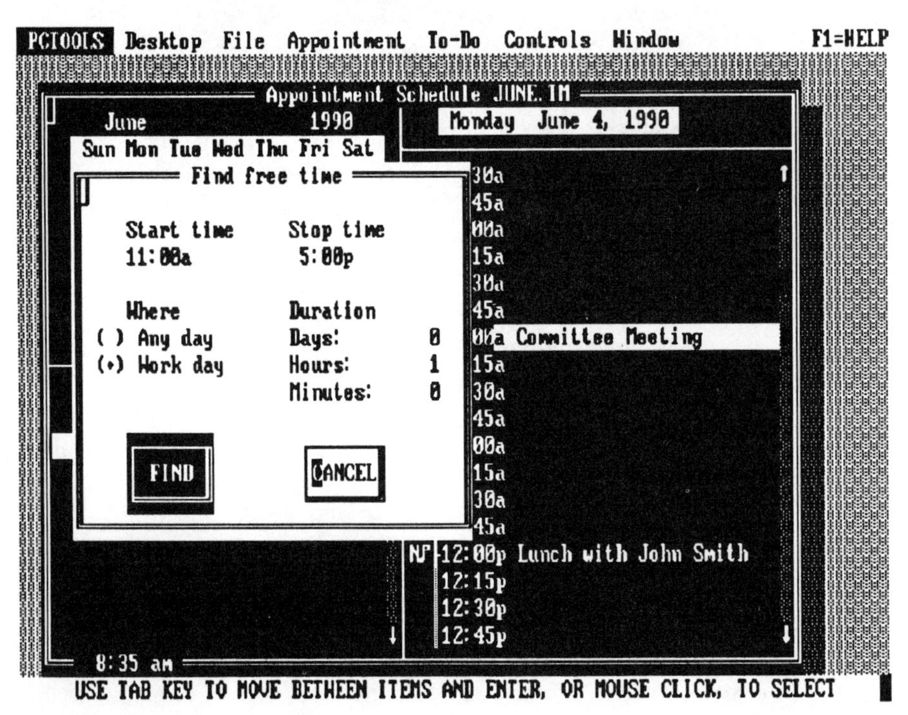

5-9 Finding free time

work day, and time duration. It will search in the future to find the needed time and stop in the daily scheduler when it does.

The Show time usage command shown in Fig. 5-10, and also selected from the Appointment menu, graphically represents the appointments, free time, and any conflicts. Appointments are shown as solid dots, free time as a shaded area, and open dots as possible conflicts. The window shows five days at a time. You can look at different dates by selecting the PgUp key to advance five dates and the PgDn key to select the previous five dates. The Home key will always give you the present date.

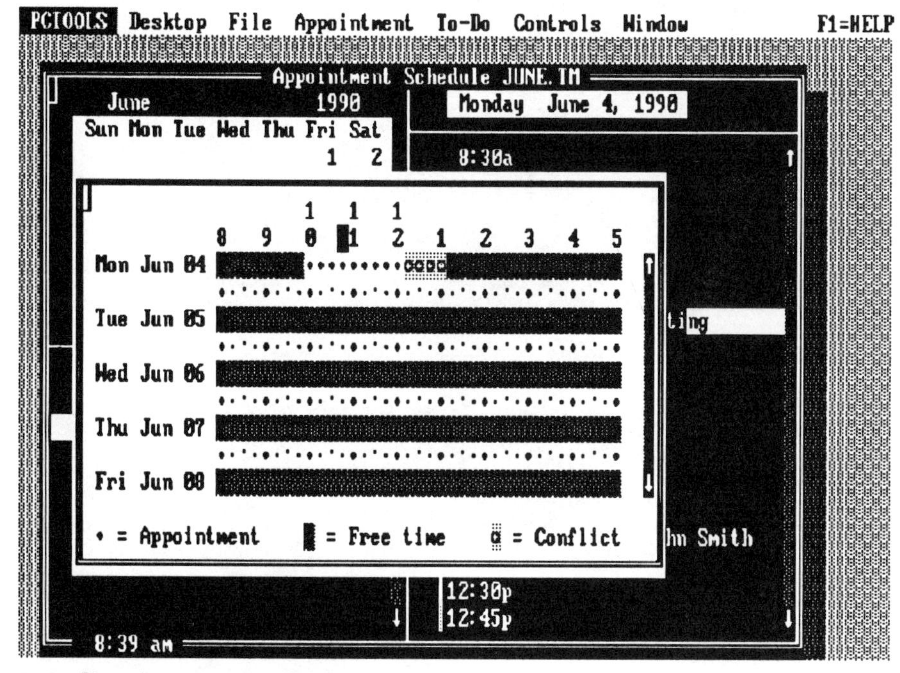

5-10 Show time usage command

The Attach note command of the Appointment menu lets you attach a note to an existing appointment. Select the appointment to which you want the appointment attached, and choose the Attach note command. If you want to edit an existing note, retrieve the appointment containing that note, press the F6 key, and select Alter note.

To-do list

To create a to-do list, select the New to-do entry command shown in the To-Do menu of Fig. 5-11. The box shown in Fig. 5-12 appears, which lets you create an entry for the to-do list. You can enter a note, the starting and ending date, and the

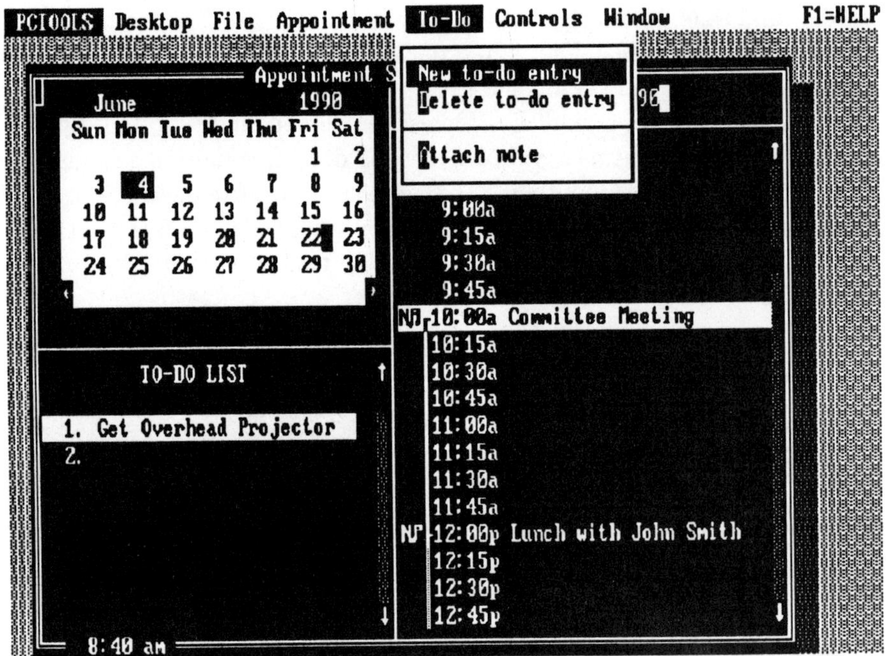

5-11 To-do menu

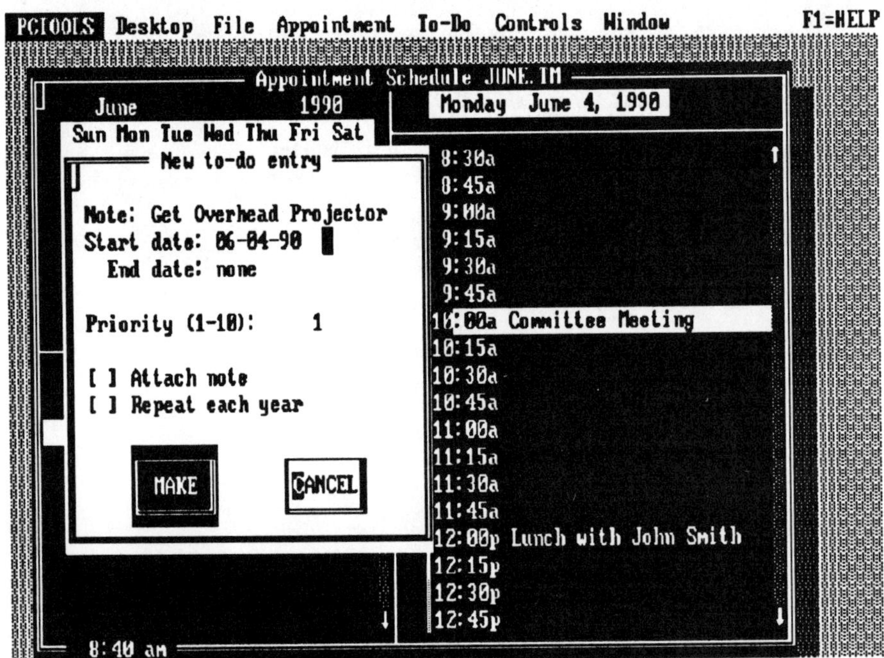

5-12 Creating a To-do entry

priority. The priority is a number from 1 to 10 that lets you select the order in which the notes are displayed. You can also attach a note to each to-do entry and have it repeat for each succeeding year. The To-Do menu of Fig. 5-11 also lets you delete a to-do entry or attach a note to an entry.

Controls menu

The Controls menu, displayed in Fig. 5-13, lets you change appointment settings, set holiday dates, delete old appointments, and change the width of the display.

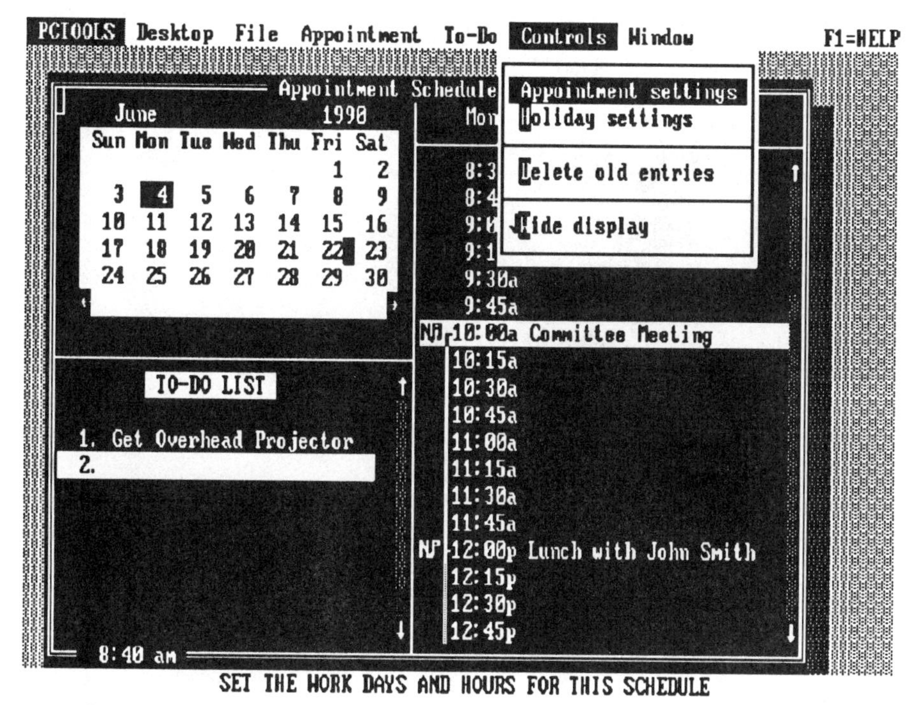

5-13 Controls menu

The Appointment settings selection generates the dialog box shown in Fig. 5-14. The default hours of the work day are from 8:00 a.m. to 5:00 p.m. If your work day is different, change the hours here. You can select the work days, the starting and ending times, and the time increments. The time increments can be either 15 or 30 minutes. You can also change the date format and the time format. When you are finished making changes to the appointment settings, select the OK button to install them, or the Cancel button to exit without changing them.

The Holiday Settings window is displayed in Fig. 5-15. To avoid conflicts between appointments and holidays, you can set the holiday dates and an asterisk

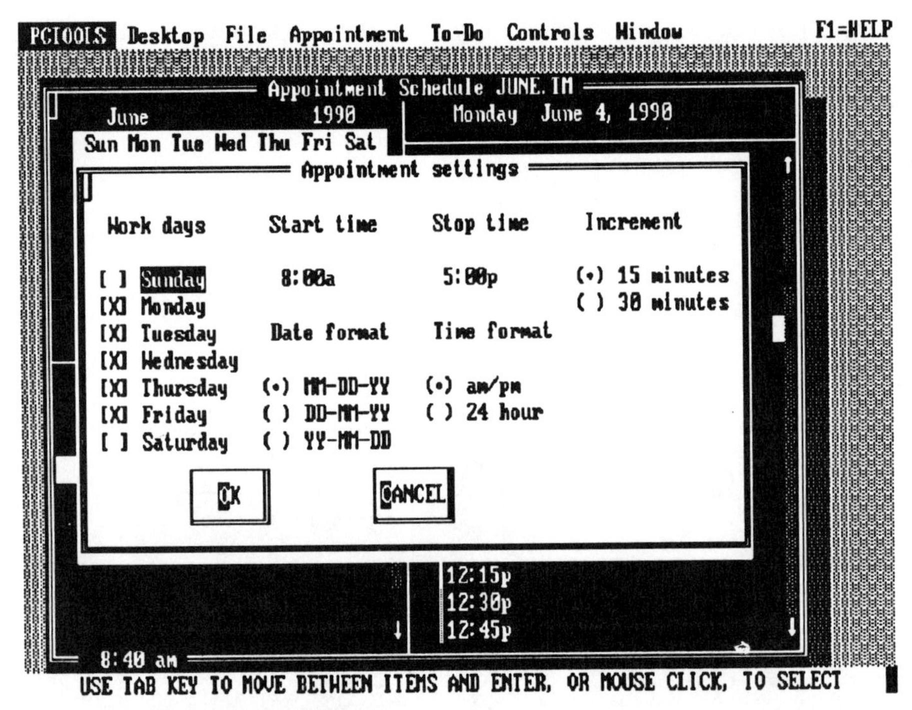

5-14 Appointment Settings menu

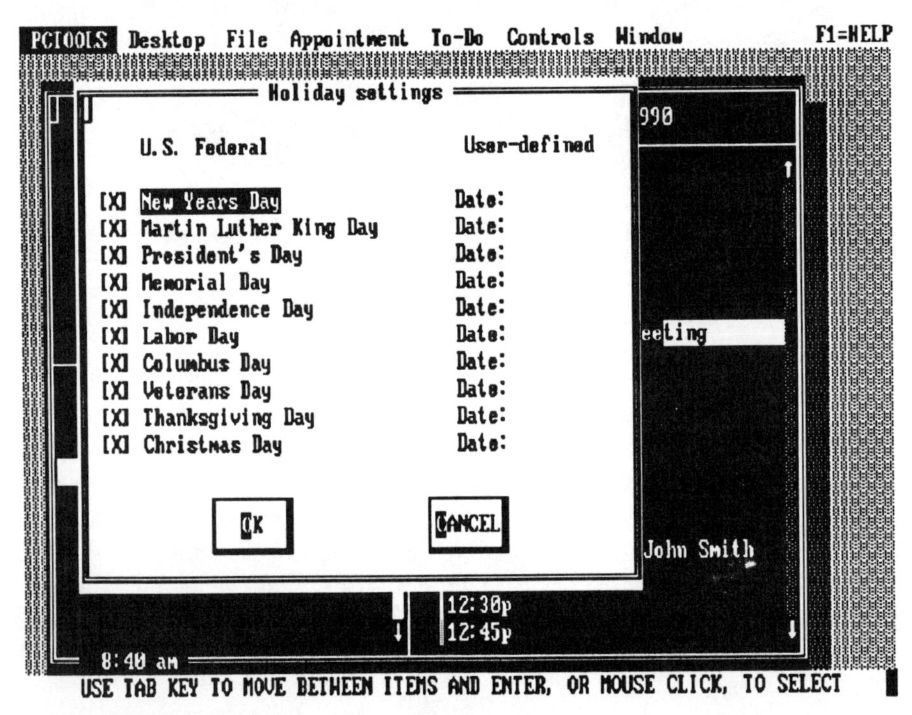

5-15 Holiday Settings window

will appear to the left of the dates in the daily scheduler. The appointment scheduler will not let you schedule an appointment or a recurring appointment on a holiday. Type in the holidays and select the OK button to implement them or the Cancel button to exit without saving them.

The Delete old entries command, shown in Fig. 5-16, lets you remove old appointments from the appointment scheduler. It also lets you select a cutoff date. Any appointments scheduled before that date are removed.

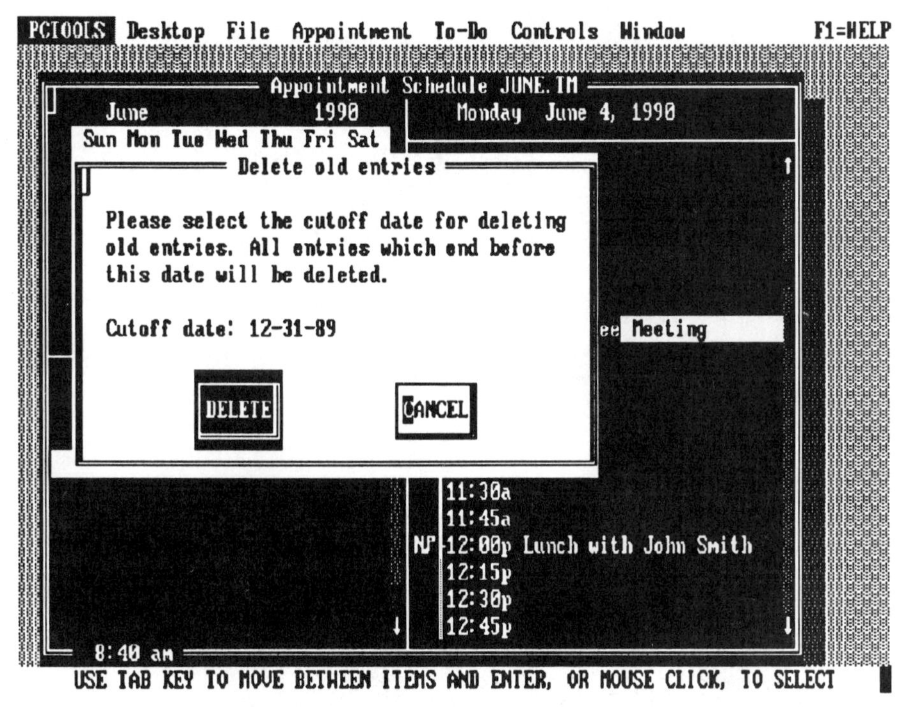

5-16 Delete old entries command

The Wide display command is a toggle switch that lets you change from a three part display to the single window display of Fig. 5-17. The vertical brackets to the left of the time slots show the length of time for the appointment. Figure 5-13 displays a checkmark next to the Wide display command to show that the feature is on.

Window menu

The Window menu is basically the same as the other utilities. The Change colors selection, which generates the box in Fig. 5-18, lets you change the colors of the background, window border, appointment, text, and time usage graph. The Move command, shown in Fig. 5-19, lets you move the appointment scheduler window around the screen using the four arrow keys.

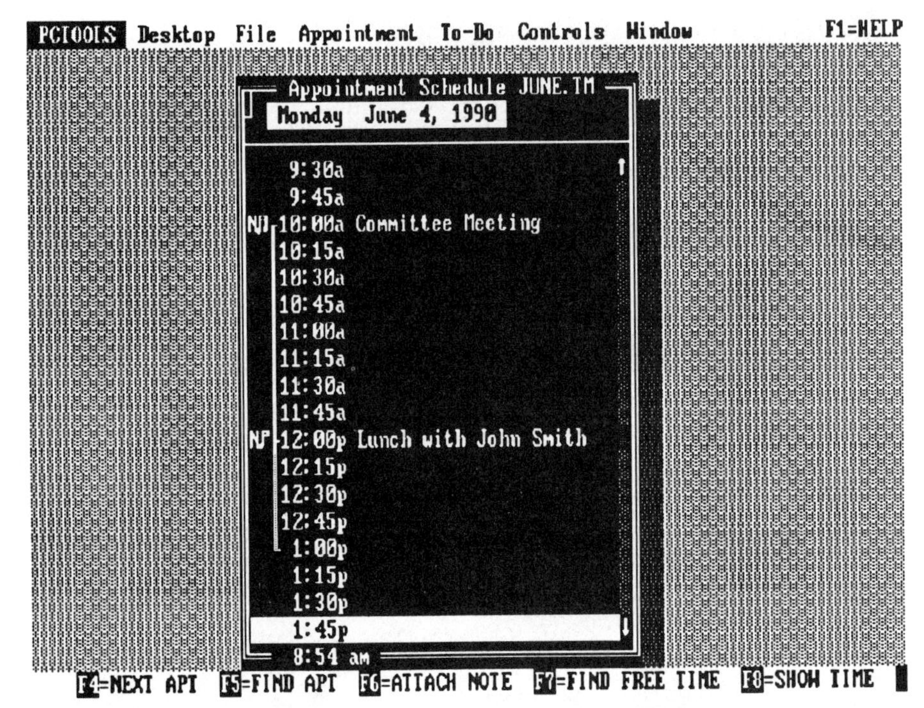

5-17 Wide display command

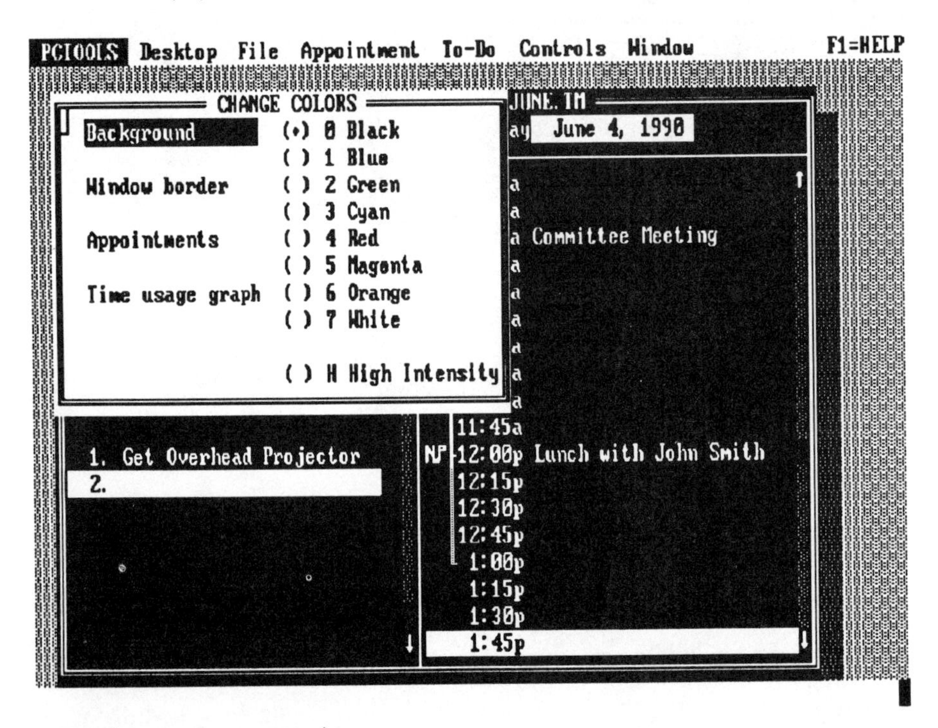

5-18 Change colors command

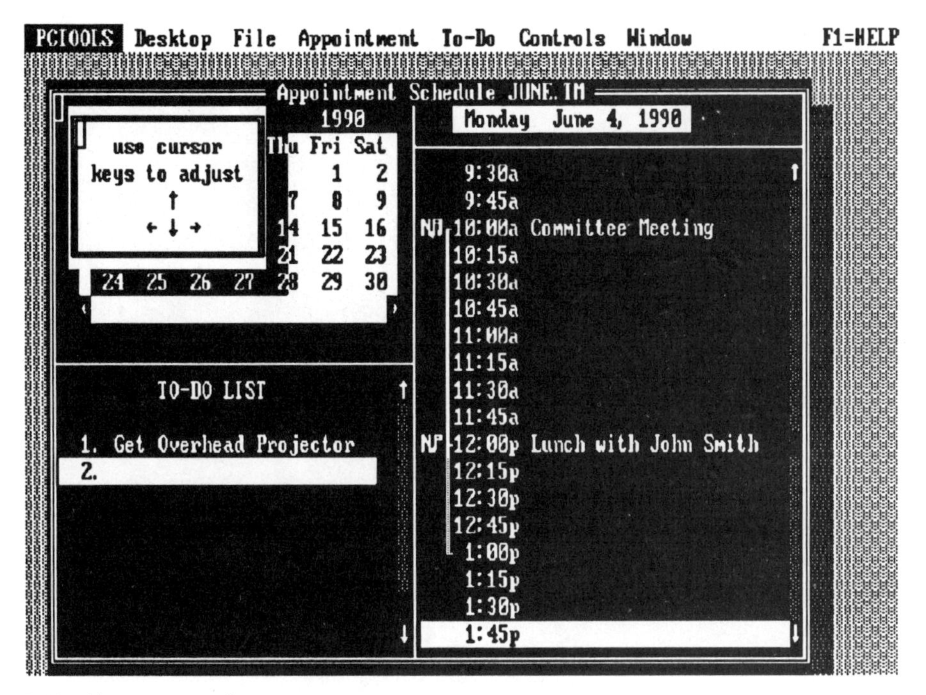

5-19 Move command

File menu

The File menu, shown in Fig. 5-20, lets you load an already-saved appointment, save an appointment, print an appointment, utilize the autosave feature, and exit without saving. All the file commands are identical to the other desktop utilities, except the Print command shown in Fig. 5-21. The Autosave feature lets you set the number of minutes between automatic saves.

The Print command lets you print out today's schedule and to-do list, weekly schedule, weekly to-do list, monthly schedule, and monthly to-do list. The list will always start with today's date.

The musical notes that indicate an alarm cannot be printed by most printers. The Translate graphics characters command lets your printer print out the musical notes using different symbols. A single musical note will be represented by the pound sign, while a double musical note will be printed with a percent sign.

Setting an automatic alarm

As mentioned earlier in this chapter, alarms can announce upcoming appointments. They can also run programs at pre-determined times. Figure 5-22 shows how to set up an alarm to run a program at a certain time after being prompted to do so. The first thing to do is select the program and time to run it. Next, write a message in the note box.

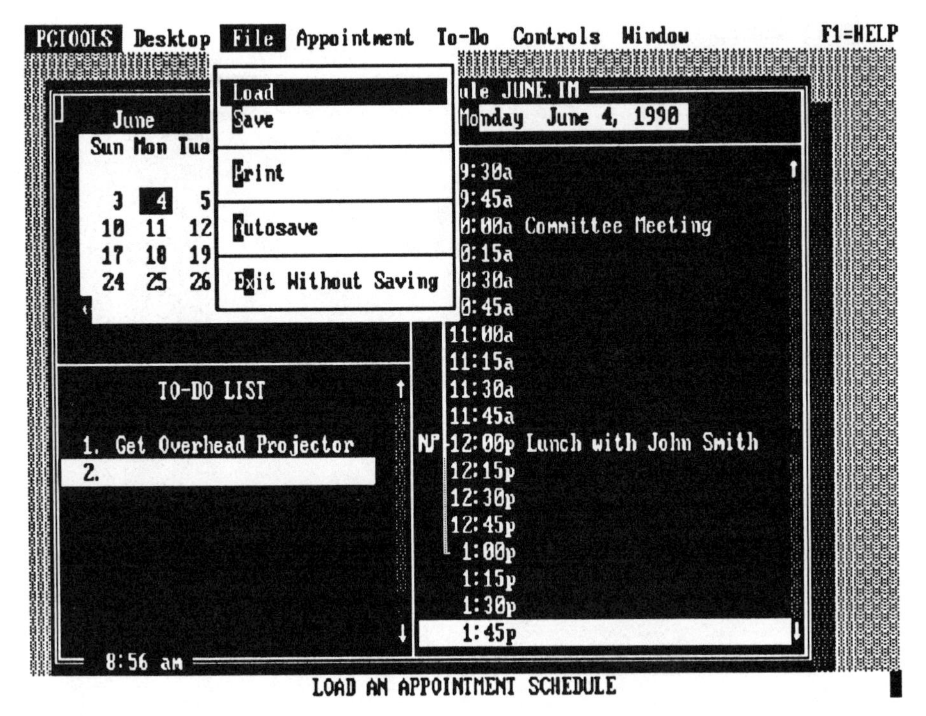

5-20 File menu

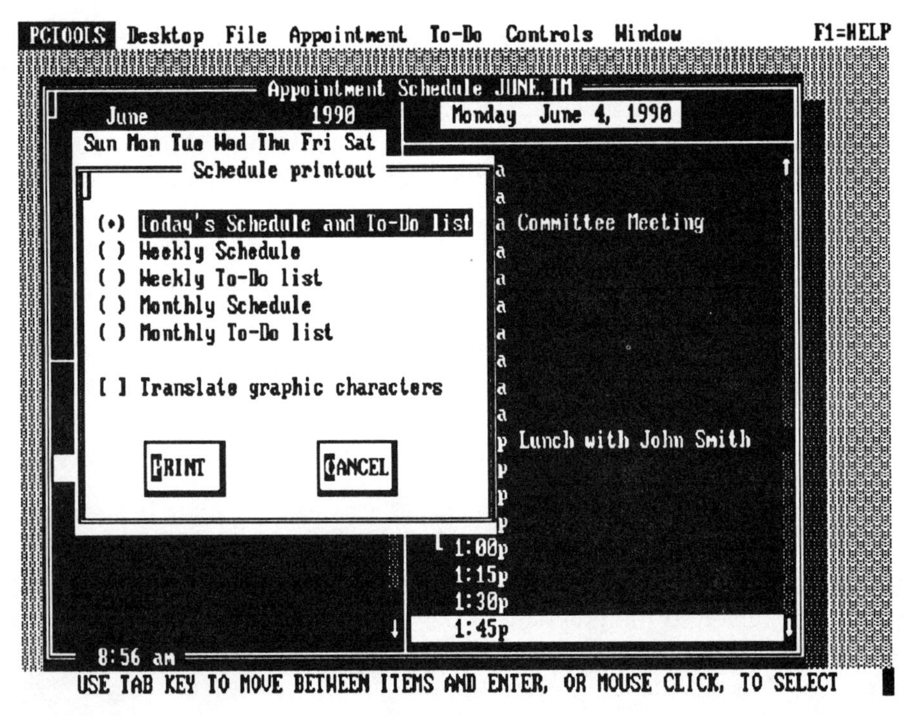

5-21 The schedule printout

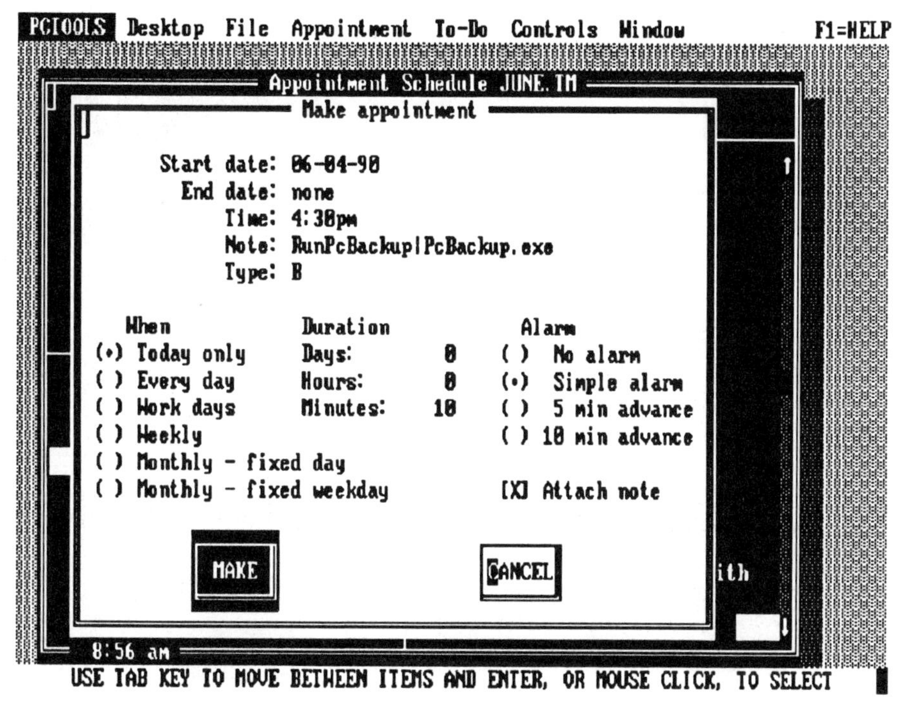

5-22 Scheduling a preset backup

The message in Fig. 5-22 is RunPCBackup ¦ PCBackup.Exe. The first part of the note, RunPCBackup, is the text of the prompt that will be displayed when the alarm is run. It will ask you if you want to run the program. The second part of the statement is the name of the file that will be executed, in this case, PCBackup.Exe. They must be separated by the pipe symbol (¦). The program must be an executable program having an extension of .BAT, .COM, or .EXE.

If you don't want to verify running the program, type ¦ Backup.Exe, and no prompt will be shown. The backup program will simply run at the designated time.

Notepads can also be loaded at the desired time, using the same technique. Figure 5-23 shows how text is loaded into a notepad. The prompt, "Board Meeting," will be displayed on the screen asking you if you want to load the notepad text (Notes.TXT) into the notepad. If you select OK, the notes for the board meeting will be displayed on the screen.

Figure 5-24 shows how the two preset alarms appear on the daily schedule. Macros, discussed in Chapter 7, can also be used with the appointment scheduler. Using macros to run a program and automatically dial a phone number will be discussed at that time.

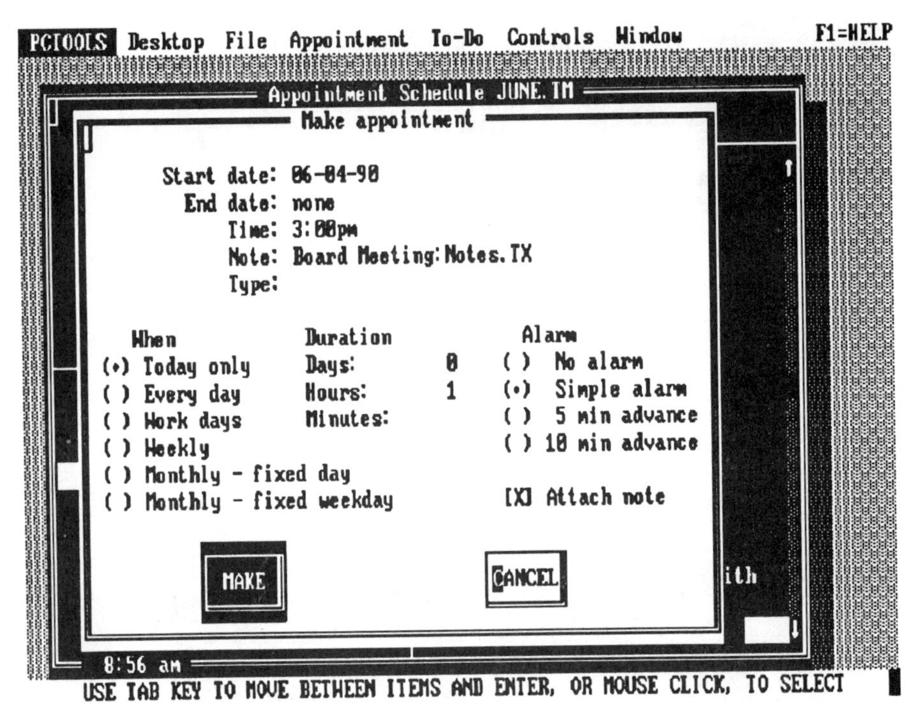

5-23 Automatically loading a notepad

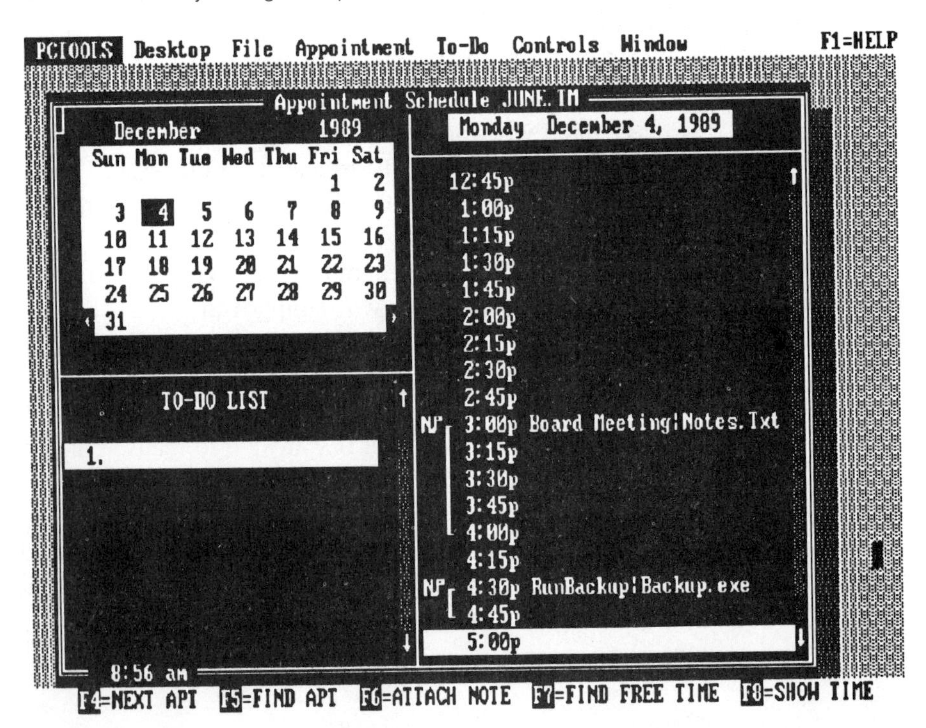

5-24 Preset alarms in daily scheduling

6
CHAPTER

Telecommunications

In this chapter, you will learn what data communications is, and how to use telecommunications to access a service or another computer. You will learn about electronic mail, bulletin boards, networking, information services, and sending and receiving text files.

Writing reports at home or in a business can be very frustrating. You always seem to be missing pieces of information that are located at a library, office, or friend's home, and there is never time to get it. *Communications software*, programs that electronically link one computer to another, provide a way to quickly and accurately communicate with distant sources of information. These software packages can link your computer to a friend's personal computer so you can exchange messages, programs, or data. You can also easily link a personal computer to a larger computer system in order to transfer, or *download*, data or programs stored on the larger computer to the personal computer.

Your microcomputer is capable of communications, as well as information processing. It can link through a telephone system in order to send and receive data to and from virtually anywhere in the world. To implement this capability, you must first equip your computer microcomputer with a modem and a communications program.

A *modem* is a special hardware device that acts as a link between the computer and a telephone system. It enables you to connect to a communications system or another computer. Both the sending and receiving machines must be able to handle identical signals. Establishing a connection is a relatively easy task with the help of the modem and the communications utility of PC Tools Desktop.

The communications tool will introduce you to the world of information services. From it you can obtain stock quotations, use banking services, obtain airline and restaurant reservations, use news services and consumer buying services, and participate in software exchange.

Electronic mail

Electronic mailboxes are areas within computer-maintained files that are reserved for the reception of user-addressed messages. Messages addressed to specific users can be sent directly to the appropriate "mailboxes." These messages are received automatically. Users can check their electronic mailboxes for messages on a regular basis.

The telephone is limited because you must find the right person at the right time. Much time is lost trying to get two people on the line simultaneously. People usually end up talking to an answering machine. A letter might solve the problem, but might take several days to reach its destination. Electronic mail is the solution; it is as speedy as a telephone, and guarantees message delivery like the postal system.

Another advantage of electronic mail is that messages can be sent to more than one mailbox. The owner of your VCR shop might send one message to a top salesperson, and another message to several salespeople to inform them of a meeting. Messages are addressed to certain recipients through codes embedded within a message file. The person who receives these messages can save them to disk or transfer them to a printer or other output device.

There are many forms of electronic mail services. Many subscription services are offered by commercial carriers, such as MCI, CompuServe, Western Union, ITT, and Prodigy. Some of these will be discussed later in this chapter. Many of the services, such as CompuServe, offer bulletin boards and information services.

Bulletin boards

An electronic *bulletin board* can be thought of as a large mailbox shared by many people. Messages posted on bulletin boards can be read by any user accessing the board. Bulletin boards are often set up for special interest groups or organizations. There are user groups for specific brands of microcomputers, where hints, upgrade tips, and other information are exchanged over bulletin boards. Some bulletin boards have electronic mailboxes that are available for private message exchanges.

Bulletin boards can also be an excellent source of free and inexpensive software. Many bulletin boards provide collections of *public domain* software. This software can be copied by any user. These programs are written by individuals and made available to anyone accessing the bulletin board. In return, the originators of the software might ask for minimal or voluntary payments from users. Variations of almost every commercially packaged program can be found in some type of public domain software.

Bulletin boards are set up and maintained on some type of computer. Depending upon the size of the bulletin board and the number of users, microcomputers might handle incoming calls and messages. Once the computer is set up, most of the work is performed by the computer without human intervention. Some bulletin boards provide a telephone number so users can reach a person to answer questions about the program. Some bulletin boards are too large to run on microcomputers, but all are designed to be accessed by them.

Networks

Within some organizations or databases, microcomputers are linked to share data and information. These structures of connected computers are known as *networks*. Microcomputer terminals can be linked to minicomputers or mainframes. Under this arrangement, users can access many or all of the resources maintained by the larger machines. These networks make use of uploading and downloading.

Uploading refers to the capability of sending data from remote terminals to the minicomputer or mainframe, using remote terminals to manipulate data that is then stored on larger machines. *Downloading* is the reverse. Files or processing capabilities are transferred from the central computer to microcomputer terminals. Networking provides virtually limitless opportunities for sharing data resources.

Information services

Information service organizations maintain files for public access, and provide callers with special access. Subscribers pay fees for the services they use. Information services provide data on virtually any subject, including personal finance, entertainment, education, health, and business. Two of the more popular information services are described below.

CompuServe

CompuServe is based in Columbus, Ohio and is the country's largest information service, with over 200,000 subscribers. It was purchased by H&R Block in 1980 and is available 24 hours per day, anywhere in the world.

CompuServe is used by both the consumer and the business person through its Consumer Information Service and Executive Information Service. CompuServe provides stories and features from newspapers, wire services, and organizational profiles; financial news; electronic banking and fund transfer; airline scheduling and reservation services; movie, theater, book, and restaurant reviews; medical and health information; electronic games; mail service and shopping; online encyclopedias; personal computing software; and many other features and software.

CompuServe's CB function lets you converse on its channels, as on a citizen's band radio. You assume a handle or name so nobody will know who you are. Enter GO CB at CompuServe's main menu to access this service.

CompuServe also provides games and related software. To enter the games section, type GO IBMNEW from the main menu of CompuServe. You'll find games, personal opinions, and free utility programs.

The Comp-U-Store is an electronic mail-order catalog from which you can order merchandise, from clothing and travel to food and liquor. Anything you order will arrive by UPS or Federal Express. Most of the prices are higher than in local stores, but it eliminates the time spent shopping and traveling. As you buy things, CompuServe charges your credit card or withdraws money directly from your bank account. Don't go wild or you'll wind up in the poorhouse.

A subscriber to the Executive Information Service has access to all the options just discussed. In addition, he or she can access a range of financial, demographic, and editorial information. Stock quotations and commodity prices are available. Historical market information, market and industry indexes, and national and international news wire services are also available.

A subscriber to CompuServe can do research for a presentation, make airline reservations to a distant city, transfer funds among bank accounts to pay for the flight, send messages to arrange for airport pick-up, check reviews of restaurants in that city, and even consult weather reports to help in packing the appropriate clothing for the trip.

Recently, CompuServe has acquired the Source, a business-oriented informational network.

Dow Jones News

The Dow Jones News Retrieval service provides business and financial services to users. These services include stock market quotations, profiles of business organizations, financial news, and financial forecasts, as well as general news and stories. The Dow Jones News Retrieval was started in 1974 to offer financial news to stockbrokers via the ticker tape. In 1979, the financial service was opened to anyone who had access to a personal computer. Quotations on stocks could be easily accessed any time of the day. The service also carries the text of the *Wall Street Journal* and other well-known financial publications.

Modems

Information services are available for a minimal investment. All you need is a microcomputer equipped with a modem, and a telephone. Once this hardware is installed, establishing communications is simply a matter of traveling through the menu structure of the communications program and the services themselves.

To communicate with another computer, your personal computer has to be physically connected to the telephone line. This requires a hardware interface called a *modem*. On the transmitting end, a modem converts binary signals, produced by a computer or peripheral device, to analog signals, which can be sent over the public phone system. On the receiving end, the modem converts the analog signal back to a digital signal, which is then forwarded to the computer or associated device.

Data in its most basic form is a series of electronic pulses of two varieties. These values are represented numerically as *bits* (binary digits), 0 and 1. Every character (for example, a letter or a number) consists of eight bits, or one *byte*, according to the commonly accepted ASCII coding scheme. *ASCII* is an abbreviation for American Standard Code for Information Interchange. For example, the letter A is represented by the code 01000001, the numeral 9 by the code 00111001, and the quotation symbol by the code 01010110.

Every message transmitted by a computer, whether it is a word, number, or even a picture, is broken down into bytes. The bytes are further broken down into bits. A computer processes information one byte (eight bits) at a time; a telephone line carries a message one bit at a time. A device known as a serial port converts the parallel stream of bytes to a serial stream of bits for transmission, then back again to bytes on the receiving end.

The telephone was built for voice transmission. It is an analog device that represents information in a continuous, smoothly varying signal. A computer, on the other hand, is a digital device that uses discrete signals to represent data as the presence or absence of an electrical voltage. A modem is an abbreviation for modulator and demodulator. A *modulator* converts a digital signal to analog. A *demodulator* does the reverse; it converts an analog signal to a digital signal. A modem is a device that performs both functions when connecting personal computers to the telephone system. In short, modems convert electricity into sound and sound into electricity.

It is advisable, when buying a modem, to purchase one that is Hayes-compatible. Choosing a modem is the first step in telecommunications. Just as with every other kind of computer peripheral, there are many modems available commercially, and they differ in price and capability. The speed of a modem, the maximum rate at which it can transmit or receive data, is one of its more important distinguishing characteristics. This rate is measured in *bits per second* (bps), or *baud*. In most cases, the two terms are used interchangeably.

Three hundred bps is considered a low speed for data transfer. Medium speed is 1200 to 9600 bps. Anything over 9600 is high speed. Most high-speed modems require specially installed data communications lines and are not suitable for personal computers.

In general, as the speed of a modem increases so does the price, but the cost of all modems is dropping. A few years ago, 300 bps was the standard transfer rate for most personal computers. Now 1200 bps has become the standard, and the 2400-bps rate will become the standard in the next few years.

Even though the high-speed modems cost more, they return savings in the form of reduced communications costs. Most information services charge according to the amount of *connect time*. Therefore, the faster the modem, the less time it takes to transmit data, which results in lower communications costs. It takes approximately one minute to fill a 24 × 80 screen at 300 baud. It only takes 16 seconds to fill that screen at 1200 baud, and eight seconds at 2400 baud.

Modems are also chosen according to their physical design. Some can fit into an expansion slot inside a personal computer, while others need to stand alone on a desktop. An internal modem is generally cheaper than an external modem. It does not require additional desk space, and also does not require an outside, or extra, power supply.

External modems work with a variety of personal computers. This is useful if you expect to change from one brand of computer to another, for example, from IBM to Apple. It is also easier to set switches and control volume on an external modem. Flashing lights on the front of the unit help you monitor the status of the call. An external modem takes up additional desk space and requires its own electrical outlet.

One type of external modem is an *acoustic coupler*. An acoustic coupler plugs into a computer and connects to a standard telephone. The handset of the telephone is then inserted into a pair of rubber cups. Carrier signals are produced and received through the mouthpiece and earpiece of the telephone handset.

Telecommunications menu

When you enter the Telecommunications menu of the desktop program, Fig. 6-1 appears. It shows the name of some telecommunications services, their telephone number, availability, and baud rate. The baud rate, as previously mentioned, is the speed of a modem, the maximum rate at which it can transmit or receive data.

Duplx stands for duplex, which describes how information is sent. A full-duplex microcomputer can send and receive information simultaneously. A half-duplex computer can send and receive information, but not at the same time. Most modems are full duplex.

The P represents parity. The question of parity must be answered according to your modem manual. All data communications are subject to noise or disturbances on the line. This can alter the value of the transmitted data and produce

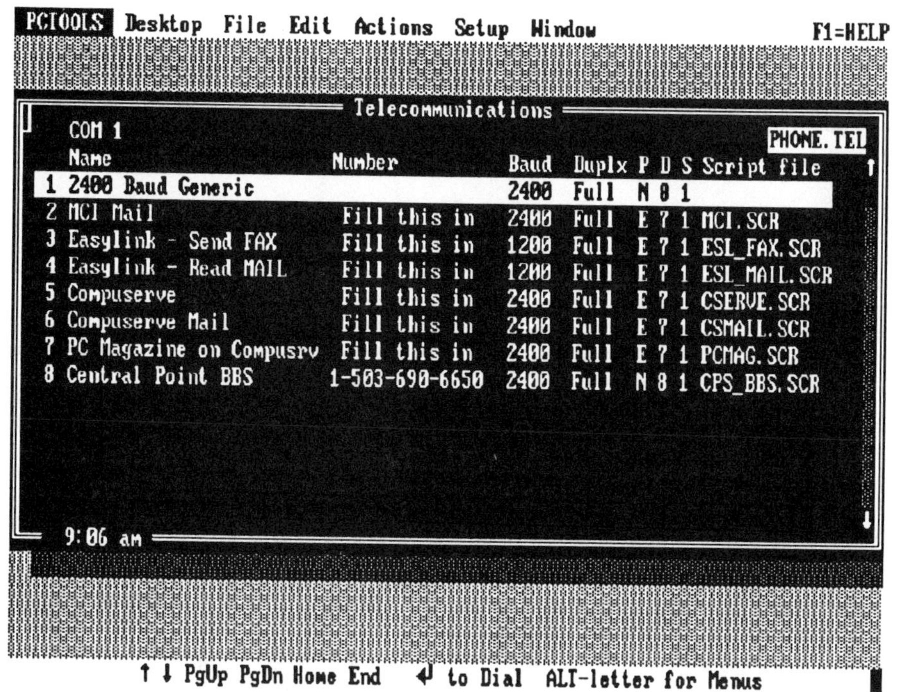

6-1 Telecommunications menu

garbled messages. *Parity checking* detects these types of errors. A parity bit is an extra bit added to every byte, which checks whether a byte has been altered in transmission. Parity is normally specified as odd or even.

The value of the parity bit depends on the other bits in the byte. If the parity is specified as even, then in each byte the total number of bits, including the parity bit, must be even. For example, the letter T is represented by the ASCII code 01010100. A parity bit is added at the end of the byte to make it even parity. It becomes 010101001. There are now four ones in the code, or even parity. If there is already an even number of ones in the code, no parity bit is added. The letter U in ASCII code is represented by 01010101. When the parity bit is added, it becomes 010101010. The parity bit is 0 because there is already an even number of ones. The telecommunications program represents even parity with an E, odd parity with an O, and no parity with an N.

The column entitled D represents the number of data bits. The default (for ASCII characters) is seven bits. Type eight bits if you are using nonASCII formats, including extended character sets or binary data that use all eight bits. If you are unsure of which setting to select, choose eight bits. Most communications are eight-bit communications.

The letter S in the column headings stands for "stop bits." The bits in a message travel in groups. Data is sent and received one character at a time. *Stop bits* are added to identify the end of a data character. The default setting is one stop bit.

The Script file logs on to a certain communications system, and contains the instructions necessary to do this. It is stored in either ASCII or the desktop text files.

The scroll bar lets you use the mouse to move through the entries while the close box in the top left corner of the window lets you close the box or leave the application.

Edit menu

The Edit menu of the desktop, shown in Fig. 6-2, lets you edit, create, or remove an entry. The Edit entry selection creates the screen shown in Fig. 6-3, which lets you change a telephone number, the baud rate, etc.

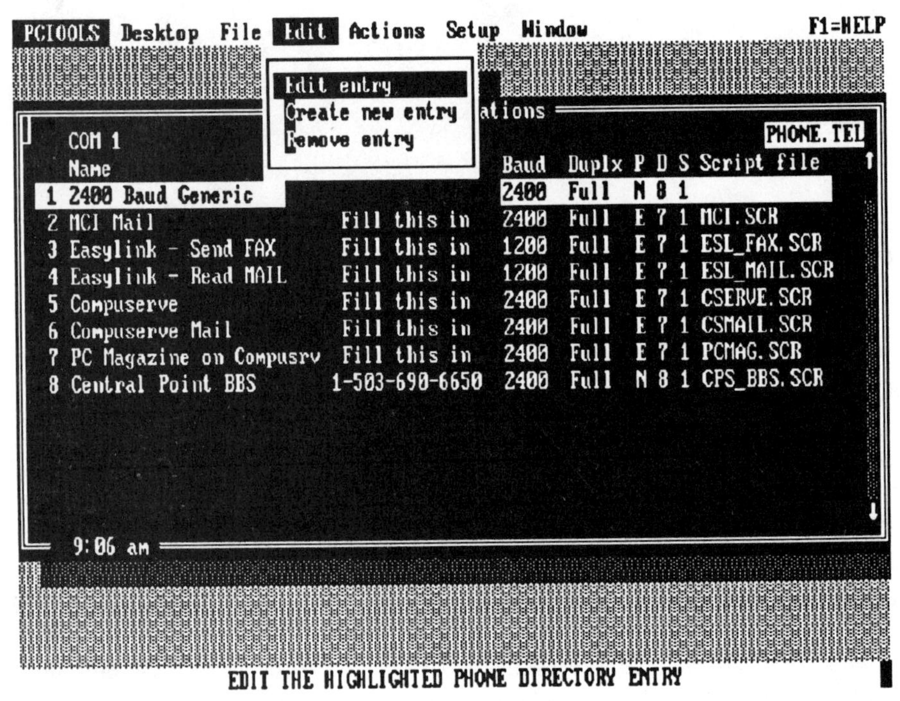

6-2 Edit menu

The Name line could be a company, computer service, or person's name. It can contain up to 50 characters.

The Phone number line lets you enter or change the telephone number. It can also be up to 50 characters long. All parentheses, hyphens, dashes, and spaces are ignored when it looks up the telephone number.

The Script file line contains the name of the file that has the information necessary to access the computer service.

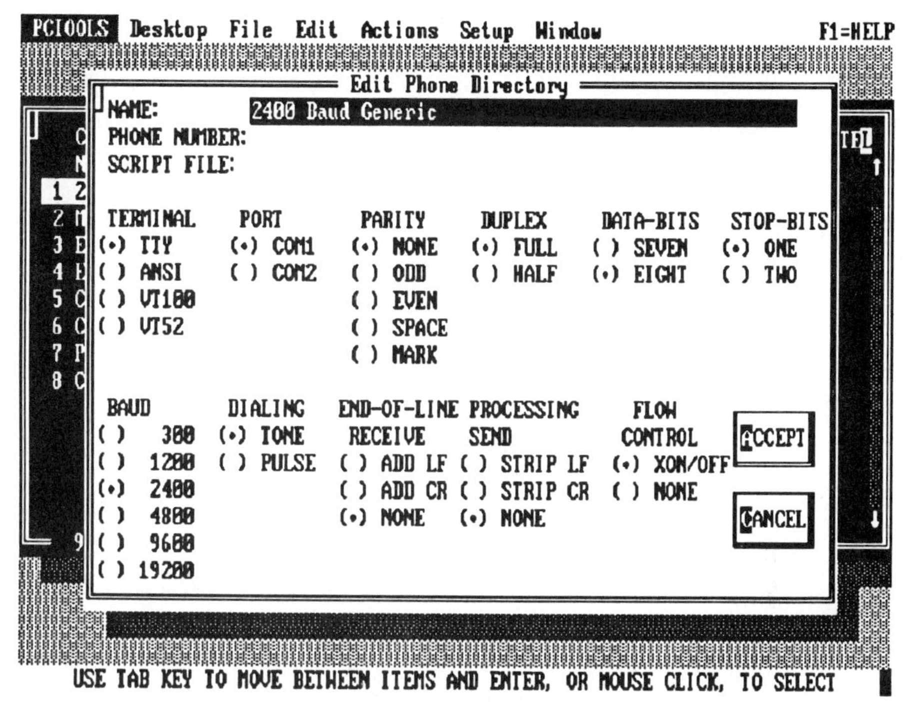

6-3 Edit phone directory command

The Terminal item shows the type of terminal being used. The default is TTY, which is the most common terminal for bulletin boards or computer services. The ANSI terminal is used with terminals that utilize the ASCII extended character sets, such as graphics and animation. The VT100 terminal emulates the DEC VT100 terminal, while the VT52 terminal emulates the DEC VT52 terminal.

The Port section determines which communications port you are using (in most cases, COM1 or COM2).

The Parity item, discussed above, checks communication errors by adding an extra bit to each byte of data sent. None will not check parity, Odd is odd parity, Even is even parity, Space means a 0 bit is always added as parity bit, and Mark means a 1 is always added as the parity bit. The parity must match the settings of the remote computer.

Duplex is the way information is transmitted. Full duplex means both ways simultaneously. Half duplex means both ways, one at a time.

The Data bits item is the number of bits in the remote computer to represent each character. Most systems use seven.

The Stop bits identifies the end of a data character transmission. One or two characters is added to the end of each character.

The Baud definition shows the speed of transmission, taken from your modem setting. Dialing specifies a touch-tone or rotary (pulse) phone.

The End-of-line processing (receive or send) item specifies Add LF (add a line feed), Add CR (add a carriage return), Strip LF (strip or remove a line feed), Strip CR (strip a carriage return), or None. The None option does not change the data being sent or received. You will have to check the settings that the subscription service wants and select them here.

Flow control checks the flow of data between the two computers. Xon/off tells the computer when to send information and stop sending information. These settings must match in both computers. Selecting None means that the two computers will automatically send data at their own time, but there is a greater chance of error. Again, you must check the manual of the computer service to see what setting to use.

Typing A for accept saves the information, while C for cancel lets you start again.

The Create new entry option from the Edit menu (Fig. 6-2) lets you add new entries to the list, while the Remove entry option lets you remove options from the list.

Setup menu

The Setup menu, shown in Fig. 6-4, shows two commands: Modem setup and Full online screen. The Full online screen option is a toggle that either lets you see a communication as it is happening (online), or lets you do other work during the communication.

The Modem setup command initializes the modem before you start communicating, as shown in Fig. 6-5. The telecommunications utility already has two strings that initial 1200- and 2400-baud modems. These strings are for Hayes-compatible modems. You might have to enter a different string for a different modem. Check your modem manual for the proper initialization string. The Connect string item contains the characters CONNECT, which tell your modem to make a phone connection. The CONNECT command initiates dialing of the phone number in the phone directory. If you try to initiate dialing while already connected to another computer, you will be asked to hang up or disconnect from the present service.

The Actions menu, shown in Fig. 6-6, lets you automatically dial a number from the entries in the list of computer services, manually dial a number, or hang up the phone. To dial, highlight the service you want to dial and select Dial from the Actions menu. Follow the on-screen instructions for your particular service to complete the communications process.

To dial a number not on the list, or a possible alternate number, select Manual from the Actions menu. A screen appears indicating that you are online (on the phone) and asks you to type in your telephone number. After you are finished typing in the number, you will be prompted for the necessary information for your particular computer service.

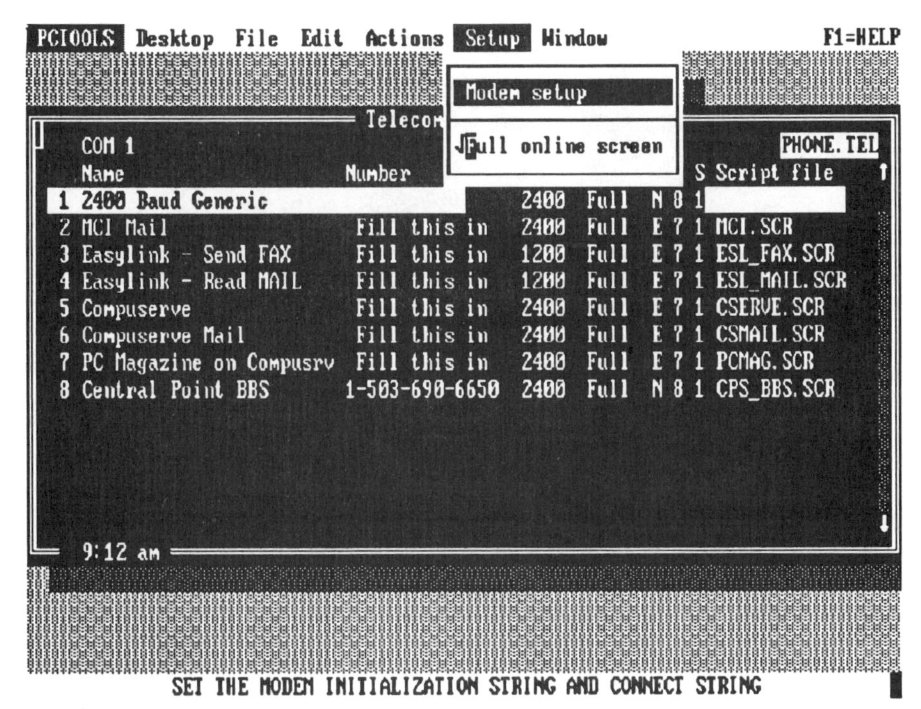

6-4 Setup menu

6-5 Modem setup command

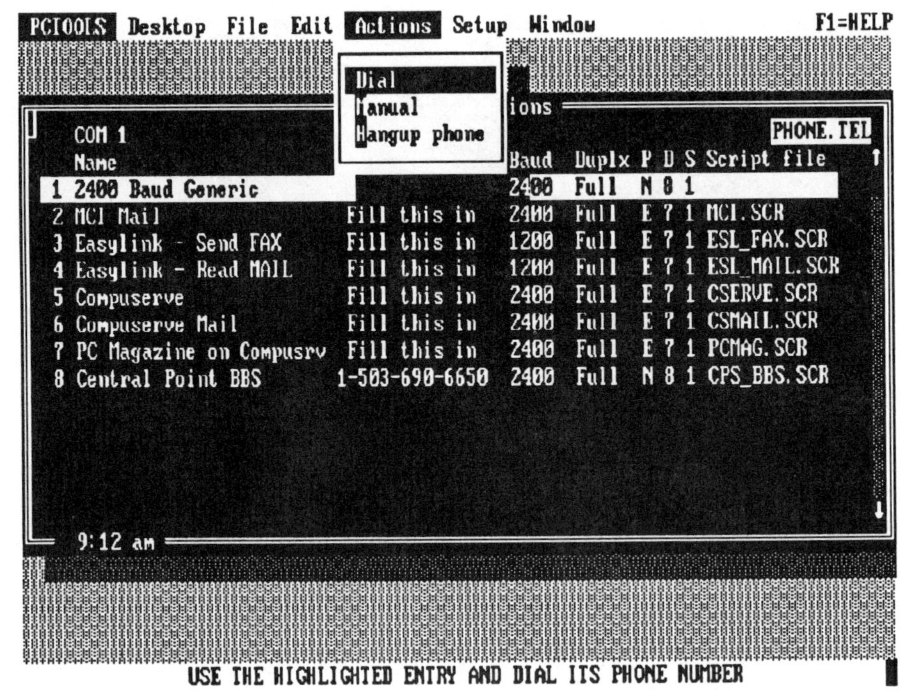

6-6 Actions menu

The Hangup phone option lets you disconnect the modem before you begin online communications. Once you are communicating with the online service, you will still be allowed to disconnect the phone with the Hangup phone option. This ends the communications session and returns you to the phone directory window. When online, you will also be given an End transfer option, which lets you stop sending or receiving information but lets you remain connected to the computer service. This is useful if you want to utilize another area of the computer service.

Sending and receiving files

Handshaking is the way two computers determine when to start and when to stop sending information. You must specify whether or not you want to use Xon/off. This is the form of protocol, or handshaking, used when two devices initially establish communications with one another. *Protocol* is the set of rules used for communicating. The communications settings are stored in a file so that re-establishing a broken connection is much easier.

The two protocols used for microcomputer communication are ASCII and XMODEM. ASCII protocol is for sending files between computers, but doesn't check for transmission errors. XMODEM protocol is used by most microcomputers and includes a lot of error checking. XMODEM is slower than ASCII, but is more accurate and should be used to send and receive files.

7

CHAPTER

Macros, clipboard, utilities, and autodialer

In this chapter, you will learn what a macro is and how to create, edit, save, and delete one. You will learn how to cut and paste information to and from the clipboard, and how to use the desktop utilities and the autodialer.

Macros

You sometimes need to enter repetitive sequences of keystrokes in order to communicate commands in a program. Some examples are the keystrokes needed to print and address, save a worksheet to disk, and print a document to the printer.

Whenever you find that you are repeatedly using either the same set of keystrokes or a piece of identical data, you should think about creating a macro. A macro is a sequence of keystrokes or block of information that you assign to a two-key combination, using the Ctrl or Alt keys. A macro is like a recorded message; once you've "taped" a piece of information and given it a name (a Ctrl-key or Alt-key combination of keystrokes), entering that name automatically plays back the tape.

You save time and effort by creating macros for repetitive typing, like commonly used paragraphs, your name and address for the top of correspondence, the time and date for insertion into a document, and routine tasks, such as printing standard documents.

Creating a macro

To create a macro, select Macro from the Desktop menu and Fig. 7-1 appears. You can either load an old macro or create a new one. All macro files end with the extension .PRO. The macro utility comes with several built-in macros that you can examine to see how they are created.

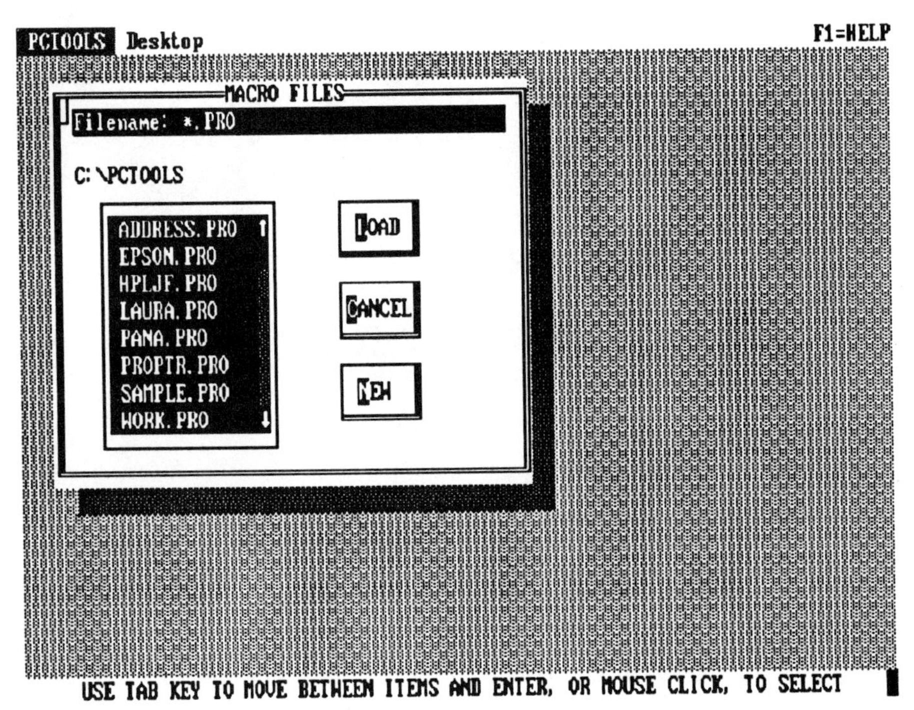

7-1 Creating a macro file

Figure 7-2 shows some of these sample macros, from the file SAMPLE-.PRO. The first macro shown in Fig. 7-2 lets you load the PC Tools Desktop by pressing Ctrl−F1. The macro is:

<begdef> <ctrlf1> <DESK> <enddef>

Let's look at the different parts of the macro, and then create one from scratch.

The first thing you notice in the string is that all macro commands are enclosed within left and right brackets. The command <begdef> denotes the beginning of a macro. The next part, <ctrlf1>, indicates the keystrokes that access the macro. Try to pick a sequence of keys that can be easily remembered.

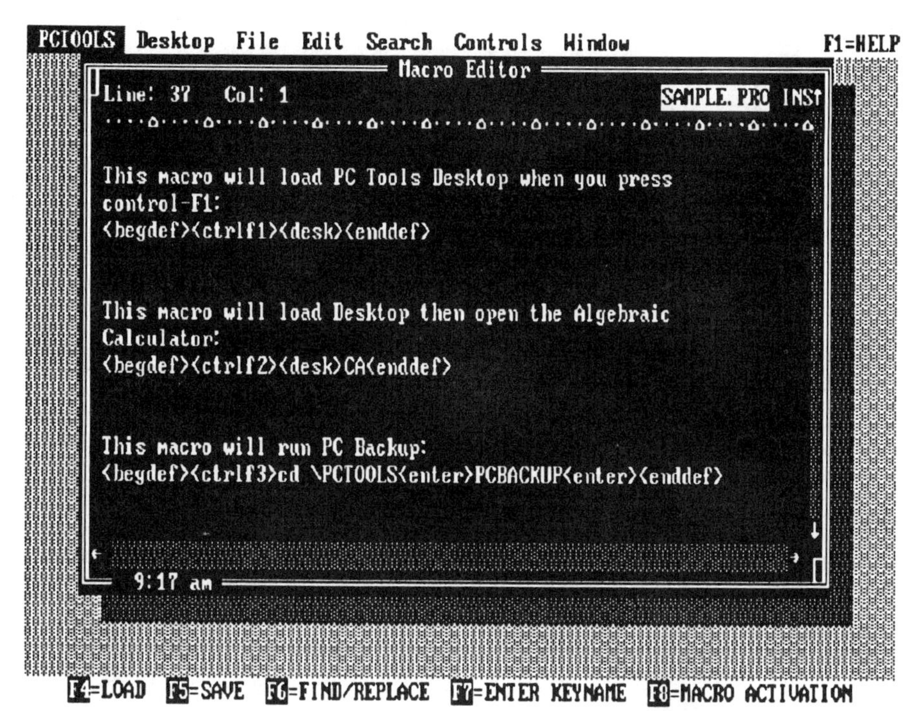

7-2 Macros from the sample file

The third part <desk> is called the script, and contains the commands or characters to be recorded. The last part, <enddef>, signifies the end of the macro definition. All macros contain these four parts.

You can use macros from the desktop, or while using other applications such as dBASE or Lotus. You can also use them in the appointment scheduler, and to run backups and compress the disk, as shown in Fig. 7-3. The sample file also contains macros that you can use with the appointment scheduler to run a program at a preset time, attach a document to an alarm, dial a phone number automatically, and dial a remote computer service to transfer files.

The PC Tools sample files also contain macros that can print files using the HP Laserjet printer (shown in Fig. 7-4), Epson printers, IBM Proprinters, and Panasonic printers. These particular macros are available only with the notepad, outline, and database utilities. Commands that specify boldface, italics, subscript, superscript, and other print styles can also be inserted into letters and other document files.

If you have a printer that is not listed in the sample macros, you will have to create your own printer macros. The following macro will turn boldface print on:

<begdef><Ctrlb>¦BOLDON¦<esc>E<enddef>

To turn the boldface print off, you could use:

<begdef><Ctrlo>¦BOLDOFF¦<esc>F<enddef>

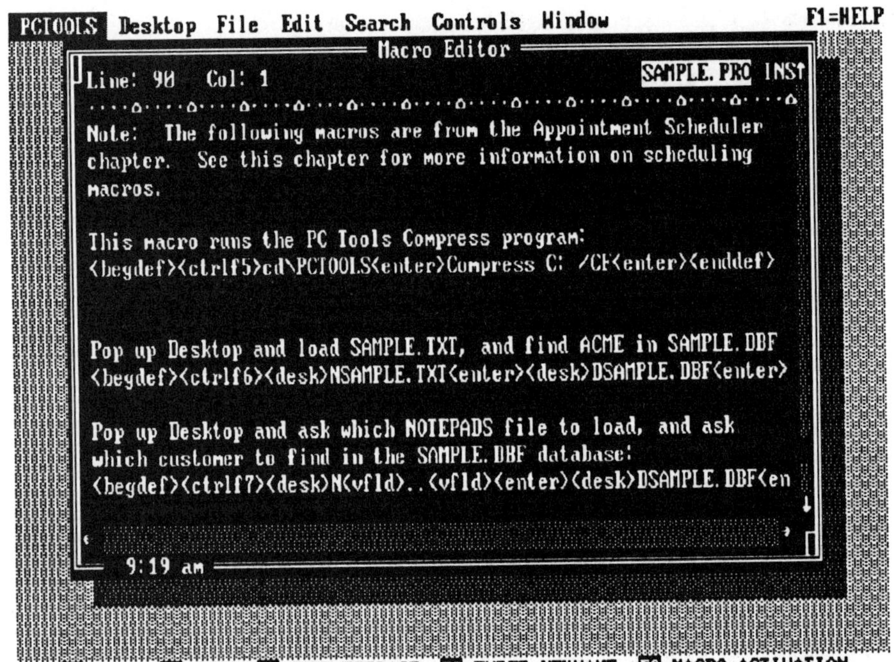

7-3 Appointment scheduler macros

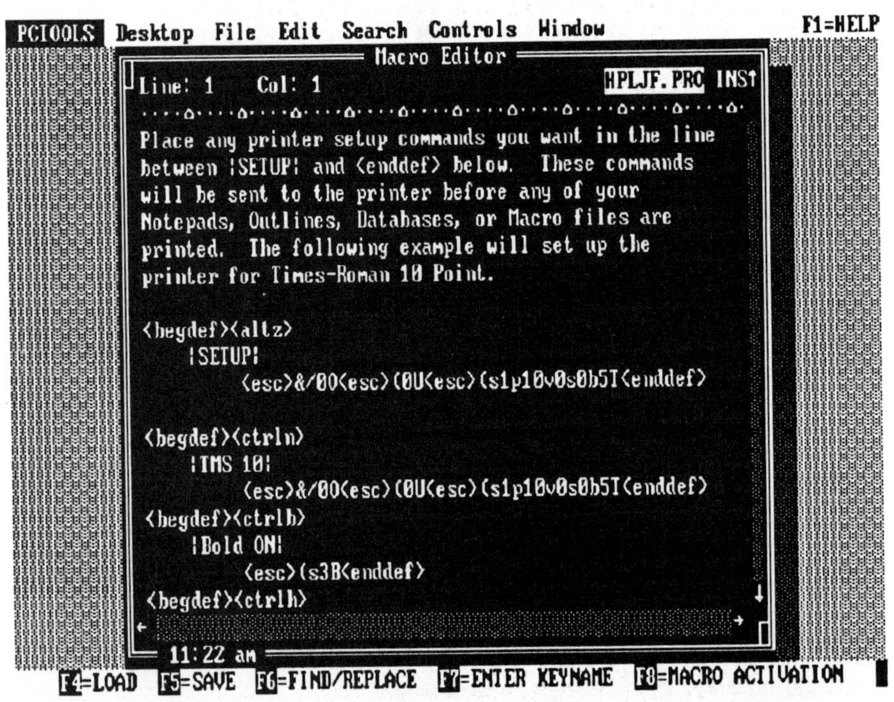

7-4 A laserjet print macro

Instead of BOLDON and BOLDOFF, you will have to substitute the printer codes that are located in your printer manual. Although these printer codes appear in the text of the notepad, they will *not* be printed. They only tell the printer to turn on or off various type styles. For example, the type style command:

¦ITALICON¦PC TOOLS¦ITALICOFF¦

would print out *PC TOOLS* in italic print.

Now, create a macro that will print out your name and address on the top of any piece of any correspondence. Create a new file and give it an appropriate name, like ADDRESS.PRO (see Fig. 7-5). You can use the same keys to enter and edit the macro as those shown in the notepad utility. You can insert a comment before the macro that describes what it is and how it can be used. The macro that will be executed is all the script between the <begdef> and <enddef> commands.

7-5 An address macro

After you have finished typing the comments, if any, press the Alt and + keys simultaneously. This will automatically display the <begdef> command on the screen. This command *must* start in the first column.

Now press the keys that will execute the macro. There is no need to insert the left and right brackets; they are automatically inserted when you press the key combinations. The definition <Ctrla> stands for the key combination Ctrl – A, representing Ctrl–Address. As previously mentioned, try to pick key combinations that represent the macro being executed. Since PC Tools uses a lot of Alt key combinations, try to stay away from redefining ones that are already established.

The third part of the macro is the script (the name and address text). In the script, you can use Tabs, spaces, etc. to format the information. To end the macro, press the Alt and – keys. This will generate the <enddef> command. If you want to enter a function key, Alt key, Arrow key, Backspace, Ins, or Del key into a macro script, you must precede them with the F7 key.

The macro is not activated until you press the F8, macro activation, key or select Macro activation from the File menu, as shown in Fig. 7-6. The macro File menu has the same menu as the other desktop tools, except for this command.

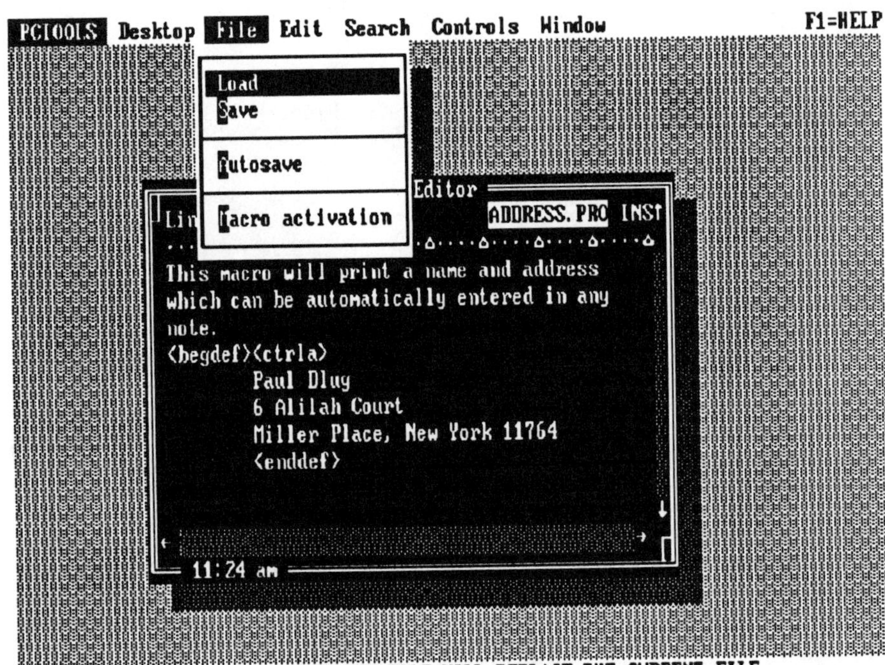

7-6 Macro File menu

When you select the Macro activation command, Fig. 7-7 appears. You can select Not active, active when in PC Tools Desktop, Active when not in PC Tools Desktop, or Active everywhere (including the DOS prompt). Select the option you want and press the OK button to implement it.

To test the macro, exit the macro editor, place the cursor where you want the macro executed, and press the programmed keystrokes. If the macro does not run properly, press the Esc key to stop the macro from running.

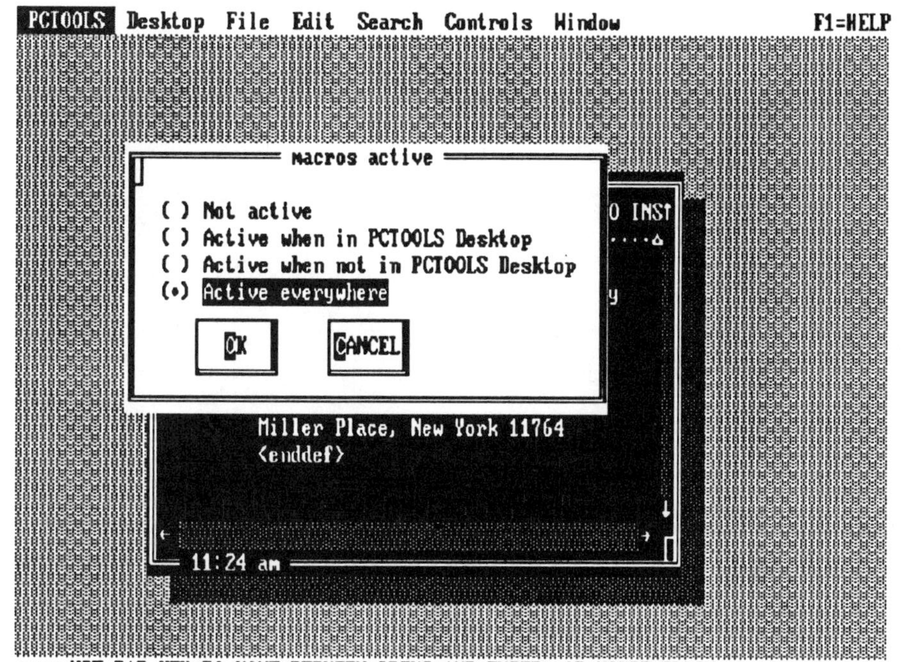

7-7 Macros Active dialog box

Macro Controls menu

The Learn mode of the Controls menu, shown in Fig. 7-8, lets you create a macro, but you must be using the desktop as a resident program. Instead of typing the keystrokes into the macro editor, simply press the keystrokes and they will be recorded as part of the macro. Macros created in this manner are saved in a file called LEARN.PRO.

To create such a macro, select Learn mode, select the application in which you want to create the macro, press the Alt and + keys to begin recording, enter the macro keys you want to use, enter the macro script, and, finally, press the Alt and − keys to end the macro. One disadvantage of creating macros in the learn mode, compared to entering it in the macro editor, is that all keystrokes (including the correction of mistakes) are entered. If you have a tendency to make a lot of mistakes, it is best to use the macro editor where they can be easily erased.

The Controls menu also has an Erase all macros command, which deactivates all the current macros but doesn't erase them from disk. Selecting the command displays the dialog box shown in Fig. 7-9.

The Playback delay command displays the screen shown in Fig. 7-10. This command controls the speed of the macro execution. In some cases, if the macro contains a lot of keystrokes, the computer will not be able to handle it because it is too fast. This will cause the computer to beep. You can set a time delay by enter-

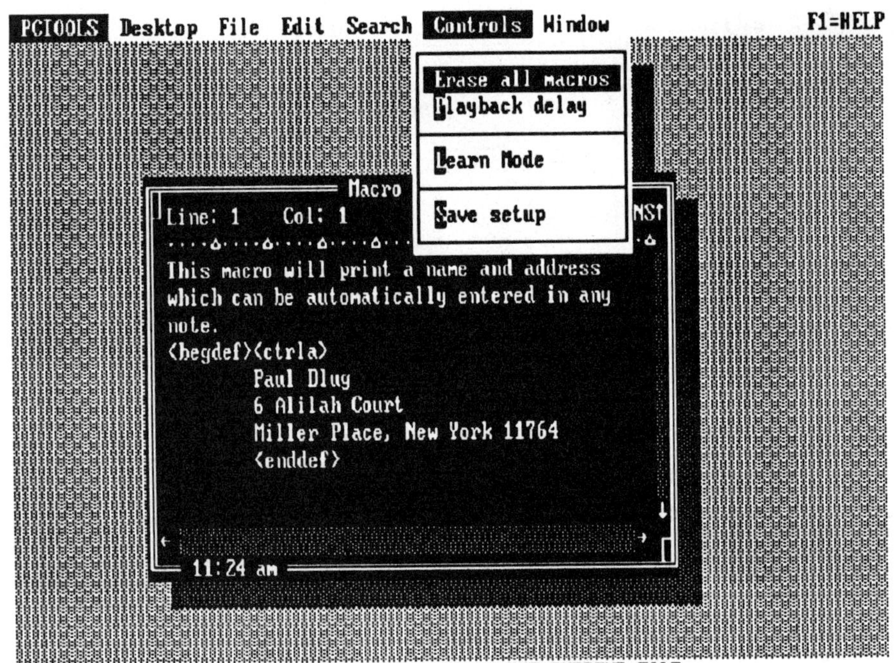

7-8 Macros Controls menu

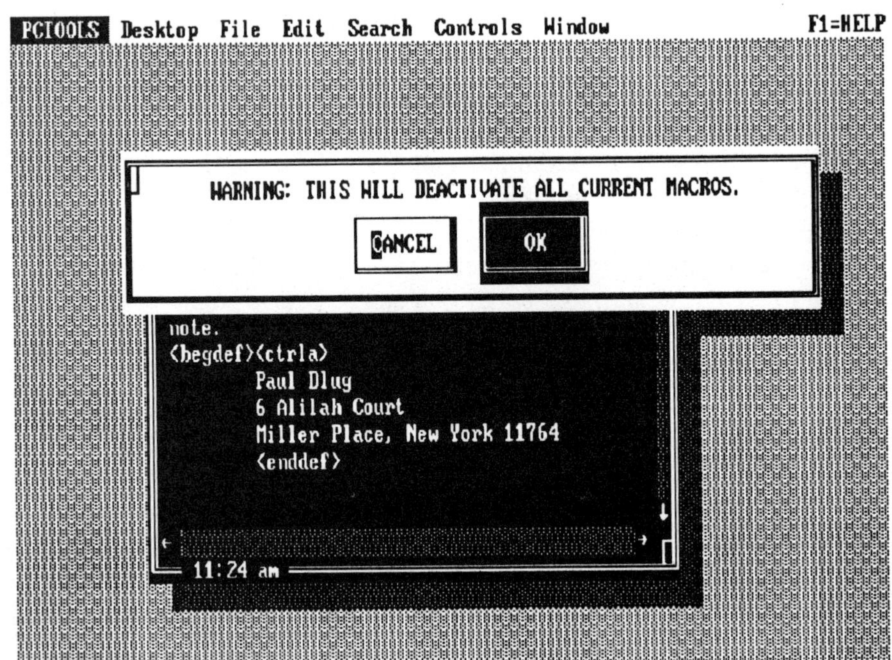

7-9 Deactivating macros

7-10 Macro playback delay command

ing a number that represents the number of eighteenths of a second to be inserted between each keystroke played in a macro. A zero delay means each keystroke of a macro is delayed by one eighteenth of a second. Select On to turn on the delay and OK to execute it.

The Save setup command generates the box shown in Fig. 7-11, which lets you save the selections you made in the Controls and Windows menus.

Macro Edit menu

The Edit menu, shown in Fig. 7-12 lets you cut, copy, and paste to the clipboard, mark and unmark blocks, delete all text, insert a file, and go to a certain line of text. It is very similar to the Edit menu of the notepad.

Clipboard

The clipboard is a temporary storage area that holds text to be copied or cut from one application to another. The clipboard can be used in any application, but is most effective in the notepad utility for moving and copying text.

The purpose of the clipboard is to hold text for later processing. The clipboard can handle about 40 to 50 lines of text. When Clipboard is selected from the Desktop menu, Fig. 7-13 appears. Any text that appears in the clipboard can now be edited or printed.

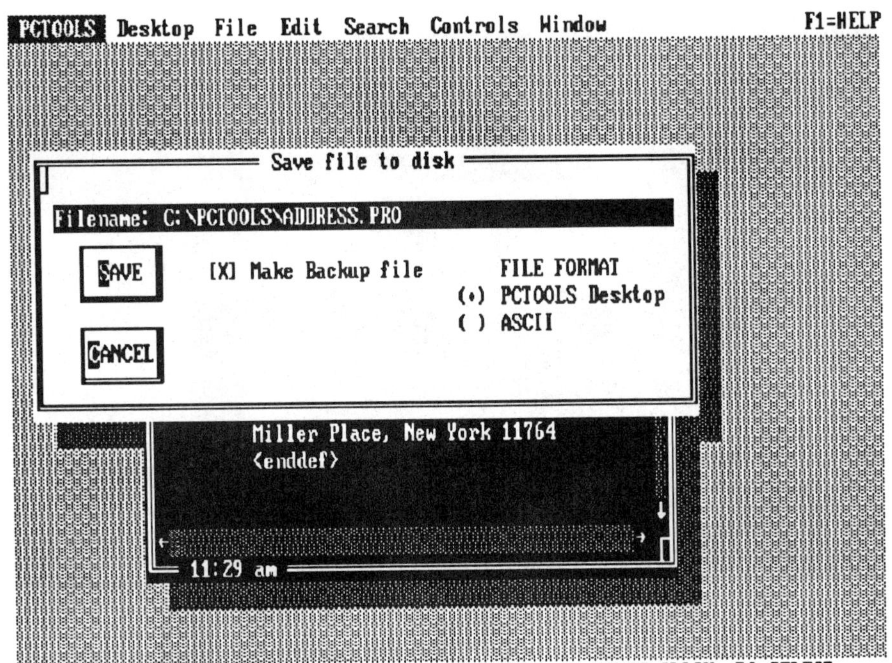

7-11 Saving macros to disk

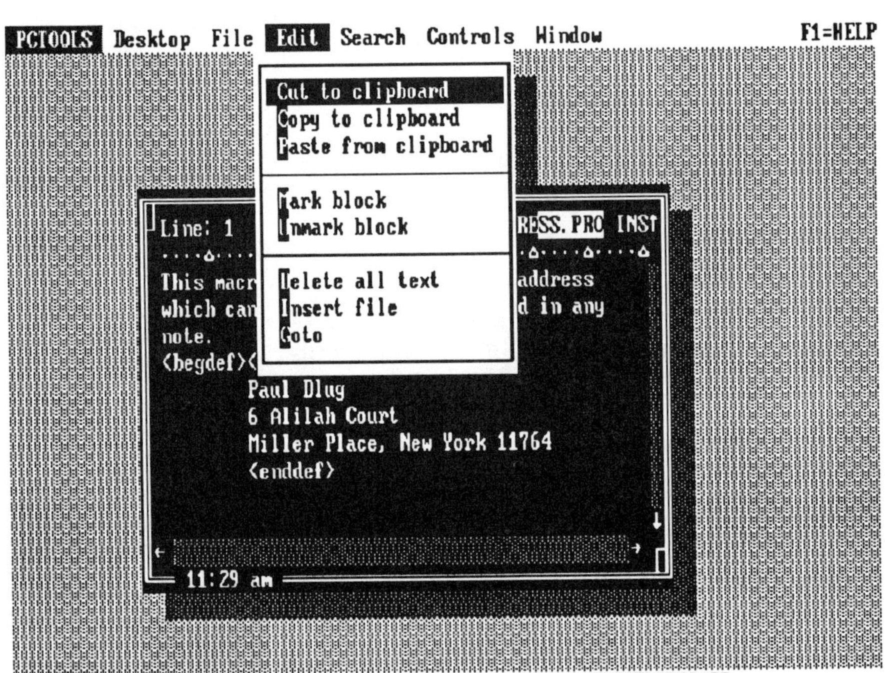

7-12 Macro Edit menu

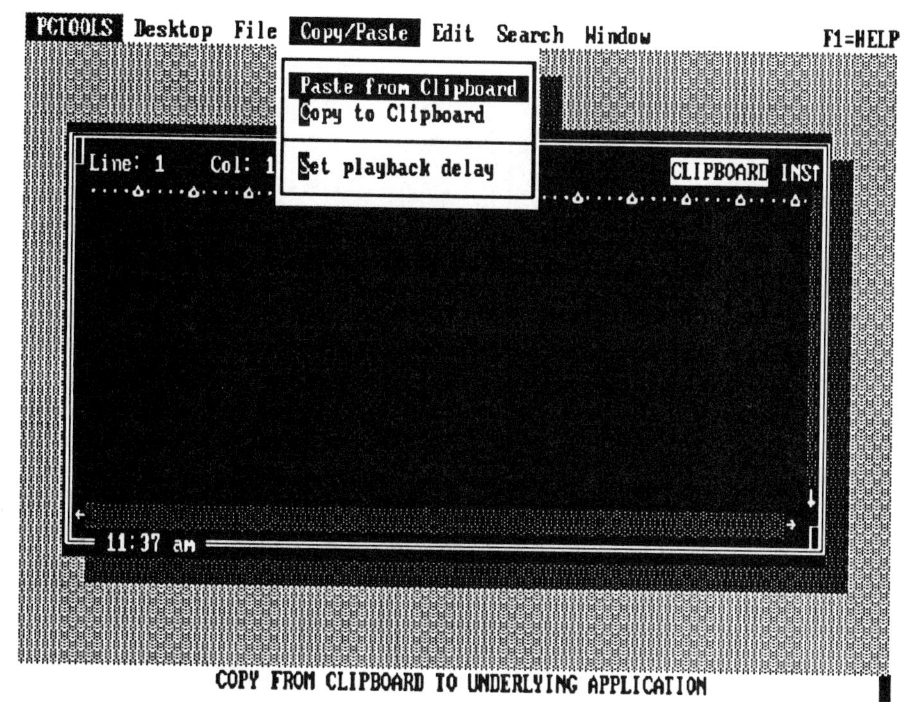

7-13 Copy/Paste menu

The Copy/Paste pull-down menu, shown in Fig. 7-13, lets you Paste from clipboard and Copy to clipboard. PC Tools Desktop must be memory-resident for you to use this menu. It allows you to cut and paste information between a word processing document, a database, and/or a spreadsheet. When you select the Paste from clipboard command, the information or text in the clipboard is placed in the current application, at the position of the cursor.

When you are in an application and want to transfer text into the clipboard, press Ctrl–Space to enter the Desktop menu. Select Copy to clipboard from the Copy/Paste menu. Position the cursor at the beginning of the text you want to copy and press the Enter key. Position the cursor at the end of the text. The block of text to be copied will now be highlighted. To copy the selected text to the clipboard, press the Enter key.

You can use a mouse to highlight the selected text by positioning the mouse pointer on the beginning of the text you want to copy and dragging the mouse to the end of the text.

Use the Set playback delay with macros or with the clipboard.

Once text or information is pasted to the clipboard, you can use the Edit menu, shown in Fig. 7-14, to edit it. The Edit menu of the clipboard is the same as the Edit menu for the notepads, discussed in Chapter 2.

Information can be copied and pasted to the clipboard with hotkeys, which are keys that have a certain function and let you bypass the menus. The hotkeys

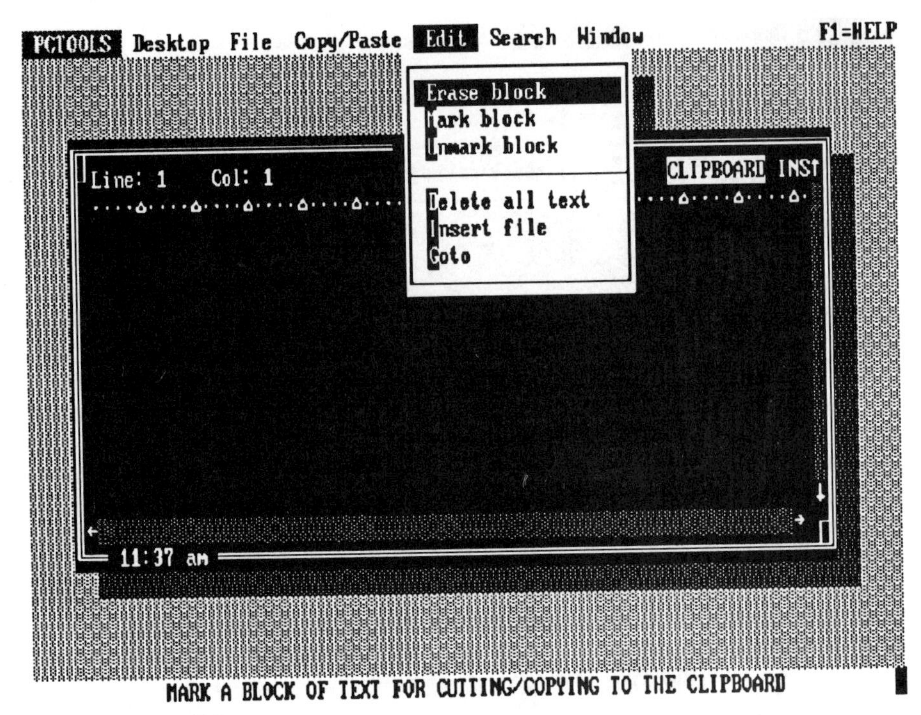

7-14 Clipboard Edit menu

that paste to the clipboard are Ctrl–Del. They only work if the desktop is memory resident. To copy text from the screen to the clipboard, exit from PC Tools Desktop, press the Ctrl–Del keys, and position the cursor where you wish to start coping text. Press the Enter key. Move the cursor to the end of the block of text you want to copy, and press the Enter key. The text is now copied to the desktop, where it can be edited, if necessary. You can also use the mouse to copy text by clicking on the starting text and dragging the cursor to the end of the text you want copied. When you release the mouse button, the text is copied to the clipboard.

Text can be "hotkeyed" from the clipboard by using the Ctrl–Ins keys. Position the cursor where you want the text to be inserted and press the Ctrl–Ins keys. The information from the clipboard is pasted wherever the cursor is positioned.

Utilities

The Utilities menu of the desktop application is shown in Fig. 7-15. The first command in the menu is Hotkey selection, which generates the screen shown in Fig. 7-16. As you can see from the screen, there are four sets of hotkeys, two of which were discussed in the last selection. PC Tools chooses the four hotkey selections, but you can change them if you want. The Clipboard Paste (Ctrl–Ins) and the Clipboard Copy (Ctrl–Del) hotkeys were just discussed. These can be specified by whatever key combination you like. The two other set of hotkeys are

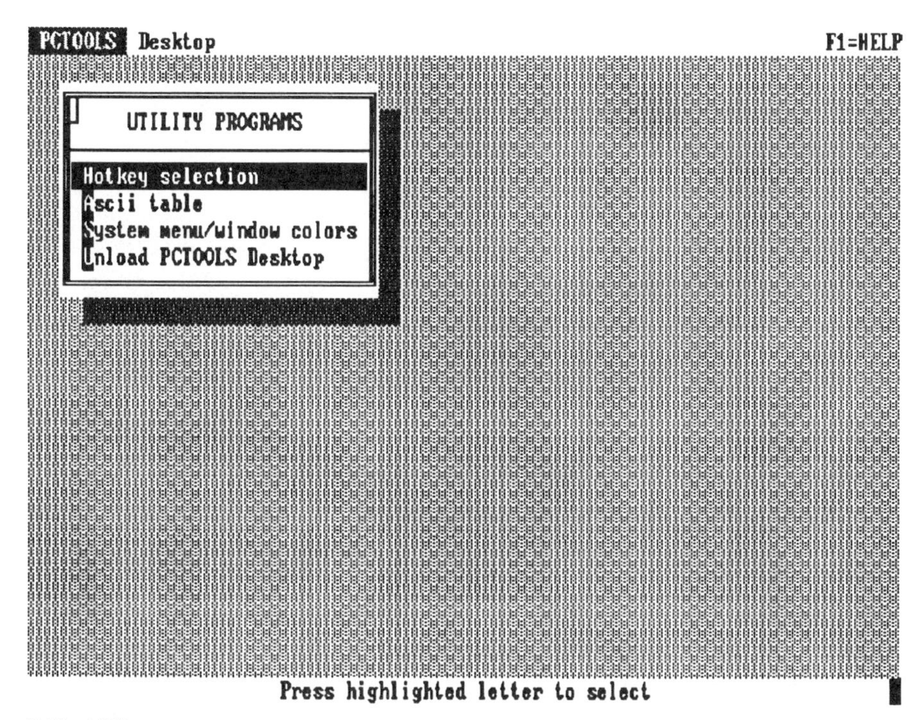

7-15 Utility programs

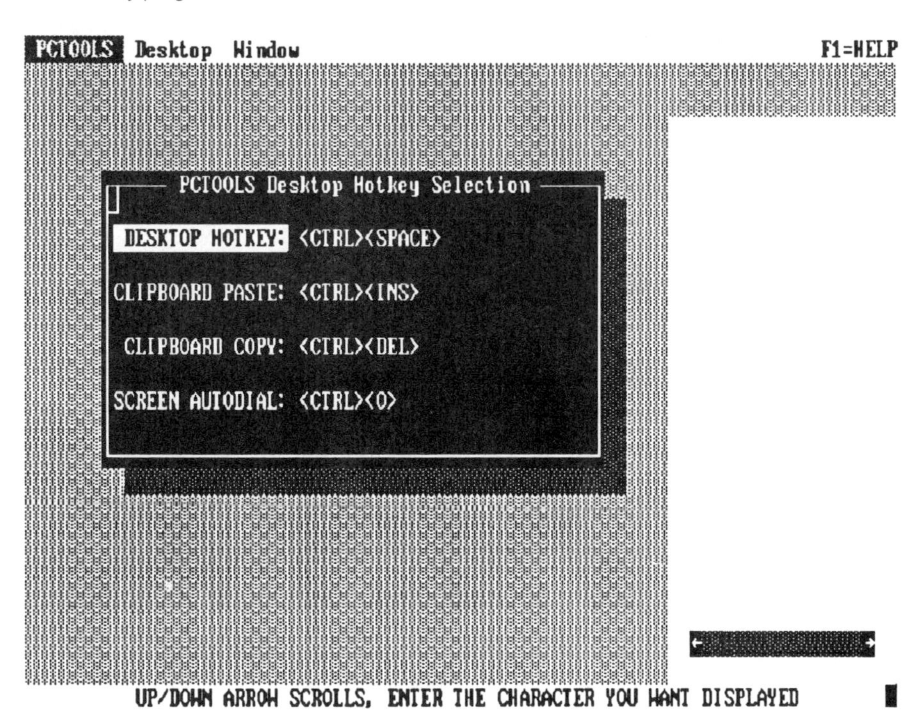

7-16 Desktop hotkeys

the Desktop hotkey (Ctrl−space) and the Screen Autodial (Ctrl−O). The Screen autodial lets you dial the phone number that appears on the screen.

To change the hotkey key-combination selection, highlight the appropriate function and type in the new key combination. The new keys appear in the dialog box, and you can use them in future applications.

The next menu selection in the Utilities menu is the ASCII table command, which provides you with the chart shown in Fig. 7-17. Three different lists of ASCII codes are displayed. ASCII (American Standard Code for Information Interchange) code is the way in which computers represent each letter, character, and number in its memory.

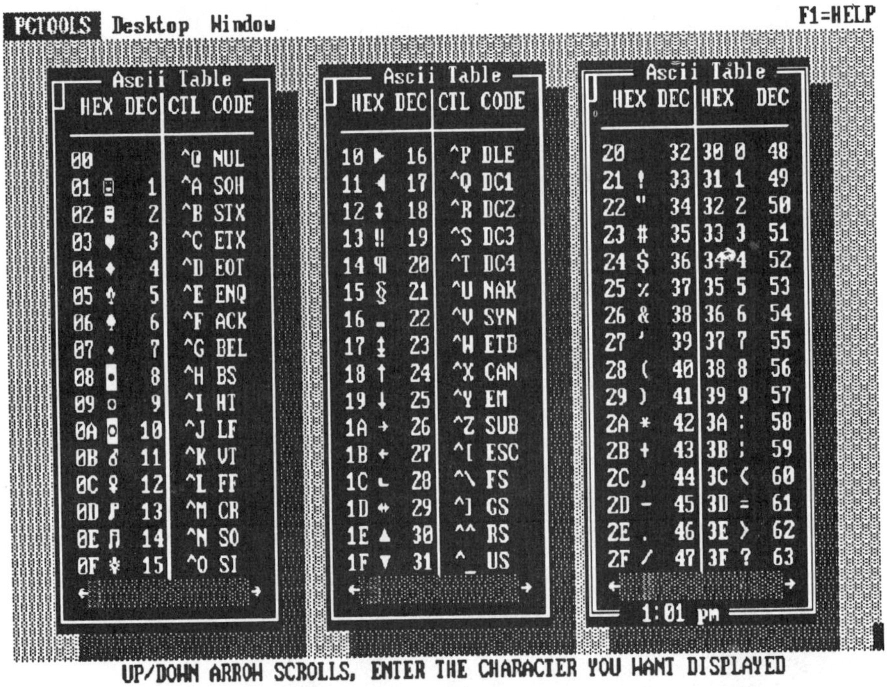

PCTOOLS Desktop Window F1=HELP

Ascii Table

HEX		DEC	CTL CODE
00			^@ NUL
01	☺	1	^A SOH
02	☻	2	^B STX
03	♥	3	^C ETX
04	♦	4	^D EOT
05	♣	5	^E ENQ
06	♠	6	^F ACK
07	•	7	^G BEL
08	◘	8	^H BS
09	○	9	^I HT
0A	◙	10	^J LF
0B	♂	11	^K VT
0C	♀	12	^L FF
0D	♪	13	^M CR
0E	♫	14	^N SO
0F	☼	15	^O SI

Ascii Table

HEX		DEC	CTL CODE
10	►	16	^P DLE
11	◄	17	^Q DC1
12	↕	18	^R DC2
13	‼	19	^S DC3
14	¶	20	^T DC4
15	§	21	^U NAK
16	▬	22	^V SYN
17	↨	23	^W ETB
18	↑	24	^X CAN
19	↓	25	^Y EM
1A	→	26	^Z SUB
1B	←	27	^[ESC
1C	∟	28	^\ FS
1D	↔	29	^] GS
1E	▲	30	^^ RS
1F	▼	31	^_ US

Ascii Table

HEX		DEC	HEX		DEC
20		32	30	0	48
21	!	33	31	1	49
22	"	34	32	2	50
23	#	35	33	3	51
24	$	36	34	4	52
25	%	37	35	5	53
26	&	38	36	6	54
27	'	39	37	7	55
28	(	40	38	8	56
29	)	41	39	9	57
2A	*	42	3A	:	58
2B	+	43	3B	;	59
2C	,	44	3C	<	60
2D	-	45	3D	=	61
2E	.	46	3E	>	62
2F	/	47	3F	?	63

1:01 pm

UP/DOWN ARROW SCROLLS, ENTER THE CHARACTER YOU WANT DISPLAYED

7-17 The ASCII table

The Hex column shows the ASCII code for the character, in hexadecimal format. Next to the hexadecimal column, the second column from the left, is the character displayed on the screen when the ASCII code is generated. The Dec column is the decimal equivalent of the ASCII code for a particular character. The Ctl column lists the control key combination that represent the character. For example, Ctrl Z (^Z), will display the left arrow on the screen. Notice that not all characters are represented by control keys.

You can scroll through the table to find the necessary control character with the arrow keys, a mouse, or simply by typing in the character that corresponds to the ASCII equivalent you are trying to find. You can use the Window selection to

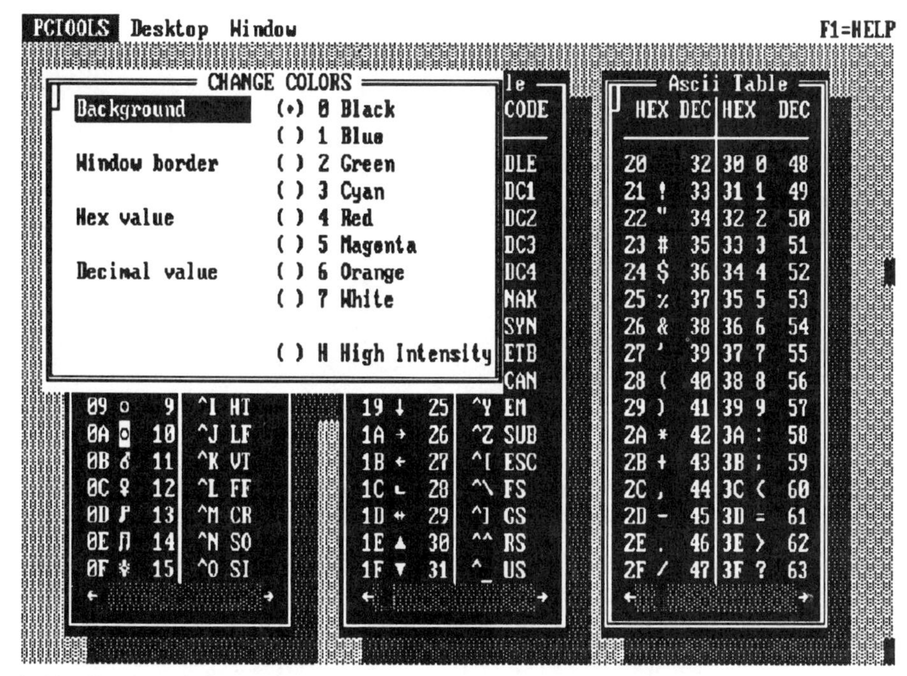

7-18 Change colors command

change the background colors, the window border, hex value, and the decimal value of the columns, as shown in Fig. 7-18.

The System menu/window colors command of the Utilities menu gives you the screen shown in Fig. 7-19. This lets you select the colors of the Desktop menu, the dialog box, and the message box for all applications. The Change Colors menu of Fig. 7-18, on the other hand, changes the colors only for the ASCII table.

The Unload PCTOOLS Desktop command lets you unload the desktop from memory, and generates the dialog box shown in Fig. 7-20. You must unload PC Tools Desktop if your application requires the memory being taken up by the desktop. This option erases the program from memory but not from the disk. If you have other memory-resident programs in addition to desktop, you must unload them before you remove the desktop. You must be at the DOS prompt to use this command. Another way of removing PC Tools Desktop from memory is to type the word KILL at the DOS prompt.

Autodialer

A phone number that is on-screen can be automatically dialed with the hotkeys Ctrl–O. A dialog box appears on the screen asking you to choose between dialing the highlighted number, dialing the next phone number on the screen, or canceling the operation. The phone number will automatically be dialed if your computer is attached to a modem.

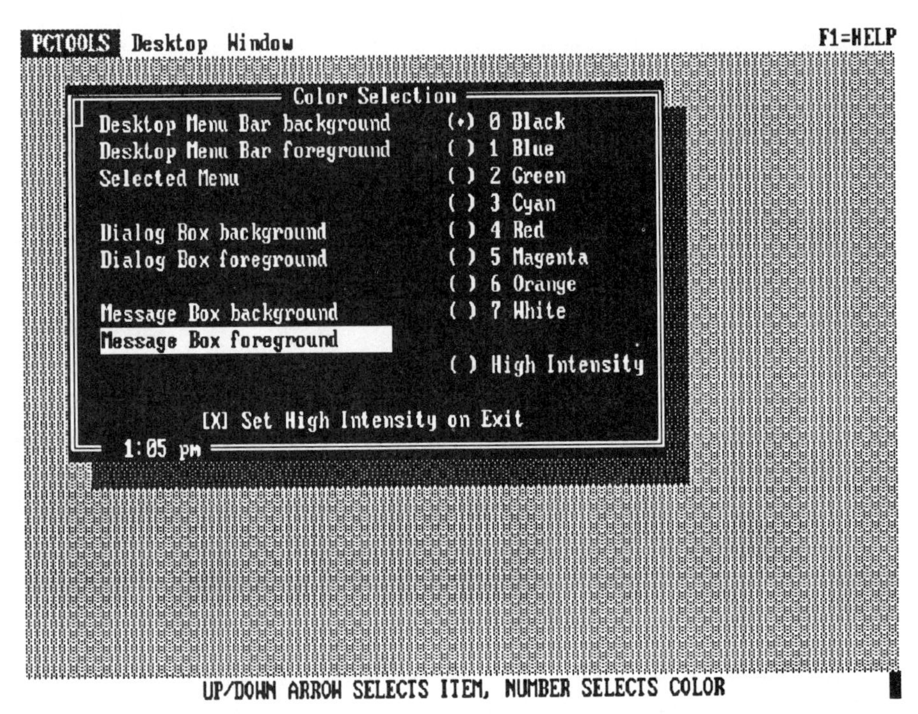

7-19 Color Selection window

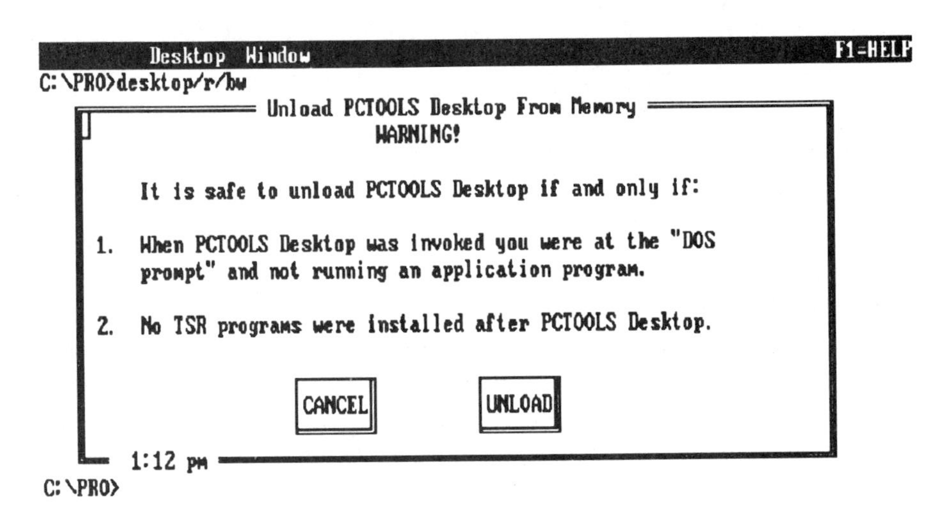

7-20 Unloading the desktop

8
CHAPTER

Calculators

In this chapter, you will learn how to use the four different calculators of PC Tools: the algebraic calculator, the financial calculator, the programmer's (hex) calculator, and the scientific calculator.

When you select Calculator from the Desktop menu, the four different calculator types appear, as shown in Fig. 8-1. These calculators can be used with any program because they are memory-resident with the desktop program.

Algebraic calculator

The algebraic calculator is similar to any arithmetic calculator. Its basic functions are shown in Fig. 8-2. It can add, multiply, subtract, divide, calculate percents, and has a memory with which you can save and retrieve positive and negative numbers. You can operate it from the keyboard or with a mouse. Simple mathematical equations, like the one shown in Fig. 8-3, are easy to solve with this calculator. The left-hand side shows the numbers and operations on the screen.

The calculation in Fig. 8-3 computes the calculation 77 * 3 / 5 + 456. It is entered by typing the following numbers and symbols: 77, *, 3, /, 5, +, 456, and =. The result, 502.2, appears both on the screen and in the window box. You can clear the result by pressing the Clr (Clear) key on the calculator, as shown in Fig. 8-4. Figure 8-4 shows that you can also edit the screen display. To edit a number in the display, position the cursor over with the up and down arrows. Type in a

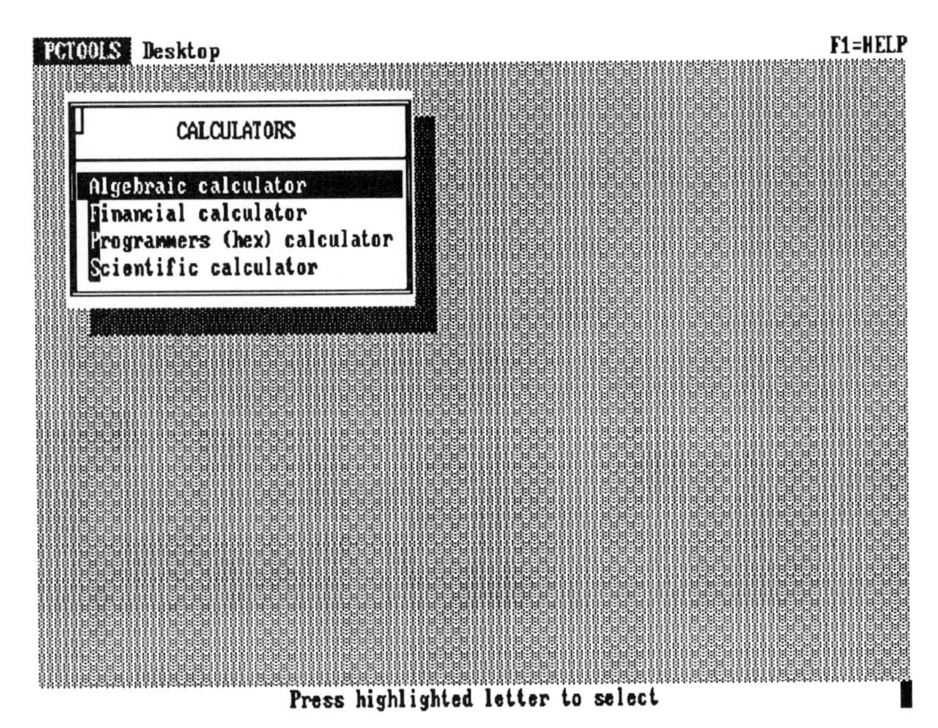

8-1 Calculator menu

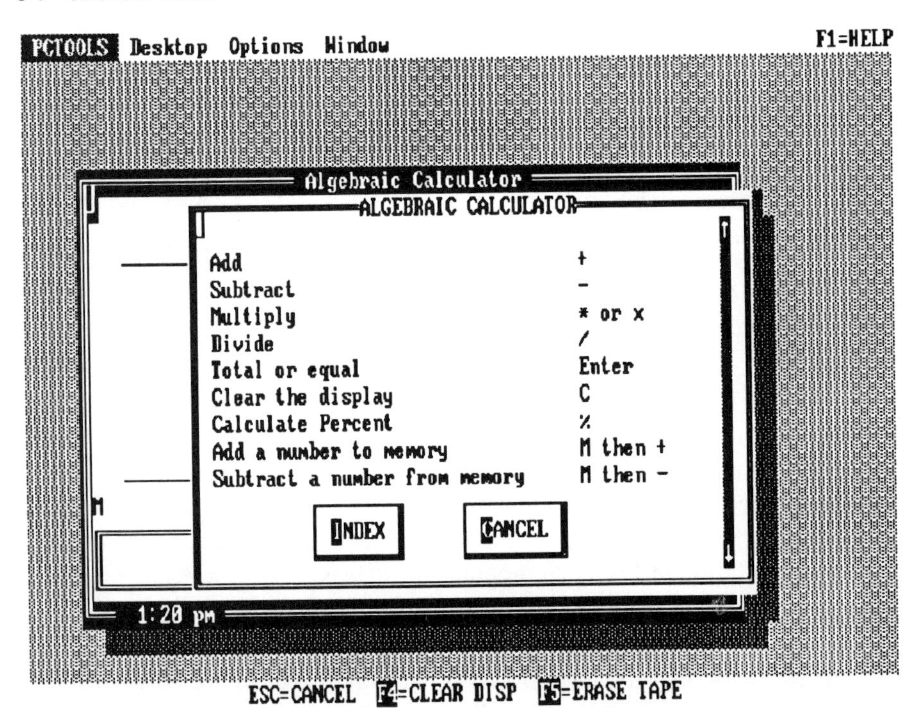

8-2 Algebraic calculator

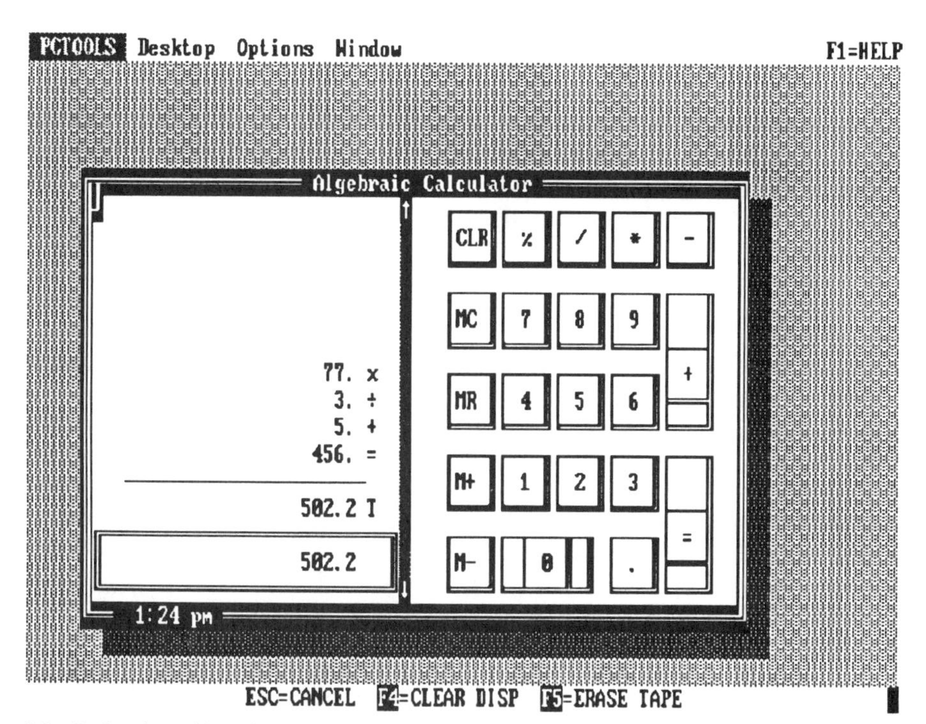

8-3 Performing arithmetic calculations

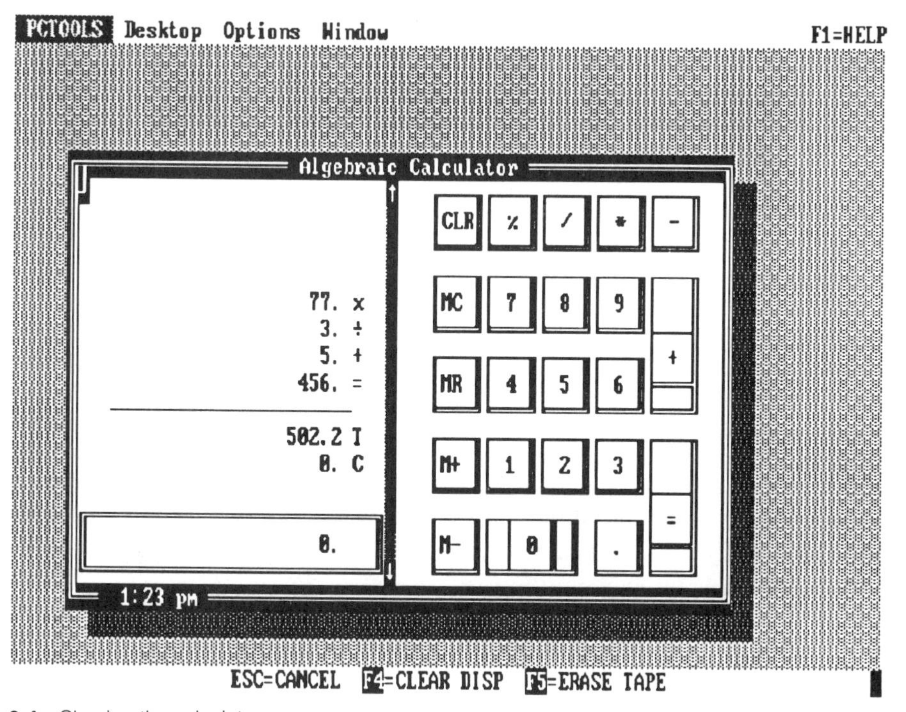

8-4 Clearing the calculator

new number to overwrite the old number. Press the End key and the screen display is recalculated with the new number.

This calculator, like most calculators, has a memory. You can save a positive number by pressing M+, and a negative number with M−. The number in memory can be recalled by pressing MR, or cleared by pressing MC. When you use a memory key, an M will appear on the screen, as shown in Fig. 8-5.

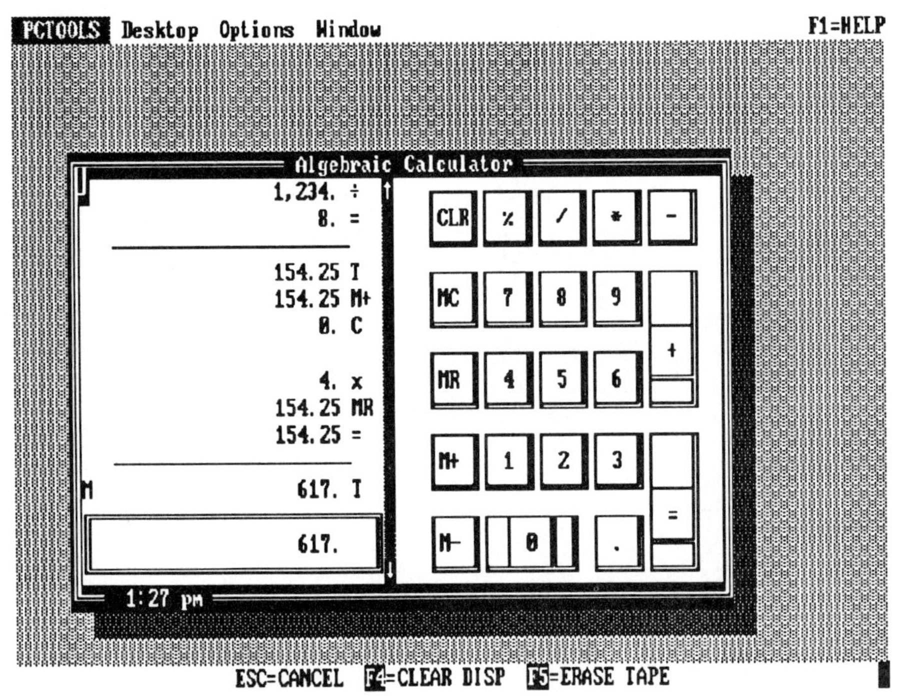

8-5 Using the memory keys

The algebraic calculator has an Options menu, shown in Fig. 8-6. The Clear display command lets you clear the calculator display. The number in the display can also be cleared by pressing the Clr key. You can also clear a display with the F4 key. The Erase tape command lets you erase all the arithmetic operations from the "tape" (screen display). You can also erase the tape by pressing the F5 key. The Copy to clipboard command lets you copy the last 100 lines of calculations you made to the clipboard, where they can be inserted into other documents. The Print tape command prints all the on-screen calculations. The Wide display command is a toggle switch that shows the calculator either with or without the keypad. A wide display is shown in Fig. 8-7.

The Window menu has a Change colors command shown in Fig. 8-8 that lets you change the colors of the background, window border, calculator keys, and tape.

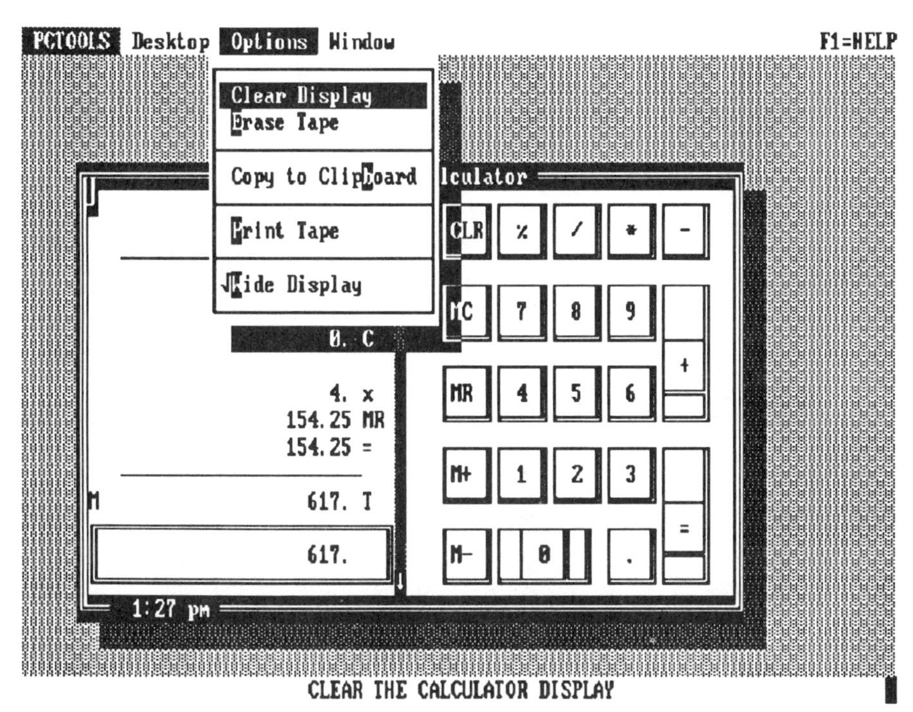

8-6 Calculator Options menu

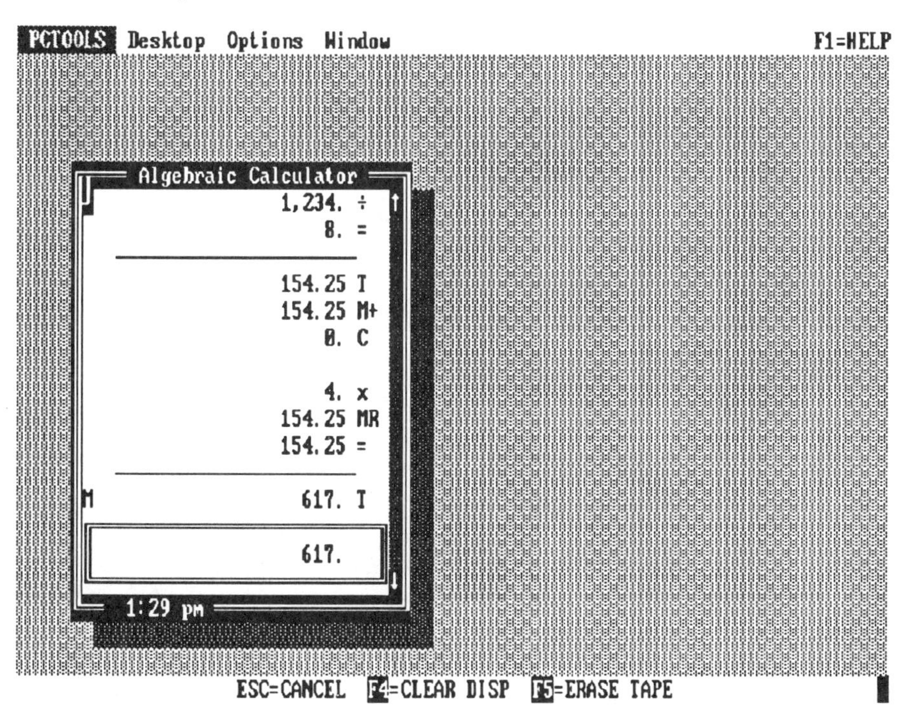

8-7 Wide window display

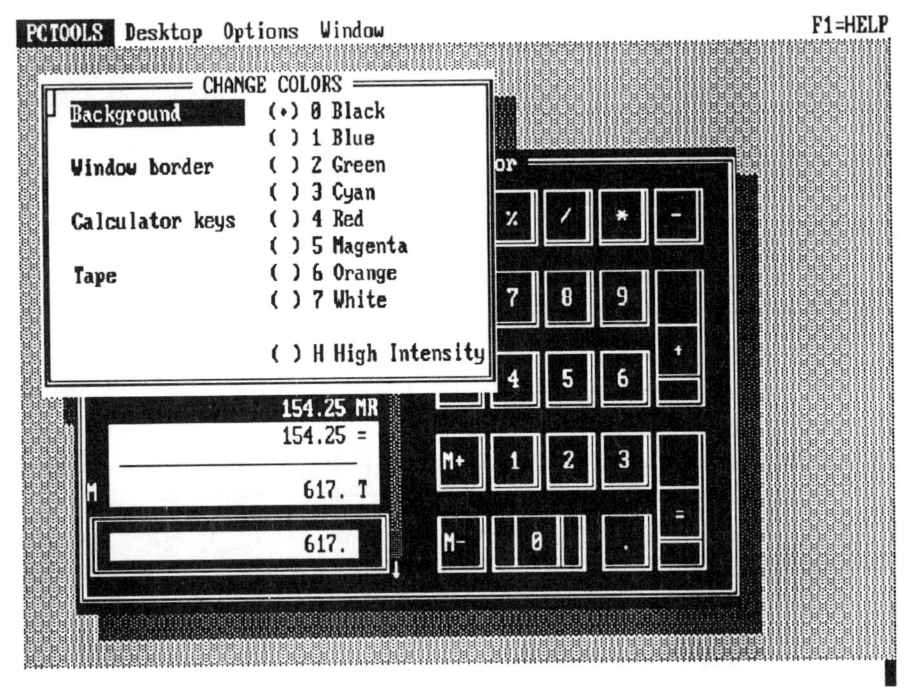

8-8 Change Colors menu

Financial calculator

The financial calculator initially appears very threatening, as shown in Fig. 8-9. It is a simulation of a Hewlett-Packard calculator. You can use it for very simple calculations or the computation of financial formulas. Simple calculations are performed the same way as on the algebraic calculator. Use the keyboard or mouse to select the number and the arithmetic operator. The numbers you select are displayed in the stack register, shown in Fig. 8-10. The stack register shows the arithmetic calculations, and displays the intermediate result of all calculations. The LSTX line holds the last number before an operation is performed.

The financial calculator has many buttons and features, all of which can be accessed with the keyboard or a mouse. By pressing the F7 key on the keyboard or highlighting the F key with a mouse, you can access all the functions on the top of each key. A small "f" will appear in the display box to indicate this. By pressing the F8 key on the keyboard or highlighting the G key with a mouse, you can access all the functions on the bottom of each key. A small g appears in the box, as shown in Fig. 8-11. If you wanted to calculate the square root of 8, you would have to highlight the square root of x key, which is the bottom part of the A key. The single letter to the left of some of the calculator keys is the key you press on the keyboard to access that key. If you have a mouse, you can move the mouse cursor to the key you want and press the mouse button.

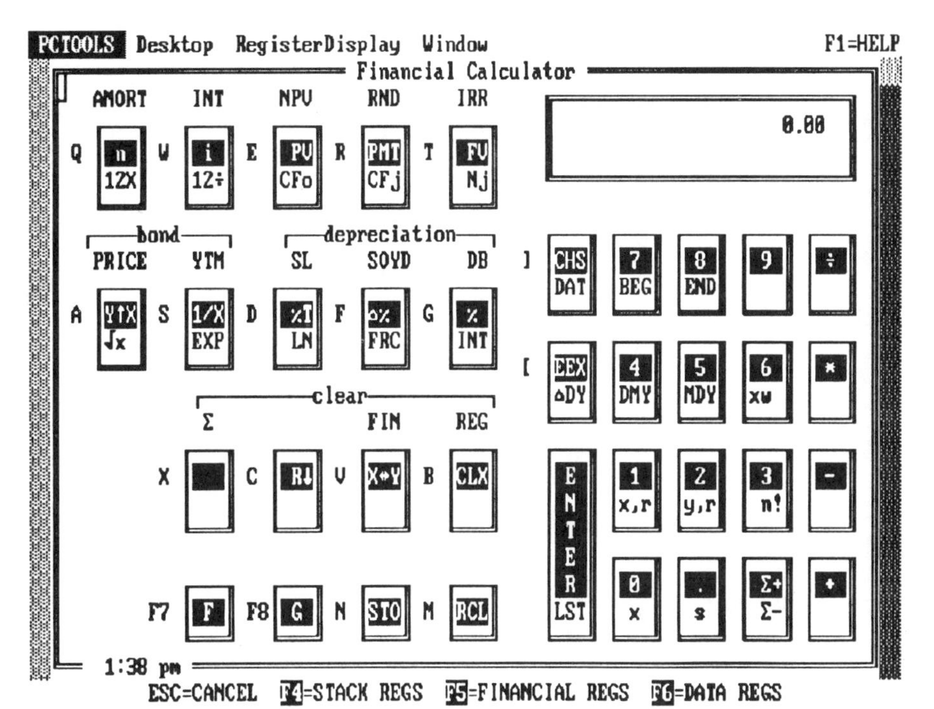

8-9 Financial calculator

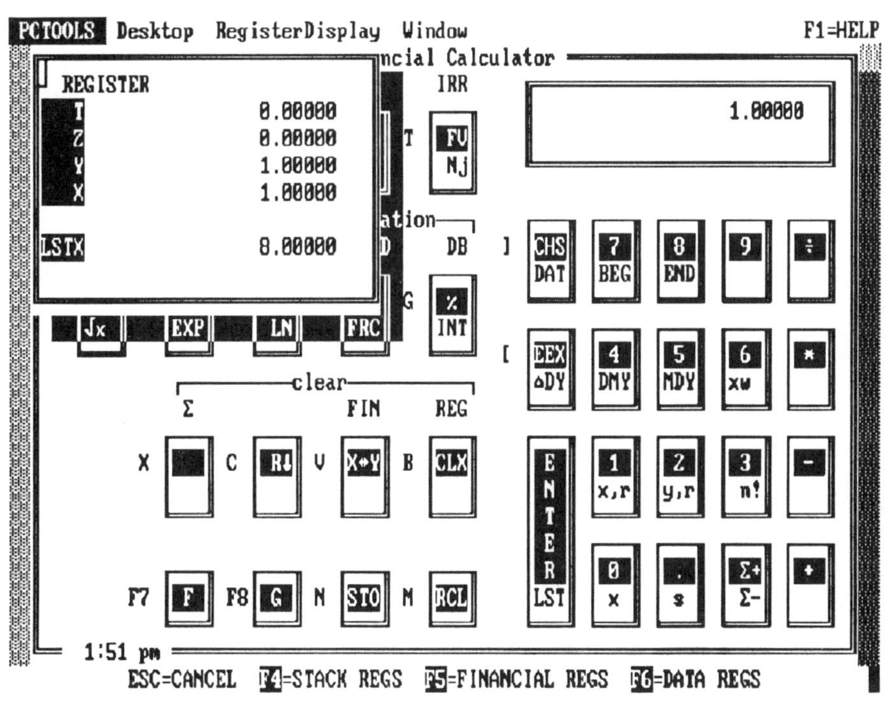

8-10 Using stack registers

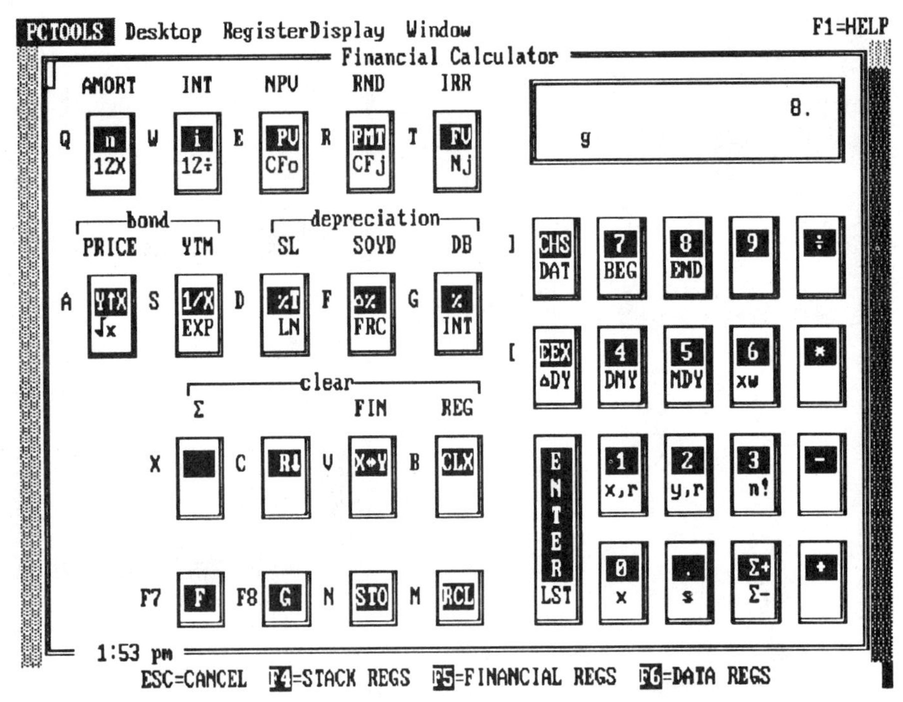

8-11 Using the "g" key

You can set the number of decimal places, up to 9, by pressing the F followed by the number of decimal places you want. For example, F2 will result in all computations having two decimal places. You can calculate continuous compounded interest with the STO/EEX toggle switch. The default date format is Month, Day, Year, but can be changed if you select DMY (number 4 on calculator pad). This will give a Day, Month, Year format.

The financial calculator has many built-in functions. When you press the F5 Financial Registers key, a box appears showing you five items that can be entered through the calculator. Entering any four items will cause the fifth to be computed. Enter information by typing the numbers in the display and highlighting the function with the keyboard and mouse.

Interest computation

Let's use the financial registers to compute interest on a loan (see Fig. 8-12). Suppose you want to take a loan out for $10,000, pay $200 per month for 6 years, and want to compute the annual percentage rate. The present value (PV) of the loan is $10,000. You are making monthly payments (PMT) of $200. It is negative because you are paying it. The number of payments (n) is 72 months, because it is for 6 years, computed monthly (6 times 12 equals 72 periods of payment). The future value (FV) of the loan is 0, which means the loan will be paid off. Interest can be computed either at the beginning of the period (notice the BEG key on the

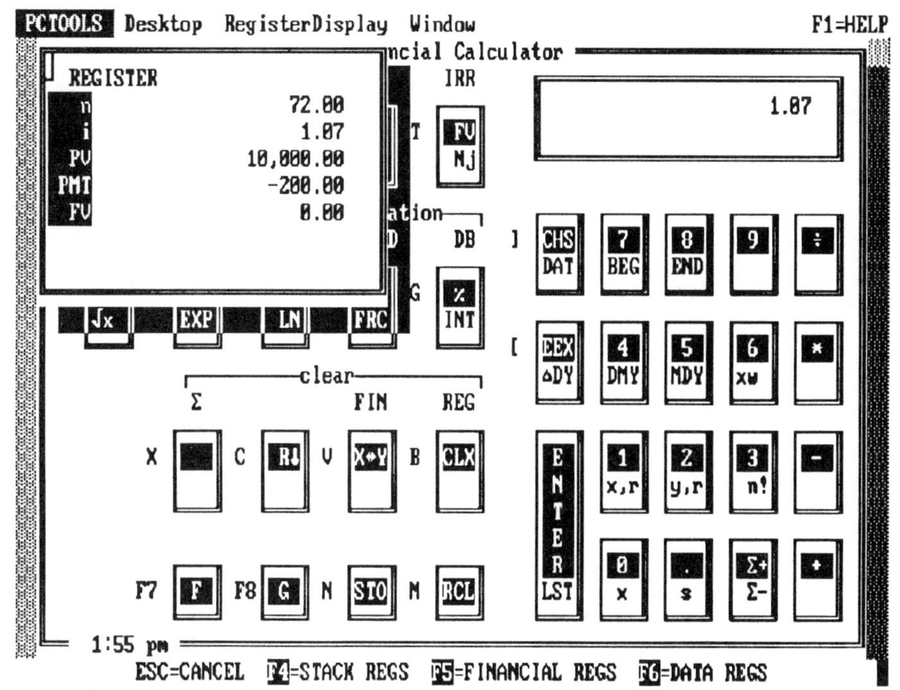

8-12 Calculating monthly interest

keypad) or at the end of the period (the END key). You should highlight the BEG or END keys before doing any calculations, to indicate which type of compounding you select. When you are finished entering all the data, press the interest rate (i) key and you will see the word *running* on the display. You will then see the number 1.07 in the display. This is the monthly interest. To get the yearly interest rate, multiply it by 12 (12.84%).

Mortgage payment

The financial calculator can also compute mortgage payments. Let's say you want to compute the monthly payment on a $100,000 mortgage, at a 10% interest compounded monthly. Figure 8-13 shows the information you need to enter into the financial register to accomplish this. The number of periods (n) is 120 (10 years × 12 months = 120 payments). The interest rate is 10% ÷ 12, or 0.83 per month. The future value (FV) of the mortgage is 0, assuming the mortgage will be paid off in full. When you highlight the PMT function, you will get a monthly payment of $1321.51. This is the monthly payment necessary to pay the mortgage.

Annuities

An *annuity* is a series of equal payments made over a specified interval of time. They can be made yearly, semiannually, quarterly, or monthly. Figure 8-14 shows

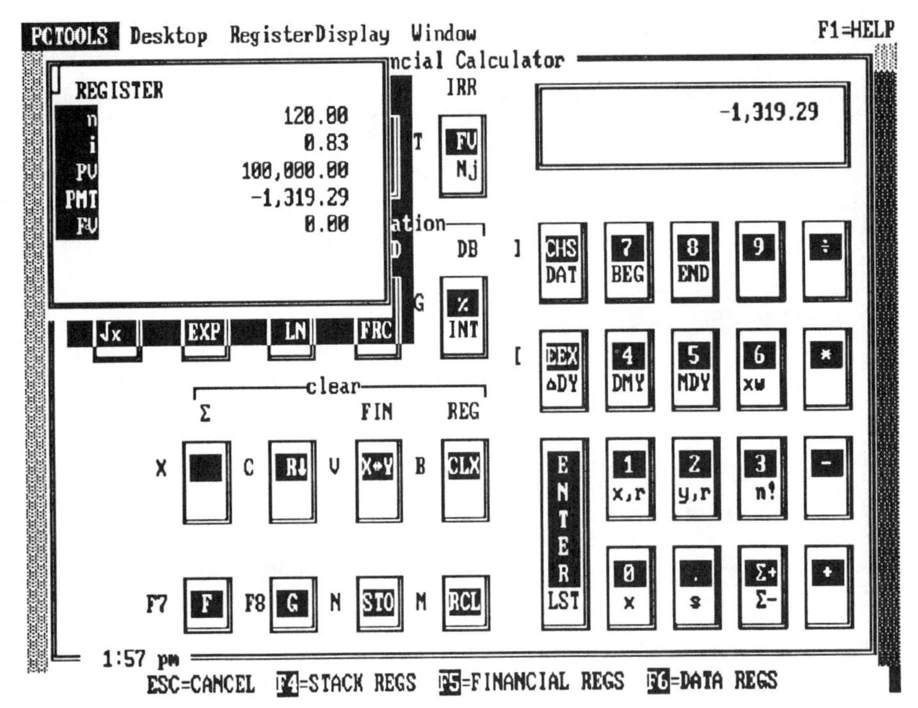

8-13 Calculating mortgage payments

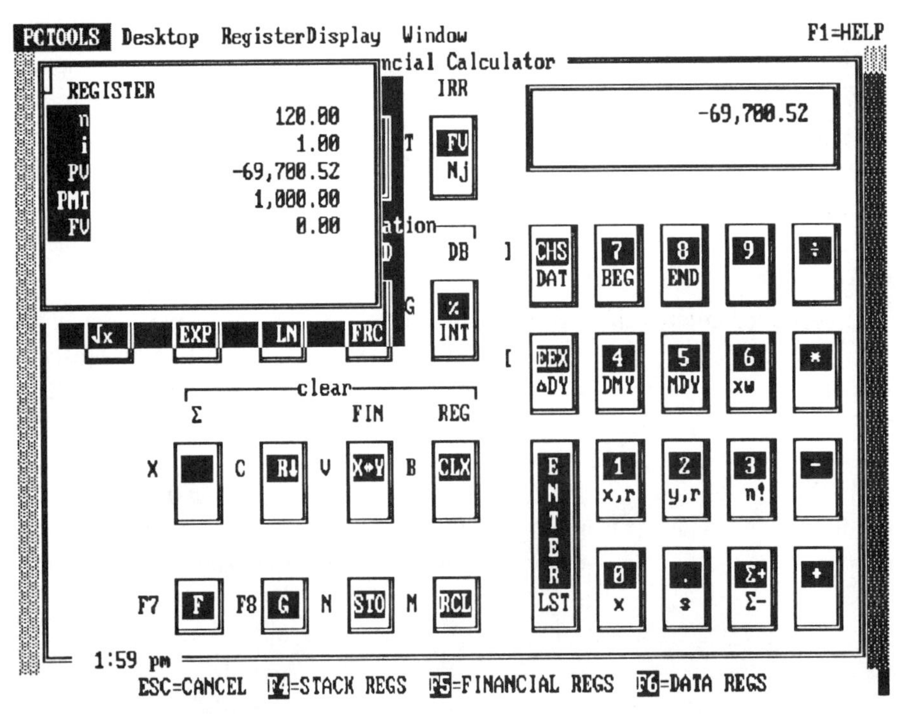

8-14 Calculating annuities

the computation for an annuity in order to pay $1,000 monthly for 10 years at a 12% annual interest rate. The number of payments is 120 (10 years × 12 months). The monthly interest rate is 1%, and the yearly rate is 12%. The present value (PV) indicates that you will need $69,700.52 to pay this annuity.

Future value

Another financial formula computes the future value of an investment. An *investment*, or commitment of cash, must provide an increase in value over time. You decide how much money you want to commit; then calculate how much that money will increase once you know the expected rate of interest. The amount of money you end up with is the investment's *future value*.

Suppose you want to invest $100 every month for three years at a 10% rate, compounded monthly. How much money will you have at the end of the three years? Enter the date into the financial register, as shown in Fig. 8-15. The number of periods is 36 (3 years × 12 months). The interest rate (i) is 0.83 (10 percent ÷ by 12). The present value (PV) is 0. The monthly payment (PMT) is $100.00. When you compute the future value, you get $4,718.18. If you invest $100 every month for three years at a 10% monthly compounded rate, therefore, you will have $4,178.18.

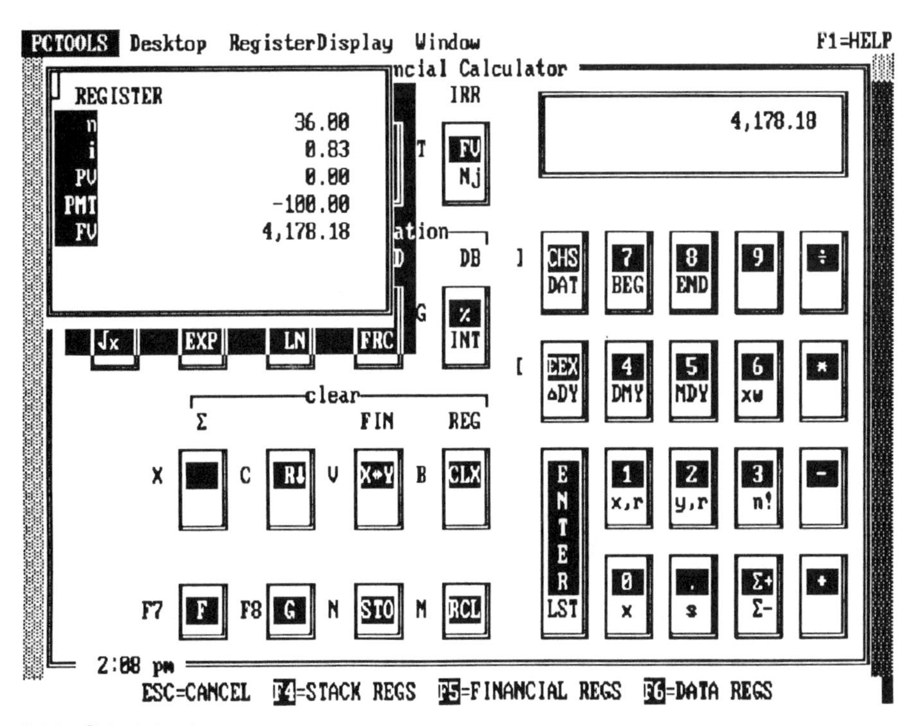

8-15 Calculating future value

Number of payments

The number of payments (n) function can determine how long it will take to reach a specific financial goal. Figure 8-16 illustrates this financial function. Suppose you have the opportunity to invest $2,500 at 10% interest, and want to determine how many years it will take to compound to $20,000.

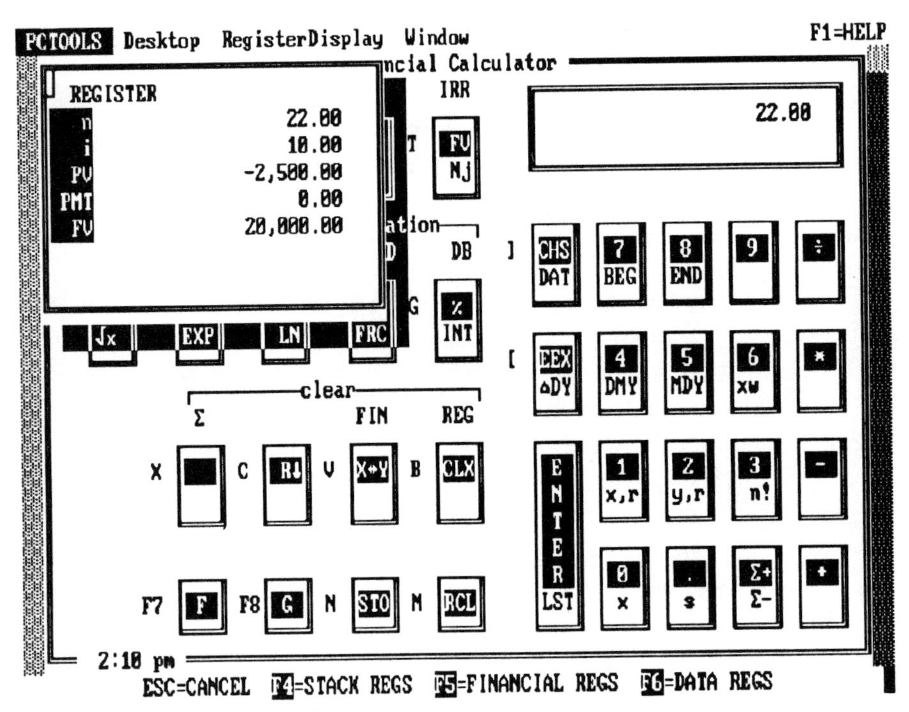

8-16 Calculating the number of payments

The interest rate (i) is the periodic interest rate (in this case, yearly), of 10%. The present value (PV), or the amount you want to invest, is $2,500. The future value (FV), the amount of money you want to accumulate, is $20,000. When you press the n key, you get 22.00. This means it will take approximately 22 years to reach the specific goal of $20,000.

Rate register

The interest rate (i) function can determine the interest rate required to reach a specific goal. Suppose you have $5,000 in cash that you want to invest for 10 years. You want to reach an investment goal of $20,000. What is the interest rate necessary to reach this goal?

Figure 8-17 shows how to perform this calculation. The goal is the future value (FV), which is $20,000. The amount invested is the lump sum of money you wish to invest, or $5,000. The number of periods (n) is how long the investment will compound. In this case, it is for 10 years. The result of this computation is

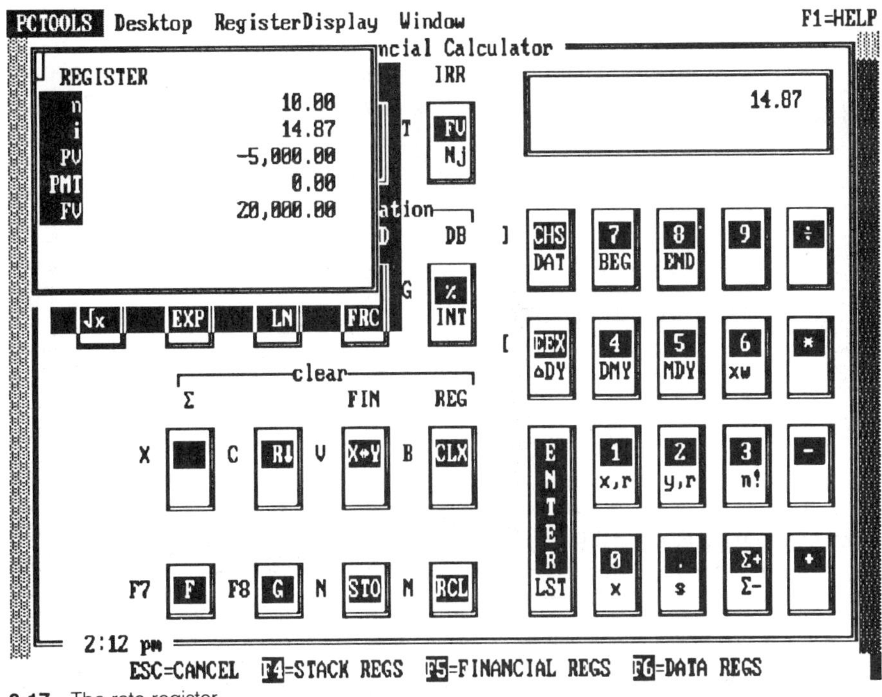

8-17 The rate register

14.87%. This means you will need an annual interest rate of 14.87% to reach a goal of $20,000, by investing $2,500 in a lump sum for 10 years.

Depreciation functions

The financial calculator has three different methods to calculate depreciation. Suppose you want to depreciate $5,000 worth of computer equipment over a period of five years. You estimate that, at the end of five years, the salvage value of the computer equipment will be $500. Let's calculate and compare the three different methods of depreciation.

Figure 8-18 depicts *straight-line depreciation* (SL). The original cost of the computer (PV) is $5,000. The salvage value of the computer, $500, is the future value (FV). The life of the item is 5 years, which is the number of periods (n). The depreciation for the computer is the same for each year with straight-line depreciation, which in this case is $900 for each of the five years. Compute this after you enter all the data into the financial register by pressing the SL key, for straight-line depreciation.

The second method of depreciation is the *sum of the years digit* (SOYD) depreciation. Figure 8-19 shows how much the computer is valued at the end of 5 years (n=5.00), which in this case is $300. If you let n equal 1, you will compute the value at the end of the first year, which is $1,500. The second year will give you $1,200, the third year $900, and the fourth year $600. You can compute the

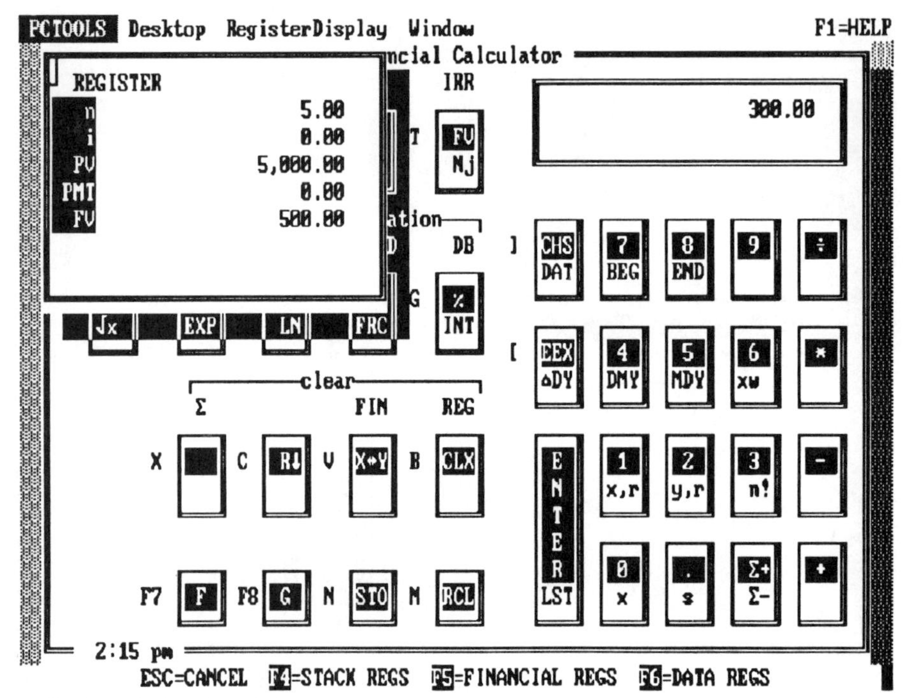

8-18 Straight line depreciation

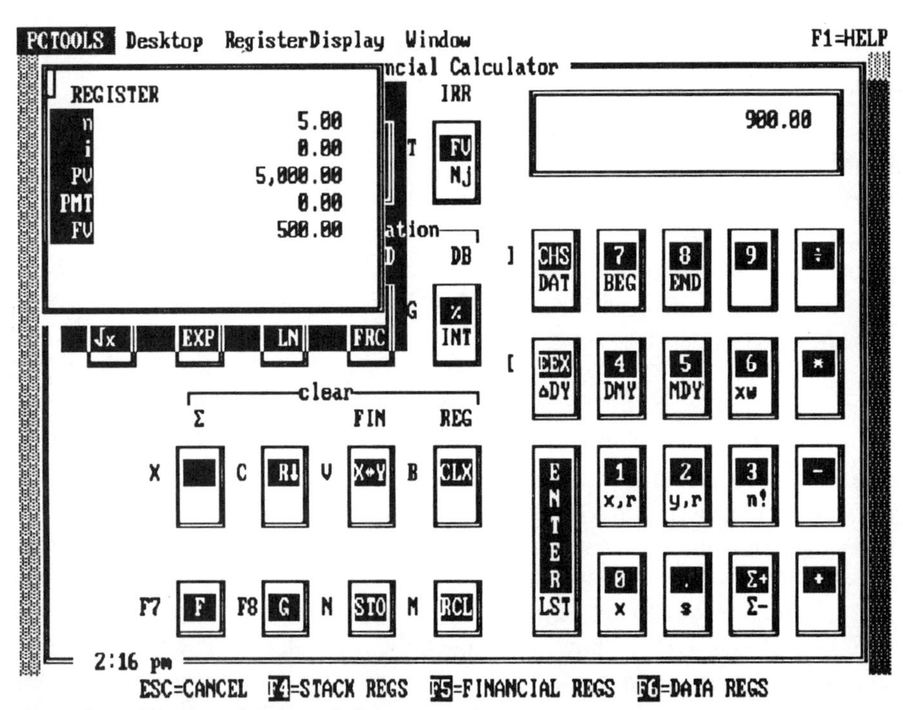

8-19 Sum-of-the-year digit depreciation

salvage value at the end of each year by changing the value of n, the number of periods.

The third depreciation method is the *double declining balance* (DB), shown in Fig. 8-20. The double declining balance method of calculating depreciation is an accelerated method. It is higher for the earlier years or periods, and lower for the later years or periods.

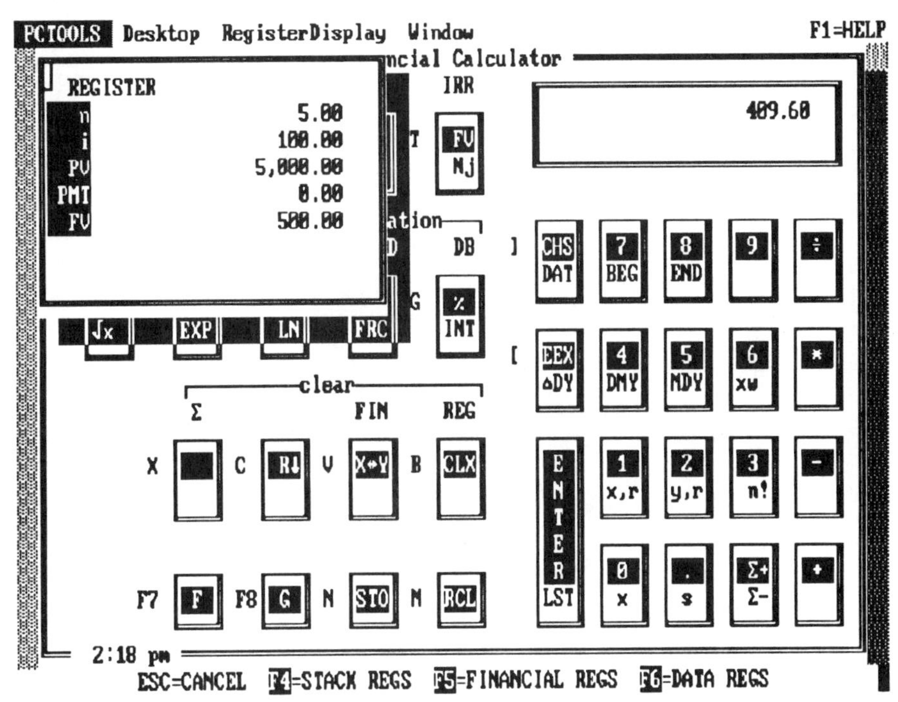

8-20 Double declining balance depreciation

Since the depreciation changes from one year to the next, the double declining balance formula requires an extra piece of information in the register, the rate (i). In this case, assume 100% depreciation each year. During the fifth year of depreciation, the computer has a salvage value of $409.60. Depending on the rate chosen, the double declining balance method of depreciation normally has the fastest depreciation of the three different methods.

Average

The financial calculator can also compute the average of any amount of numbers. Suppose you want to find the average of 98, 77, 87, 90, and 81. Type in 98 followed by the Σ+ key. Type 77 and Σ+, 87 and Σ+, 90 and Σ+, and 81 followed by pressing the × key. The × key computes the mean, or average of the number set. Don't forget to press the F8 key on the keyboard or the G key with the mouse to access the lower function of the key. Figure 8-21 shows the result, 86.60.

The complete list data registers for this average problem is shown in Fig. 8-22. It shows how the computer stores the numbers in its memory to solve the problem.

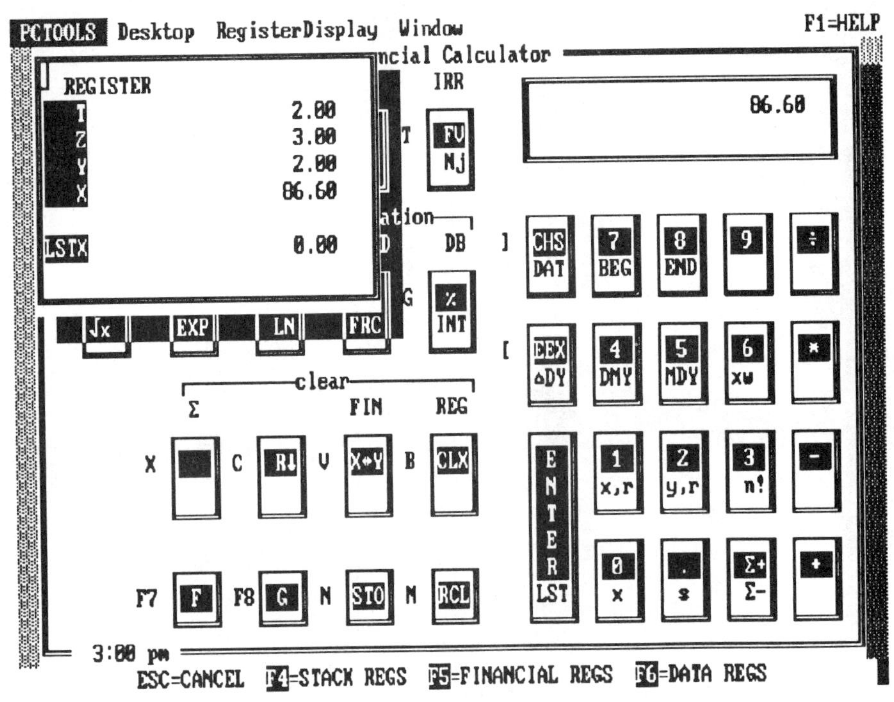

8-21 Computing the average of numbers

Window menu

The Window menu of the financial calculator, shown in Fig. 8-23, lets you change the colors of the background, window border, and f and g functions.

Programmer's calculator

The programmer's calculator lets you display numbers in different bases: hex, octal, binary, and decimal. It also displays the ASCII, or IBM graphics character that represents that number. Figure 8-24 shows the square root symbol with its associated decimal, binary, octal, and hex numbers. You can type a number in any of the four bases and it will display the matching information in the other three, including the graphic character.

You can also do arithmetic operations in this calculator by typing in a number, choosing the +, −, /, or * symbol, typing another number, and pressing the Enter key. The calculator can be cleared at any time with either the Backspace or Enter key.

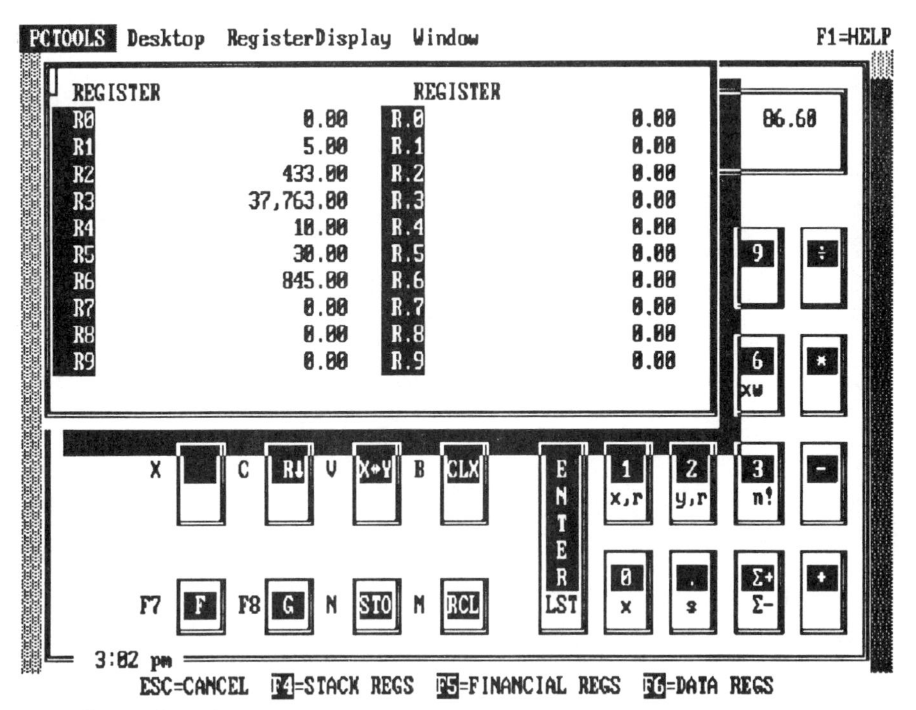

8-22 Data registers for the average problem

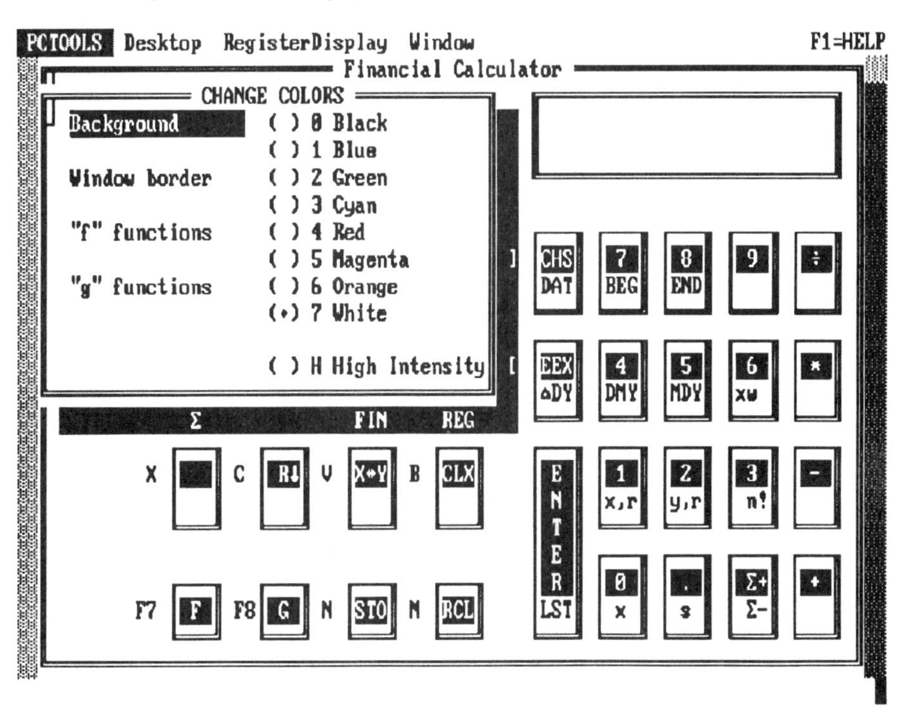

8-23 Financial calculator Change Color window

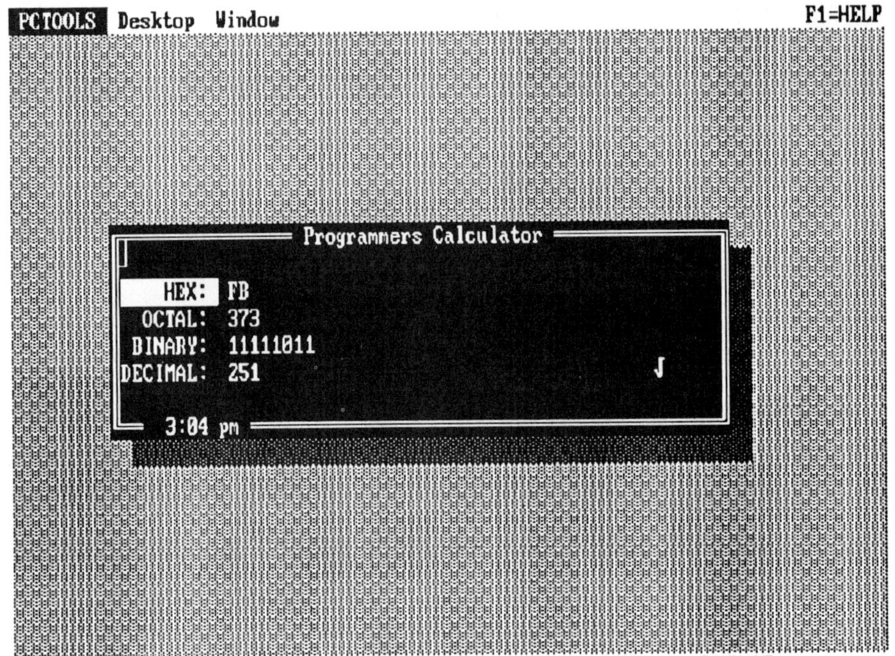

8-24 Programmer's calculator

Scientific calculator

The last calculator is the scientific calculator, shown in Fig. 8-25. It operates with reverse Polish notation. The operator (+ − * /) is entered after the number is entered. Instead of entering 7 + 9, for instance, you would type 7, Enter, 9, Enter, +. To calculate (175 − 44) * 5, type 175, Enter, 44, −, *, and 5. The result, 655, will be shown in the display box on the screen.

As Fig. 8-25 clearly shows, the scientific calculator has logarithm, trigonometric, and degree functions. It can do statistical combinations and permutations. Figure 8-26 shows the permutation of 7 items, taken 3 at a time. It also shows where the information is residing. The result, 210, means that there are 210 ordered choices to take 3 items out of a maximum of 7 items without any repetitions.

The scientific calculator can display data three different ways: Fixed (FIX), which lets you specify the number of digits after the decimal point, Scientific (SCI), which displays a number in scientific notation, and Engineering (ENG) which displays exponents in powers of three. To select the number of digits after the decimal point in fixed mode, select F FIX followed by the number you want, such as F FIX 3 for three digits after the decimal point. If you select F SCI 4, you would get scientific notation with three digits after the decimal point. If you select F ENG 4, you would be selecting engineering notation with four digits after the first significant digit.

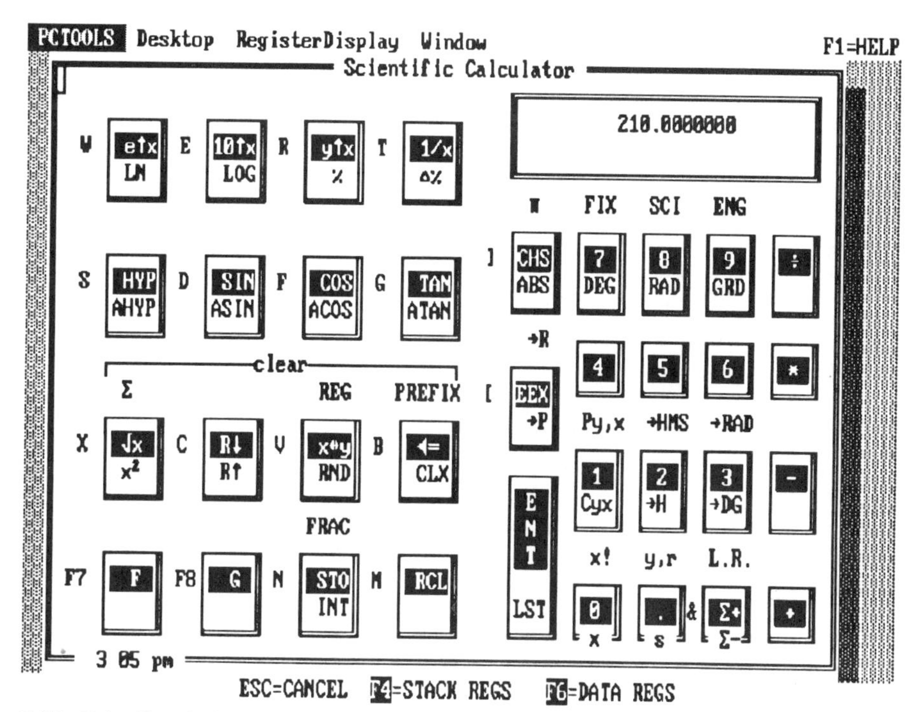

8-25 Scientific calculator

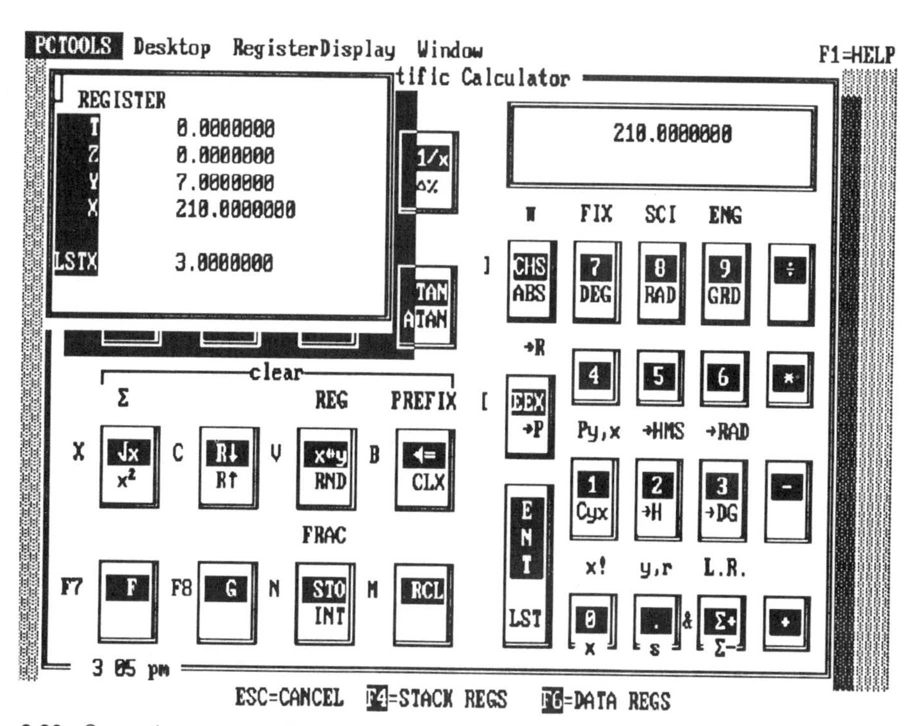

8-26 Computing a permutation

9
CHAPTER

PC Shell

In this chapter, you will learn what a shell is and how it can be used to copy, move, compare, find, rename, delete, verify, edit, run, and print files.

What is a shell?

A shell is a program that protects applications from one another. DOS is a shell that provides an interface to the operating system. It takes the commands you enter, ensures that they are carried out, and protects you from the mechanisms of DOS. The advantage of a shell such as PC Shell is that it provides extra power, flexibility, and other features beyond those found in normal DOS. It is also much easier to use than the DOS shell.

PC Shell provides you with a menu based interface for working with DOS. This means you can do your work by selecting options from a menu, rather than by typing in commands. PC Shell lets you copy, move, find, rename, delete, edit, and print files just by selecting a particular item from a menu.

PC Shell can be a resident program, which means you can access it while running other programs. It can also run as a nonresident program from the DOS prompt if you type PCSHELL. To run PC Shell as a resident program, type PCSHELL/R, as shown in Fig. 9-1. You can use the hotkey Ctrl – Esc to execute PC Shell anytime you are in another program. It can be changed to another key combination with the Utilities menu of the desktop. The screen in Fig. 9-1 also tells you how much memory is available.

```
                    PC Shell (tm)
             DOS shell and disk utilities
                     Version 5.5

  Copyright (c) 1985-1989 Central Point Software, Inc.
                 All rights reserved
                   506 Kbytes free

To activate PC Shell, press <CTRL><Esc>
```

9-1 Activating PC Shell

To remove PC Shell from memory, type KILL. This command removes both the desktop and PC Shell from memory. PC Shell can be loaded just like the desktop, but with other parameters. For example, PCSHELL/R/BW will load PC Shell as a resident program and display it in black and white if you have a color monitor. PCSHELL/LE exchanges the left and right mouse button functions to accommodate left-handed people.

Main Menu

When you press Ctrl−Esc, the Main Menu of PC Shell appears, as shown in Fig. 9-2. The top horizontal menu bar displays the pull-down menu options, including the time and the scroll lock status (either on or off). If the scroll lock is off, then the cursor can move freely in the File List window. Otherwise, the cursor usually stays in the center of the window.

The second line from the top of the screen is the drive command line. It shows the available drives for the system, and highlights the current drive that is in use. You can change from one drive to another by pressing the Ctrl key and the drive letter from the keyboard. For example, if you press Ctrl−D, you change the selected drive to D. You can also select the desired drive by clicking on it with the mouse.

The Tree window on the left side of the screen displays a graphic tree of all the directories and subdirectories on the disk. The directory you are working with is marked with a checkmark. It has its own scroll bar on the right-hand side of the window. To scroll in a window, use the up and down arrows to move one item at a time, or the PgUp and PgDn keys to move one screen at a time. You can also use the mouse to click on the arrows of the scroll bar in order to move the cursor through the subdirectory names. To select a subdirectory, click on the name of the subdirectory with the mouse or use the arrow keys to select it from the keyboard.

The File List is located on the right side of the screen. It shows all the files that are located in the subdirectory currently checked off in the left window. It too has its own scroll bar on the right-hand side of its window. Every time PC Shell is

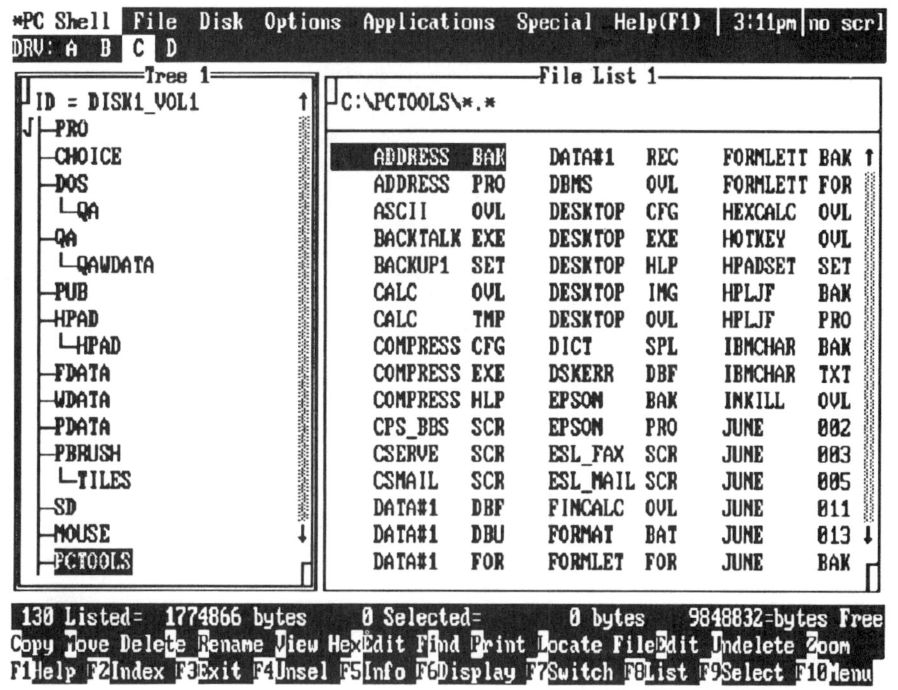

9-2 Main Menu of PC Shell

run, the disk information is read and the File List is updated to reflect any new changes. You can also select and unselect files by clicking the mouse button over the appropriate files.

Immediately below the Tree Window and File List is the status line, which shows the number of files in the directory and how many bytes they occupy, the number of files selected and how many bytes they occupy, and the number of free bytes on the disk.

The command line below the status line lists all the commonly used commands, which are also listed in the pull-down menus. You can choose any of these commands by choosing the highlighted letters.

The last line is the message bar. In the Main Menu, it provides some commonly used function keys. This line will change in PC Shell, depending on the screen.

Help key

You can always access help in PC Shell by pressing the F1 – Help key. It is context-sensitive help, meaning it gives you help specific to whatever you are doing. For example, if the cursor is on the Move file command, the F1 – Help key will bring up the help screen shown in Fig. 9-3. You always have access to the help index, in addition, shown in Fig. 9-4.

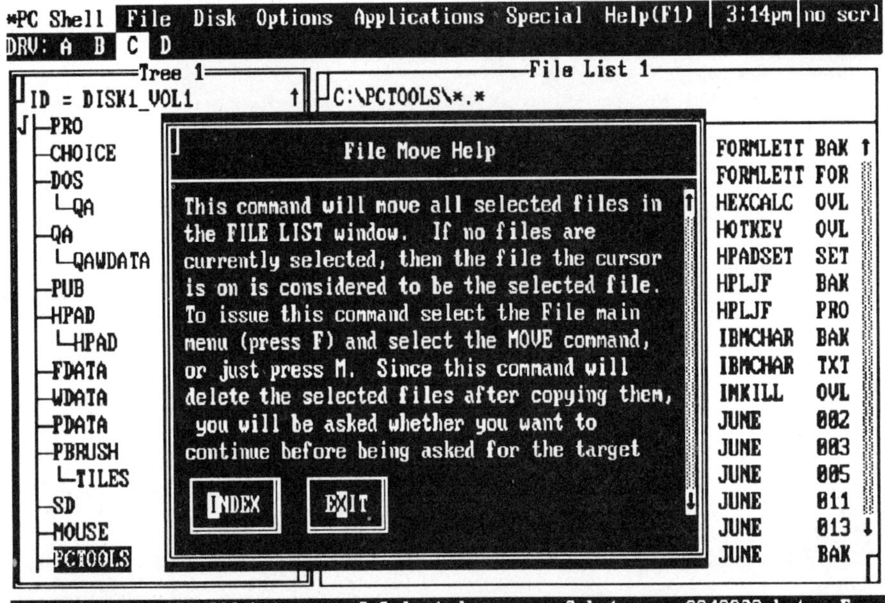

9-3 PC Shell Help screen

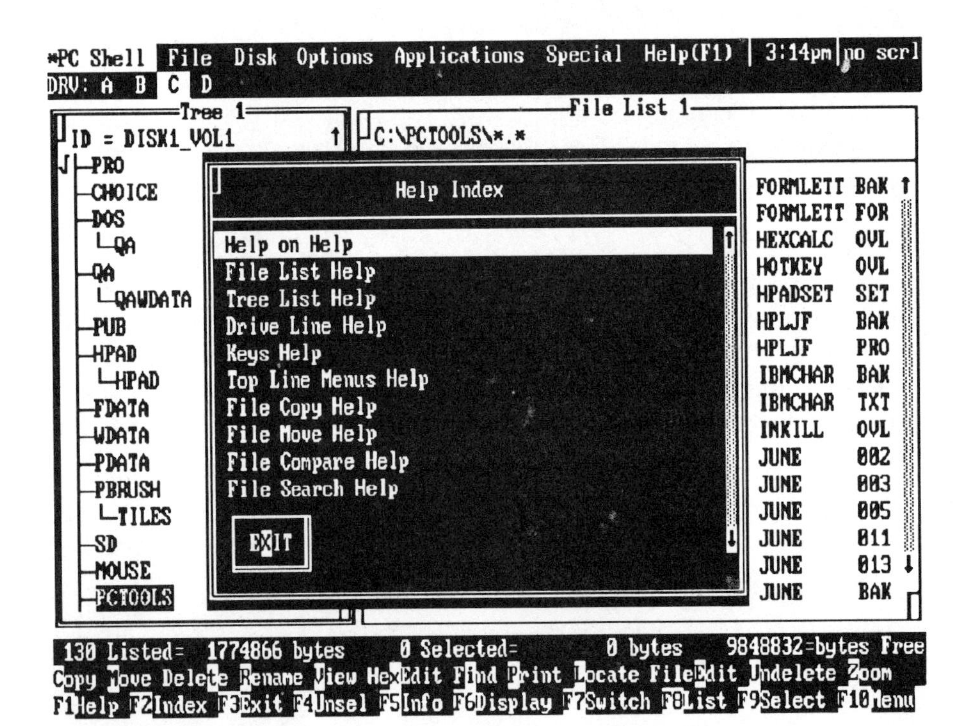

9-4 PC Shell Help index

Copy

Selecting from the pull-down File menu gives you all the commands shown in Fig. 9-5. The Copy command gives you the screen in Fig. 9-6, which lets you copy a single file or a group of files. Files can be copied from one drive to another drive, or from one drive to the same drive but a different directory. Select the files you want to copy from the File List window. You can select one or more files with either the cursor keys and the Enter key, or with a mouse. Notice, in Fig. 9-6, that the selected files have numbers to the left of them. The selected files can be in any order on the disk.

```
*PC Shell  File  Disk  Options  Applications  Special  Help(F1) | 3:19pm|no scrl
DRV: A  B
          Copy                 File List 1
 ID = DI  Move                 \PCTOOLS\*.*
  PRO     Compare
  CHOIC   Find                 ADDRESS  BAK   DATA#1   REC   FORMLETT BAK
  DOS     Rename               ADDRESS  PRO   DBMS     OVL   FORMLETT FOR
   QA     Delete               ASCII    OVL   DESKTOP  CFG   HEXCALC  OVL
  QA      Verify               BACKTALK EXE   DESKTOP  EXE   HOTKEY   OVL
   QAH    View                 BACKUP1  SET   DESKTOP  HLP   HPADSET  SET
  PUB     Hex Edit             CALC     OVL   DESKTOP  IMG   HPLJF    BAK
  HPAD    Attribute Change     CALC     TMP   DESKTOP  OVL   HPLJF    PRO
   HPA    Print File           COMPRESS CFG   DICT     SPL   IBMCHAR  BAK
  FDATA   Print Directory      COMPRESS EXE   DSKERR   DBF   IBMCHAR  TXT
  WDATA   More File Info       COMPRESS HLP   EPSON    BAK   INKILL   OVL
  PDATA   File Edit            CPS_BBS  SCR   EPSON    PRO   JUNE     002
  PBRUS                        CSERVE   SCR   ESL_FAX  SCR   JUNE     003
   TIL    Run      Ctrl+Enter  CSMAIL   SCR   ESL_MAIL SCR   JUNE     005
  SD      DOS     unavailable  DATA#1   DBF   FINCALC  OVL   JUNE     011
  MOUSE                        DATA#1   DBU   FORMAT   BAT   JUNE     013
  PCTOO   Exit PC Shell    F3  DATA#1   FOR   FORMLET  FOR   JUNE     BAK
 130 Listed=  1774866 bytes   0 Selected=      0 bytes   9848832=bytes Free

Copy selected files to specified drive:directory
```

9-5 File menu

After you select the files to be copied, choose Copy from the File menu, and the selecting target drive box will appear, as shown in Fig. 9-6. Select the drive letter with either the cursor keys or a mouse. Press the Esc key if you change your mind and want to abort the command. If you are copying to a hard drive that has directories, the program will give you a tree list of all the directories, so you can select the one you want to copy to.

If the files you are copying already exist on the new disk or directory, you will see another dialog box asking you whether you want to replace the existing file. The Replace all option lets you replace all the files in the target drive with files that have the same names in the source file. The Replace File option lets you

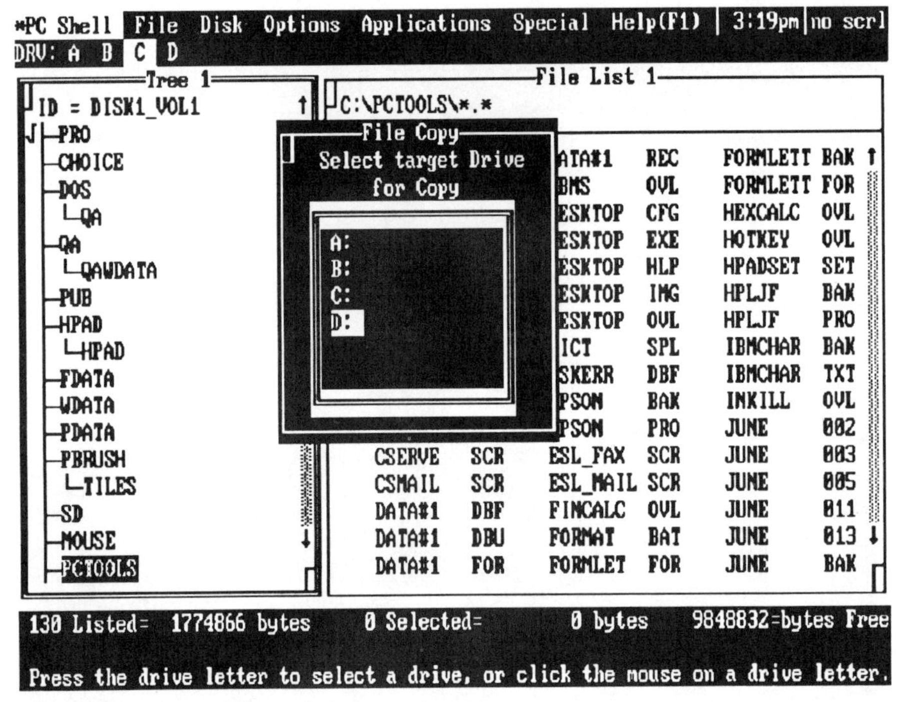

9-6 File copy command

replace one file at a time. The Next file option lets you skip the file and go on to the next. The Skip all option lets you skip all selected files and return to the PC Shell main screen. The Exit option returns you to the PC Shell main screen.

If you are copying files from one directory to another, it is easier to use the two-list display, as shown in Fig. 9-7. The Two list display command is located in the Options menu. You can also open up a second window by pressing Ctrl – Alt, plus the drive letter. For example, Ctrl – Alt – B will open a window for the B drive, letting you view and scroll through the files of two separate directories. This command is very useful if you want to copy, move, or compare two files. If you want to go back to a single window, choose the One list display command from the Options menu.

Move

The Move command lets you move a single file or group of files. The procedure is the same as that of the Copy command. Select the files you want to move with either the cursor keys and the Enter key, or a mouse. Select the Move command from the File menu, and dialog box shown in Fig. 9-8 appears. It indicates that the source or original files will be deleted. It then comes back with a second window asking you which drive you want the files moved to. Select the drive with the

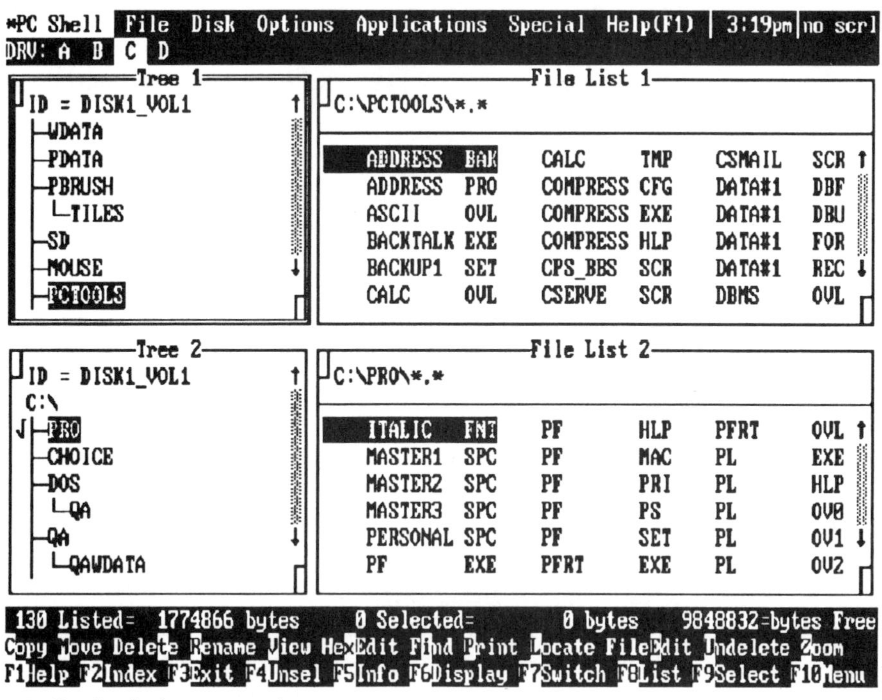

9-7 Two-list display command

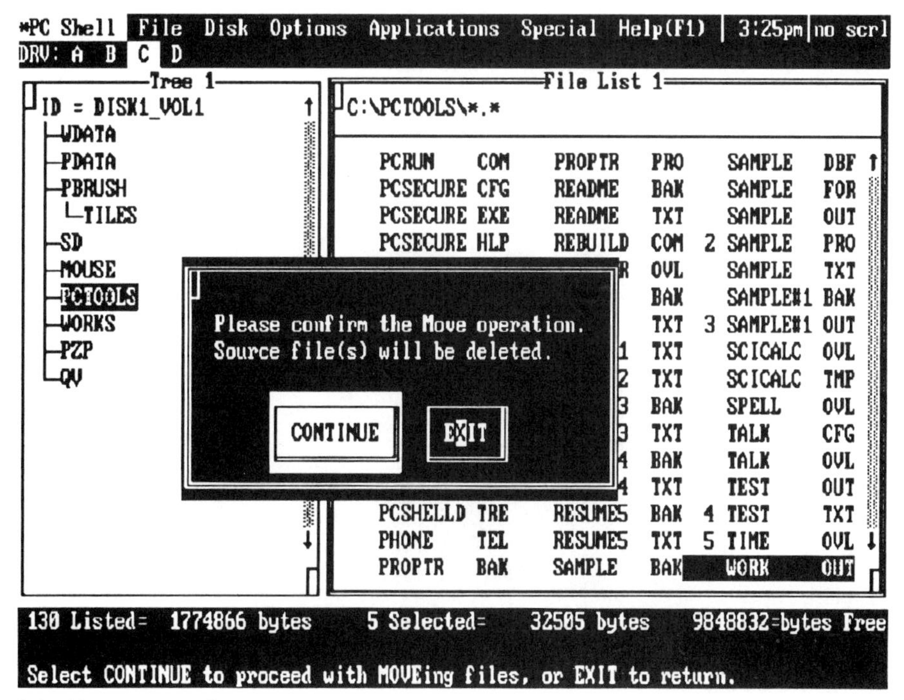

9-8 Move command

cursor keys or the mouse. If you are moving the files to a hard drive, it will display a directory tree and ask you to select where you want the files moved. The files will be moved to the new location, and the files in the original location will be deleted.

If you use the Move command in a two-list display, you will be asked if the second window is the targeted drive. If you do not want to move the files to the second window, answer No and select the correct drive.

Compare

The Compare command lets you compare two files in order to see if they are identical. This is very useful if you are copying files and want to make sure the copy program worked properly. The files can have the same name or different names, can be on the same disk or different disks, and can be in the same directory or different directories on the hard drive.

Select the files you want to compare and choose the Compare command from the Edit menu. If you are using the One list display option, Fig. 9-9 will appear asking you for the drive where the files are located. If you are using the Two disk display option, it will ask you if the files for comparison are located in the second window. If they aren't, select No. If you are comparing files on a hard drive, you will be given a directory tree and asked to select where the files are located.

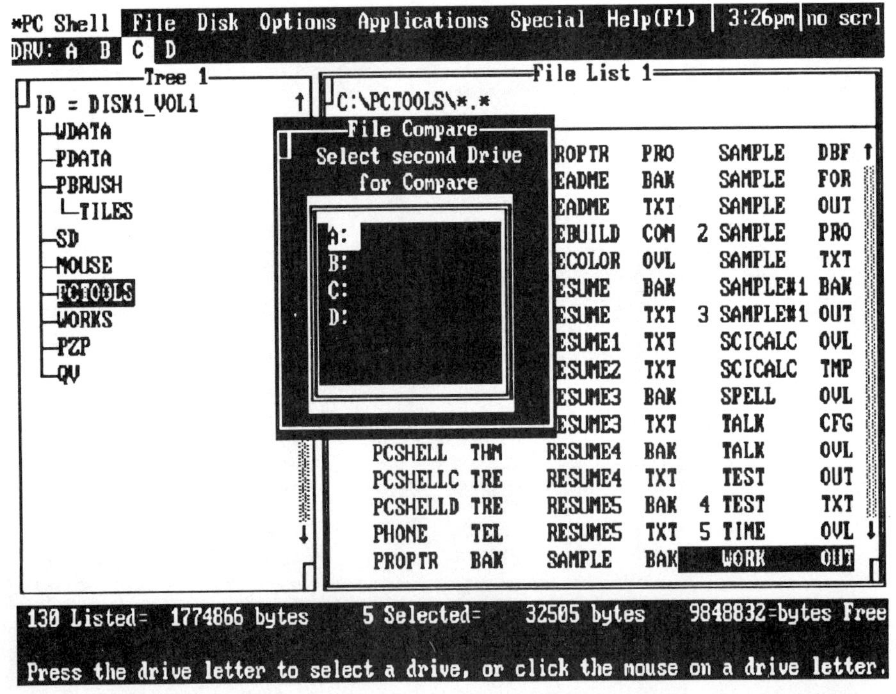

9-9 File compare command

You are asked whether you want to compare files with matching names or different names. If you select Matching names, the files are compared. If you select Different names, you are prompted for the name of the file to be compared to. If the comparison shows a difference in the files, a dialog box appears displaying the sector where the difference was found. Each file is compared separately. If you select more than one file, you are prompted to compare the files one at a time. Once the comparisons are completed, you will be returned to the PC Shell's main screen.

Find

The Find command lets you locate character strings in either all files in the window or only selected files. The character strings can be ASCII or hex character strings. The F9 key is a toggle that switches you from ASCII to hex searches.

To find a string, select the file or files you want to search for. The first file selected will be the first search, the second file selected will be the second search, etc. Choose the Find command from the Edit menu, and the File Search dialog box will appear, as shown in Fig. 9-10. To search for ASCII characters, type the search string in, up to 32 characters, in the text box. To search for hex values, press or click on the F9 key and type in the hex values you are looking for. All ASCII searches are case-sensitive, and hex searches are not. It is important to keep this in mind when executing the search.

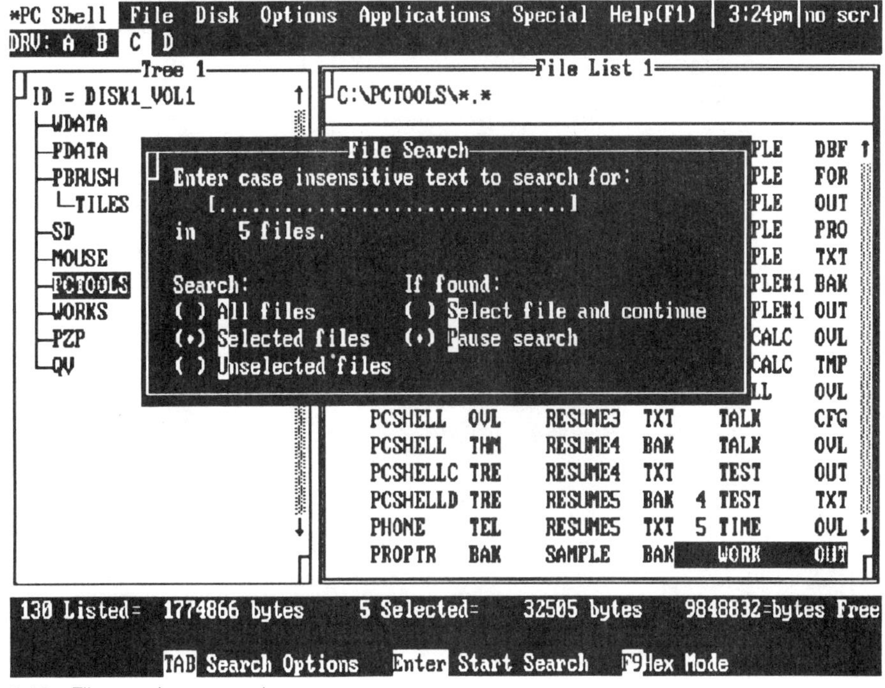

9-10 File search command

You can choose one of two options when search is completed. The Select file and continue option will mark the file and continue the search. All the files that meet the search criteria will be marked in the window after the search. The Pause search option will give you several options when the characters are found. You can choose Select file which marks the file, Continue search to continue searching, View File which displays the file, Edit file which allows you to use the hex editor on the file, and Next file which searches the next file.

Rename

The Rename command lets you rename a file. Select the files you want to rename, and choose the Rename command from the Edit menu. You can select one or more files to be renamed. If you select one file, a dialog box appears asking you for the new filename and extension. Type in the new name and extension, and select Rename to change the name.

If you select more than one file to be renamed, the box in Fig. 9-11 appears. It asks you if you wish to change all the selected filenames globally or individually. Changing a name globally is useful if you want to change all files with a certain extension to the same name but with a different extension—for example, changing all .PRO extensions to .TXT extensions. If you select Global, the screen shown in Fig. 9-12 appears. Type in the name of the program or the extensions

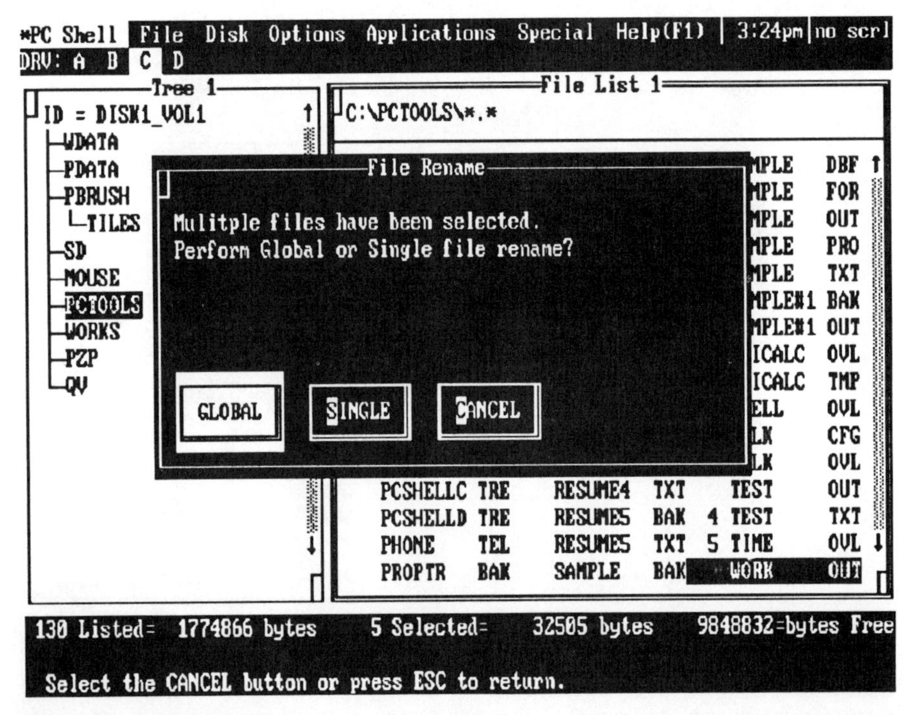

9-11 File rename command

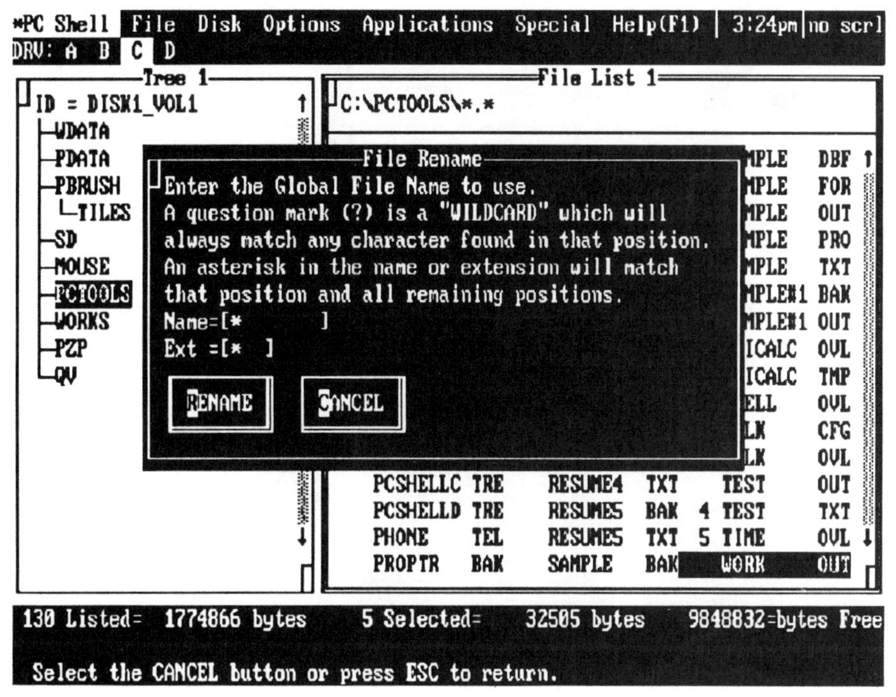

9-12 Renaming files globally

you want to change. To change from one extension to another, be sure to use the wildcard asterisk (*) in the Name field. The wildcard asterisk means the names will not be changed, only the extension.

Files will be renamed one after the other in the order you selected. After you are finished renaming one file, you will be prompted for the next file.

Delete

The Delete commands lets you remove a single file or group of files from the disk. PC Shell can remove any files from a disk, including system or hidden files. Select the file or files you want deleted from the disk. Choose the Delete command from the Edit menu, and Fig. 9-13 appears. If more than one file is selected, you will be asked whether you want to delete the listed file, skip this file and go to the next file, delete all the files without PC Shell listing them individually, or cancel the operation. If only one file is selected, you will be given only two options: to delete the file or cancel the operation.

Verify

The Verify command makes sure that a file can be read without any errors. It checks all the file sectors to see if they are good. Select the file or files you want to have verified. Select the Verify command from the Edit menu, and Fig. 9-14

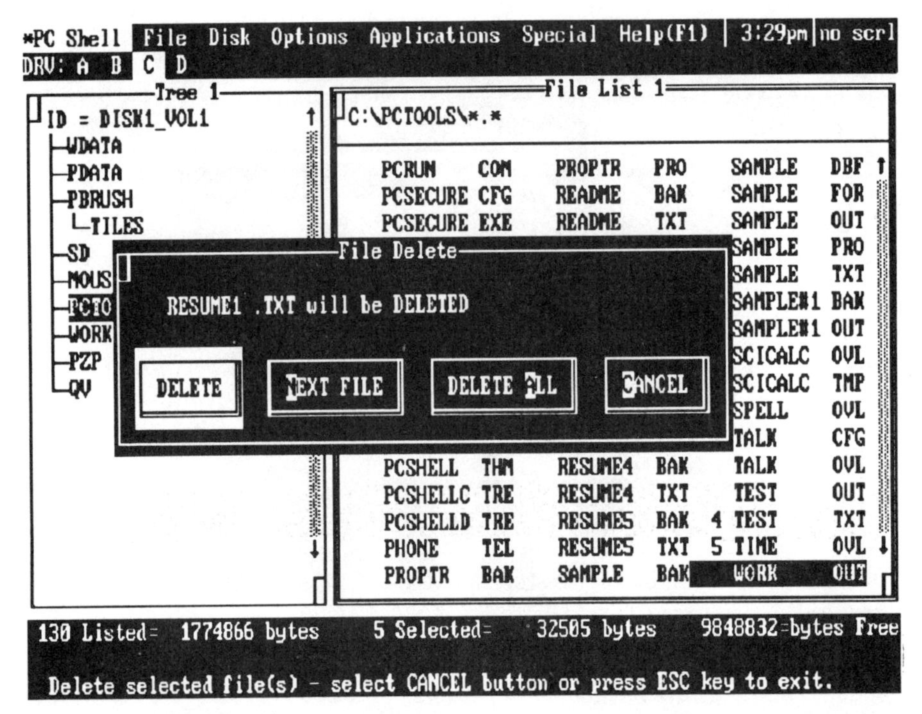

9-13 File delete command

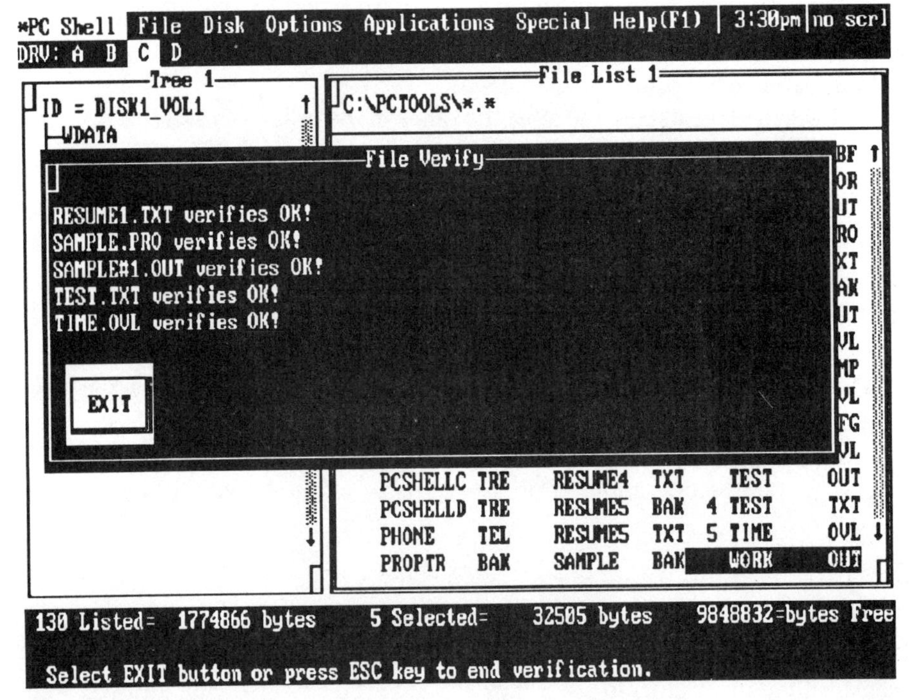

9-14 File verify command

appears. If there are no errors, you will be told the file is OK. If there is an error, the program will display the sector number where the error occurs. You can view and, hopefully, fix the error by selecting View/Edit from the File Verify dialog box. This option appears only when an error occurs.

View

The View command lets you view any ondisk file. Figure 9-15 shows the view of the notepad file RESUME4.TXT. To view a file, select the file you want to view and select View from the Edit menu. Pressing Shift−V lets you view a file that is highlighted by a cursor but not yet selected. You can use the cursor keys or the mouse to scroll through the entire document. Files that are viewed in this mode are in ASCII format.

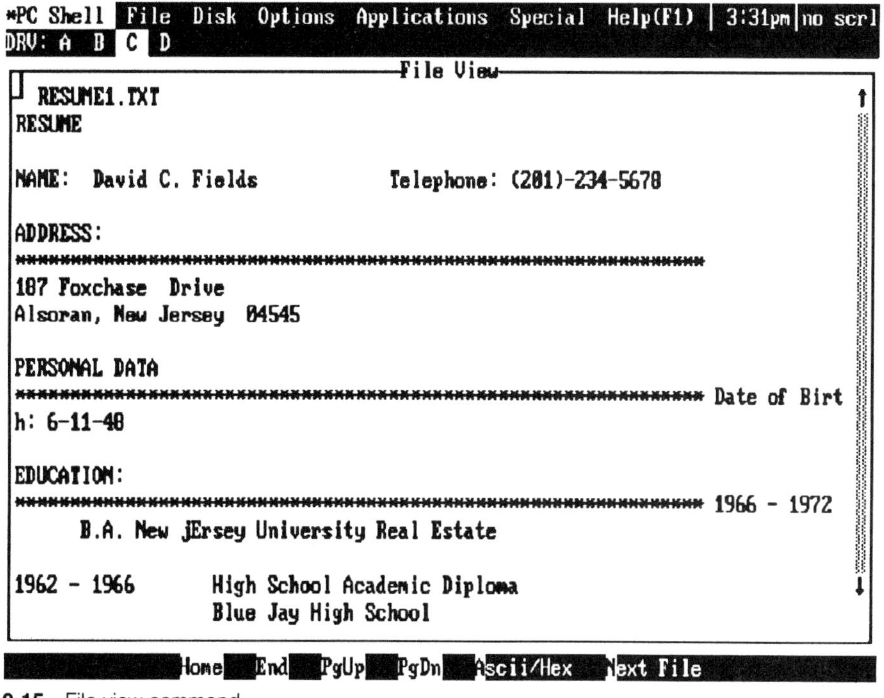

9-15 File view command

Viewing dBASE

If you want to view a dBASE or dBASE-compatible file, the file can be accessed directly by selecting it from the File List window. You can also use the Locate file command of the Disk menu to select all dBASE files. The Locate file command lets you find groups of files that belong together, like word processing documents, databases, or spreadsheets. Select the View command from the Edit menu. The database viewer appears, as shown in Fig. 9-16. The message bar at the bottom of

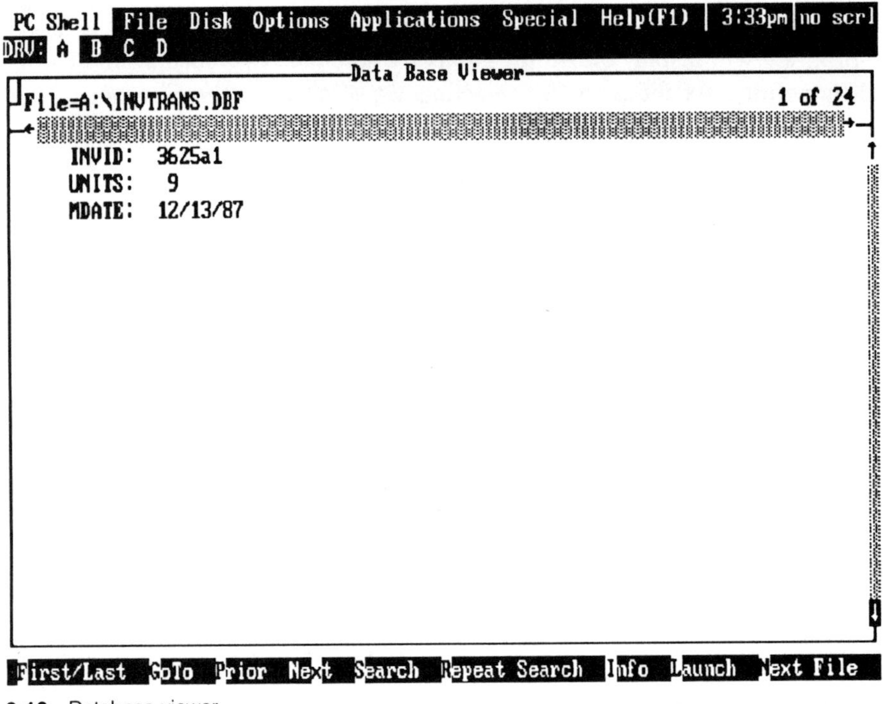

9-16 Database viewer

the screen lets you navigate through the database. The fields can be accessed by clicking on them with a mouse. If you're using the keyboard, pressing the letter X will advance you to the next record, the letter P to a previous record, and the letter F will toggle between the first and last record. Pressing the letter I will let you view the database statistics as shown in Fig. 9-17. This box gives you the name and size of the database, field names, field types, and field sizes.

Selecting Launch or pressing L will run dBASE and load the file being viewed. You can only do this if dBASE is already installed on the disk. Otherwise, you can view the database records, but you cannot run the application. To exit the database viewer, press the Esc key or use the mouse to click on the close box.

Viewing Lotus

Just as dBASE files can be automatically viewed, you can also view Lotus 1-2-3 files. Select the Lotus files you want to view. Choose the View command from the Edit menu. The 123 viewer appears, as shown in Fig. 9-18. You can scroll through the spreadsheet with the commands on the message bar.

Selecting Launch or pressing L will run Lotus and load the spreadsheet being viewed. You can only do this if Lotus is already installed on the disk. Otherwise, you can view the spreadsheet but not run the application. To exit the 123 viewer, press the Esc key or use the mouse to click on the close box.

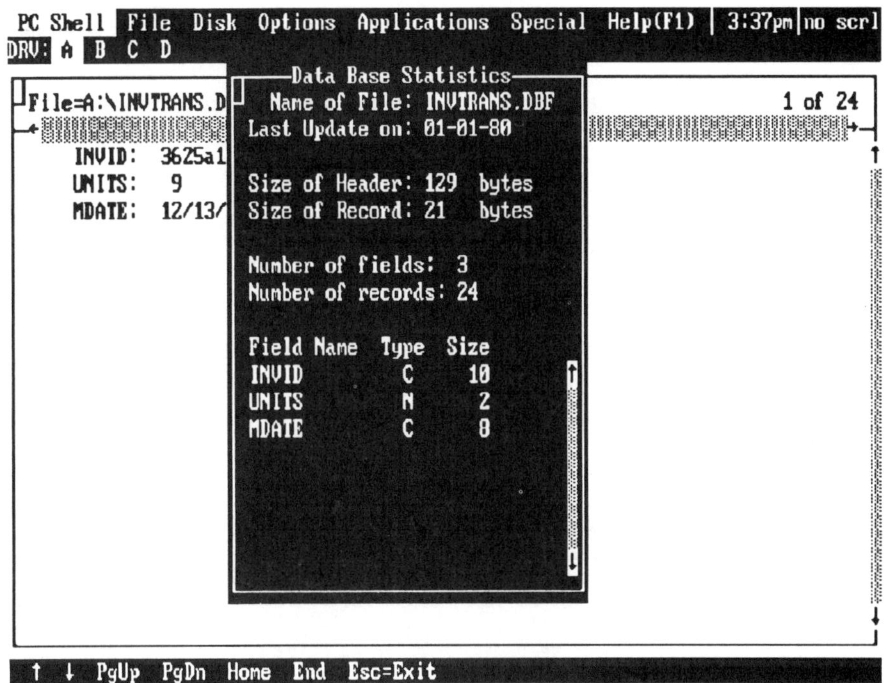

9-17 Database statistics

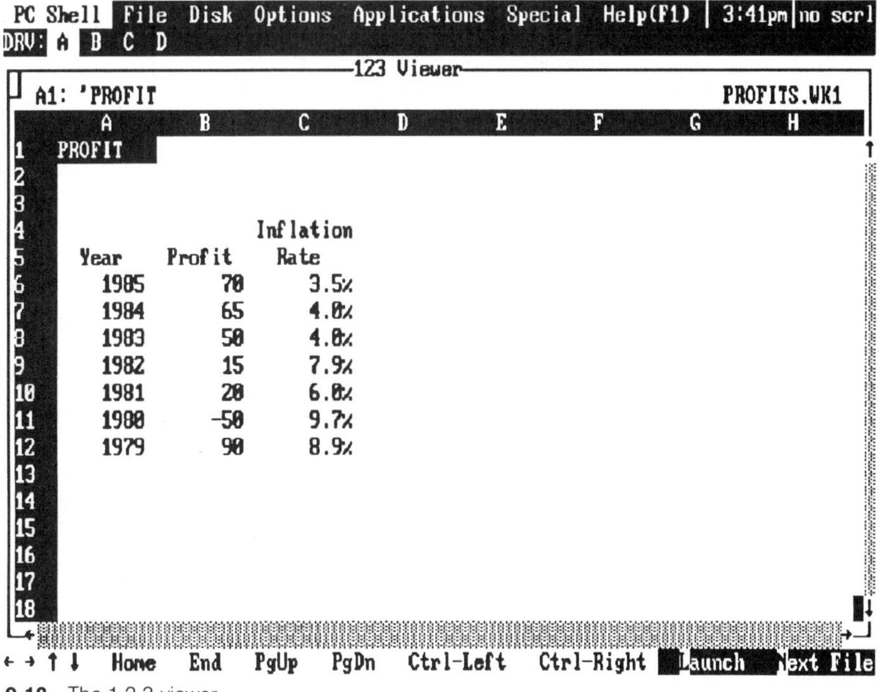

9-18 The 1-2-3 viewer

Hex edit

The Hex edit command lets you edit a file, as shown in Fig. 9-19. It can be used with the Verify or View command. The Verify command will find all errors on a disk, which you can then correct with the Hex edit command. In most cases, errors can be corrected by rewriting the same information on the disk. The top line of the dialog box shows the filename, relative sector, cluster, and absolute disk sector. The number on the left-hand side of each row is the *displacement* number, a number between zero and 511. Each sector is numbered from zero through 511. This is how the computer saves information to the disk. The number in parentheses is the same number in hexadecimal format. Following the number are the 16 bytes of information contained in that sector. Each of the bytes is a hexadecimal number. The characters on the right side of the window are the characters in ASCII format.

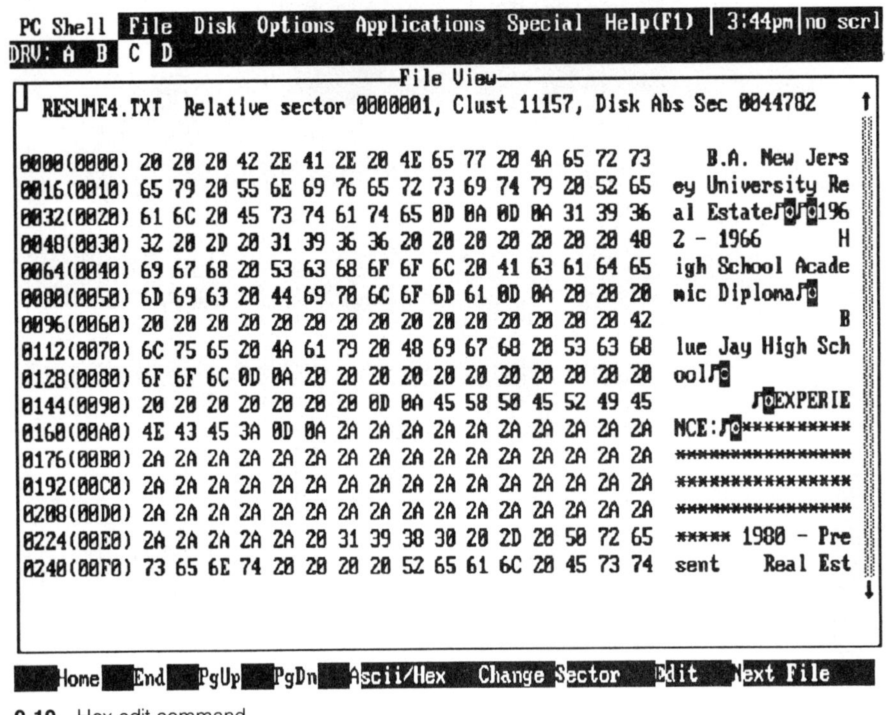

9-19 Hex edit command

Place the cursor over the appropriate byte and make whatever changes you want. You can edit both the hex and ASCII bytes by simply typing over the characters you want to replace. Changes in hex will automatically be reflected in ASCII and vice versa. The F8 key will toggle the cursor between the ASCII and hex character modes. Selecting Home will move the cursor to the beginning of the file, while selecting End will move the cursor to the end of the file.

All changes to the file will be reflected by a change in color if you are using a color monitor, or they will be highlighted if you are using a monochrome monitor. Press the F5 key or select Save to write the changes to disk, or hit Esc to exit and cancel the changes.

Attribute change

The Attribute change command lets you change the various file attributes and the time and date of the files, as shown in Fig. 9-20. Do *not* change the file attributes of the system files or copy-protected programs. There is a chance that this will prevent your hard disk from booting.

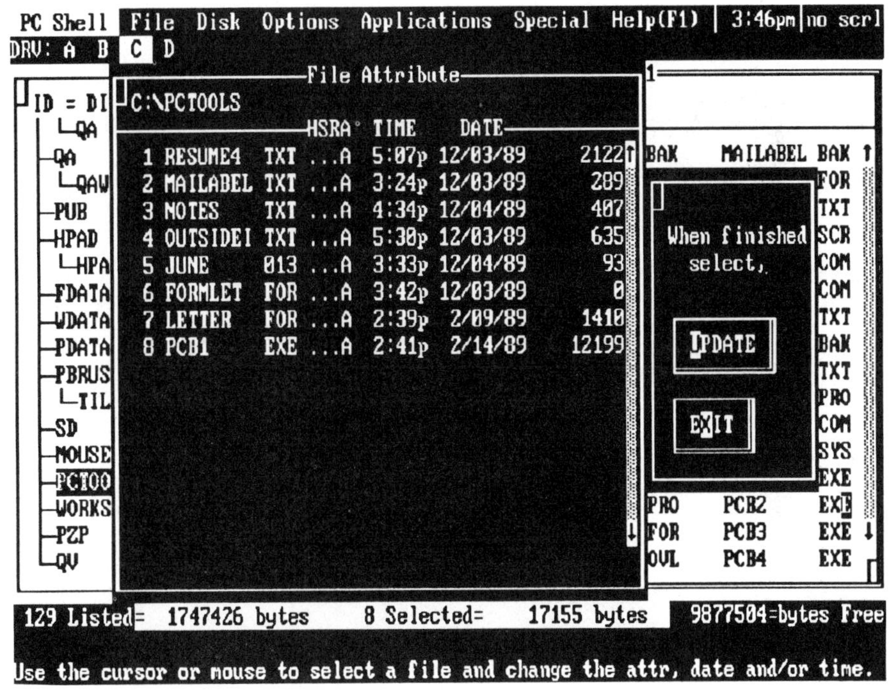

9-20 Attribute change command

The File Attribute window provides the names of the selected file or files and their attributes. The attribute is an R if the file is read-only, which means you can read it but can't change or delete it. A file is hidden (H) if it is not listed when you take a directory of the disk. A system (S) file is also hidden or invisible, and not listed in the directory. An A represents an archive file, which is used by PC Backup and DOS backup to mark those files you need to back up (files that have been changed).

The time and date fields get information from the system clock when the file is created. To change any of the attributes or the time and date, use the arrow keys or the mouse. Change an attribute by clicking on the attribute letter with the mouse or typing in the letters A, H, S, or R to get the appropriate attribute. Pressing the same letter again will turn off the attribute.

To change the time and date, use the arrow keys or the mouse to move to that field and type in the new information. All dates must be typed in the form mm/dd/yy, and each number must contain two digits. For example, Sept 5, 1990 is written as 09/05/90.

When you are finished, select Update from the window to save the changes, or Exit to cancel the changes.

Print file

The Print file command lets you print a file or series of files to the printer, as shown in Fig. 9-21. After you print one file, you can print the succeeding one with the Next command. Print as a standard text file prints the text file using ASCII characters. Dump each sector in ASCII and hex lets you print sectors in both ASCII and hexadecimal format. Print file using PC Shell print options, as shown in Fig. 9-22, lets you modify the print options and then print a standard ASCII character text file. Use the arrow keys to move the cursor to the field you want to change and type in the changes. You can modify the lines per page, top and bottom margins, left and right margins, headers and footers, and page numbers. You can stop between pages and have the last page ejected. You can specify the number of blank lines to be inserted between each printed line.

Print directory

The Print directory command prints a list of all the files in a selected directory. The filenames, size, number of disk clusters, date, time, and attributes for all the files in the selected directory are printed.

More file info

The More file info command, which generates the screen in Fig. 9-23, gives you information about the file, such as the filename and extension, file path, attributes, last time the file was accessed, file length, total clusters occupied, starting cluster number, and total files in directory.

File edit

The File edit command calls up a text editor that lets you create and edit documents while executing other programs. When this command is selected, you have the option of creating a new document or editing an existing document, as shown in Fig. 9-24. The editing commands are the same commands for the notepad utility;

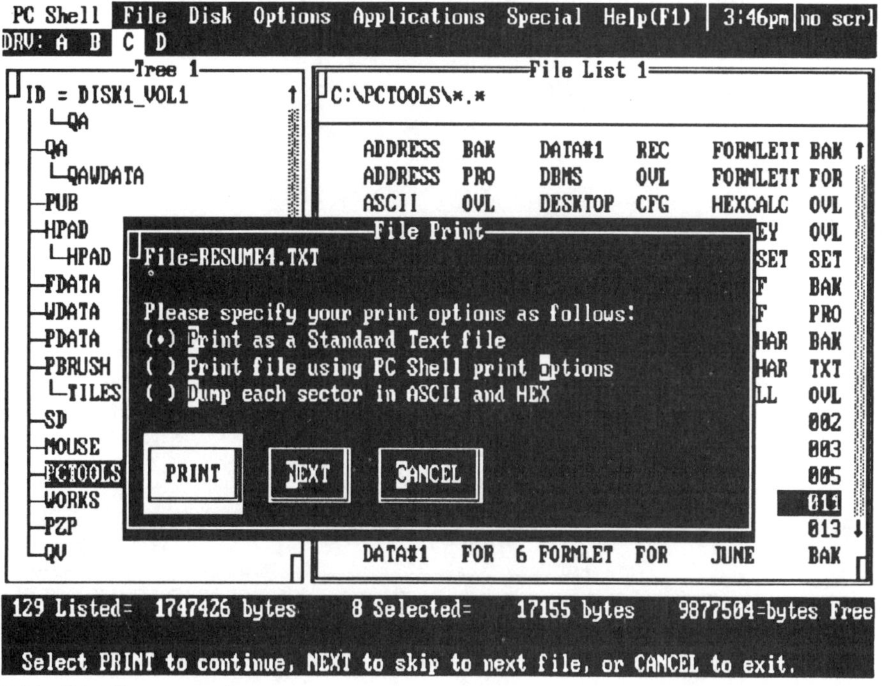

9-21 Print file command

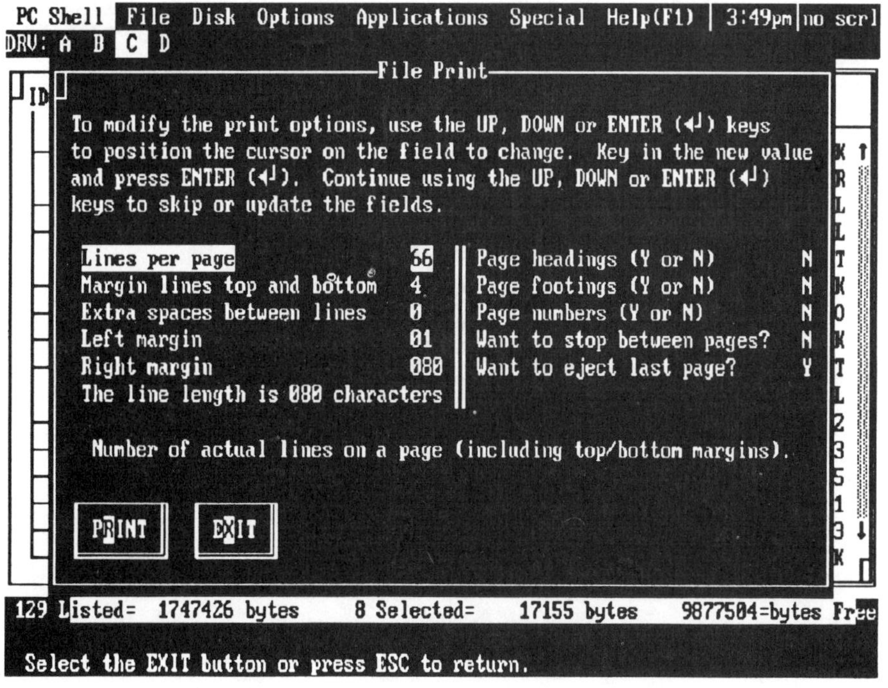

9-22 Modifying print options

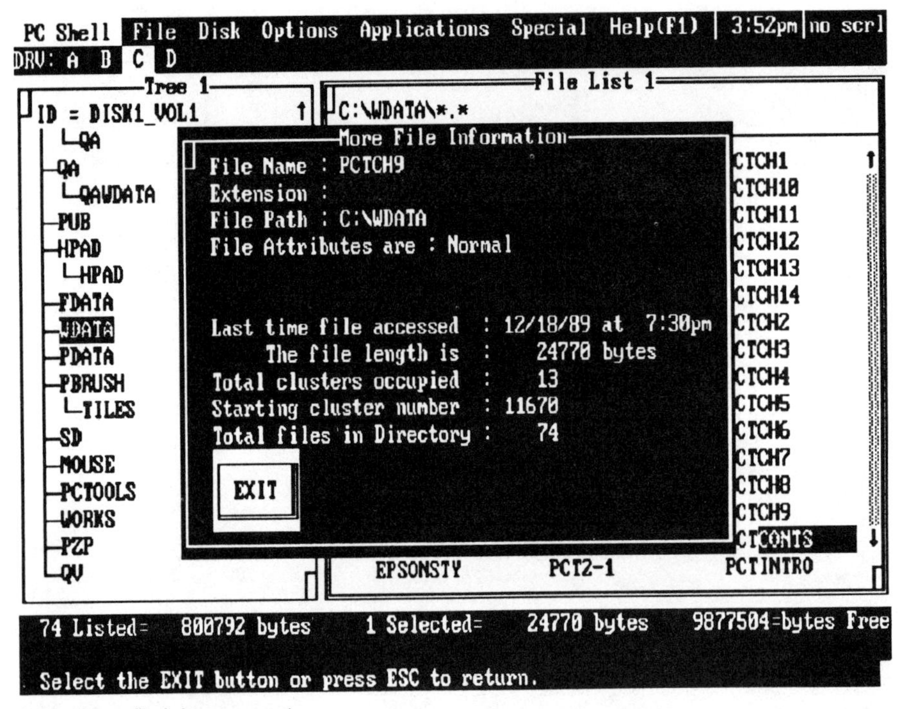

9-23 More file info command

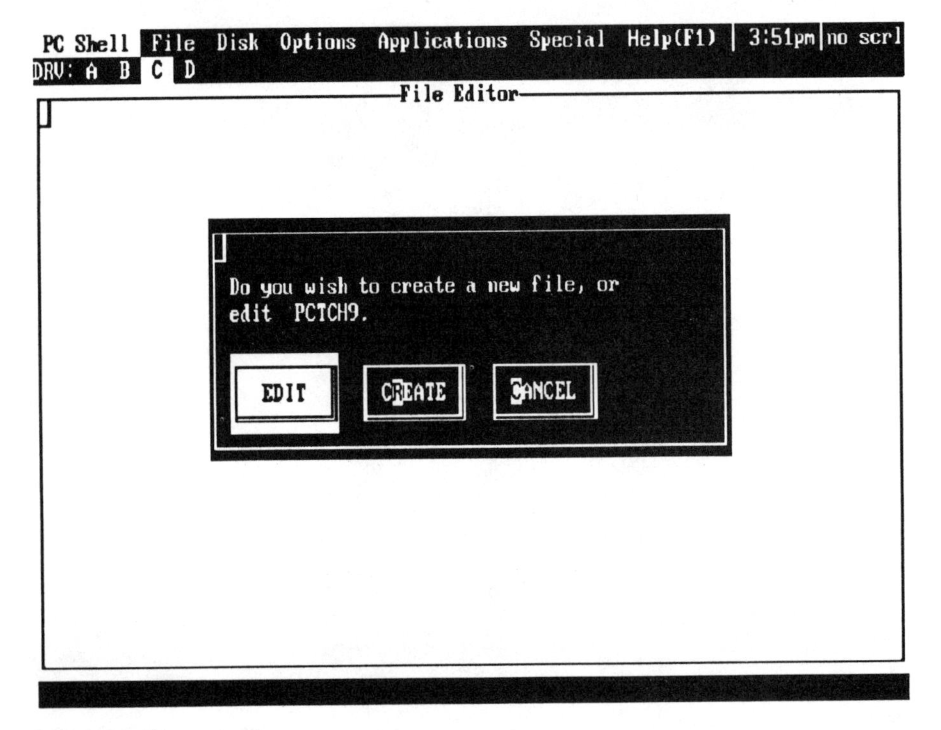

9-24 File edit command

you can cut and paste text, select a block of text, find and replace characters, and print text.

Run

The Run command lets you execute a single file or program. Programs can be run with this command only if PC Shell was loaded by typing PCSHELL at the DOS prompt. If you are running PC Shell from the DOS prompt, a Quick run command will be added to the Options menu. This command will *not* appear when you are running PC Shell as a resident program. The Quick run command frees up most of the memory of the computer, so your application can run.

To use the Run command, select a program you want to run. Select Run from the File menu, or press Ctrl–Enter. The program name will appear in a dialog box. Select Run to execute the program or Cancel to exit.

Go to DOS

The DOS command is available only if PC Shell is nonresident. If PC Shell is installed as a resident program, the word "unavailable" appears in parentheses next to "DOS." The DOS command lets you leave PC Shell to execute a DOS command. To return to PC Shell, type the word EXIT at the DOS prompt.

Exit PC Shell

The Exit PC Shell command lets you return to the DOS prompt.

10
CHAPTER

Disk activities

In this chapter, you will learn about the disk activities you can access from the Disk Menu of PC Shell. These disk activities let you copy, compare, search, rename, verify, edit, locate, format, and perform other types of disk and directory maintenance.

The Disk menu of PC Shell is displayed in Fig. 10-1. Notice that there is a small triangle to the right of "Directory Maint." This indicates that this command has a menu of its own. Chapter 9 explained how to handle single files on a disk, while this chapter deals with the whole disk itself. The commands in the File and Disk menus are very similar to each other.

Copy

The Copy command lets you copy the contents of one disk to another. It is very similar to the DISKCOPY command of DOS. The disk you want to copy is called the source disk. It contains all the files or programs to copy. The blank disk you are copying to is called the target disk. The target disk can be formatted or unformatted. If it is unformatted, the Copy command will format the target disk before

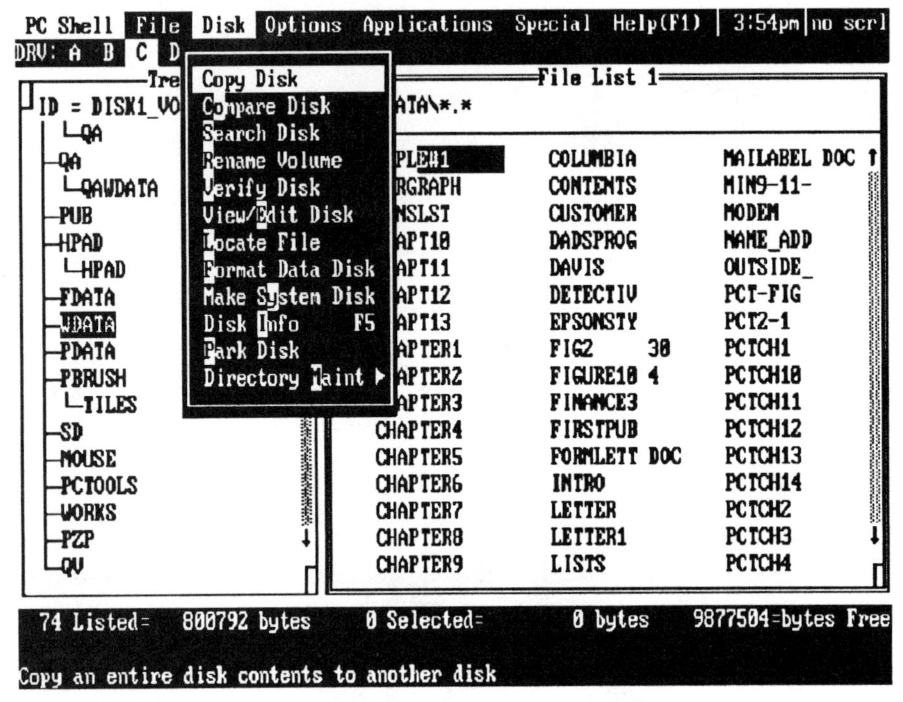

10-1 Disk menu

it starts to transfer the files. When you select Copy from the Disk menu, Fig. 10-2 appears. It asks you to select the source and target drives. Use the Tab and arrow keys to enter information in each of the windows. Both drives must be the same size and use the same density format. In other words, you cannot copy the information from a 5 1/4-inch disk to a 3 1/2-inch disk, or from a low-density disk to a high-density disk. If you want to transfer a single file, use the Copy command of the File menu.

After you select your source and target drives and press the Enter key, Fig. 10-3 appears asking you to insert the source disk. You will then be prompted to insert the target disk. Select Continue to begin the copying or Exit to end. The disk is now copied. PC Shell displays letters showing you the status of the copy: F for formatting the track, R for reading the track, W for writing a track, and a period to show that copying was successfully completed. You will be returned to PC Shell's main screen when the copying is finished.

Compare

The Compare disk command, which produces the box command in Fig. 10-4, lets you compare two disks to see if they are the same. Just like the Copy command, they must be the same size and media specification. After inserting the source and

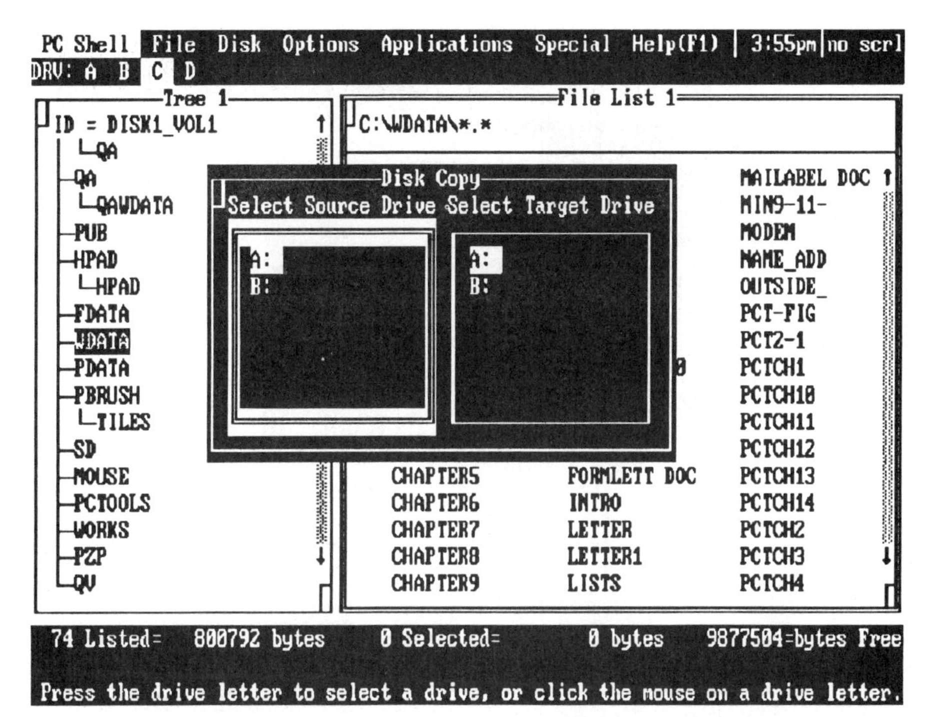

10-2 Copy command

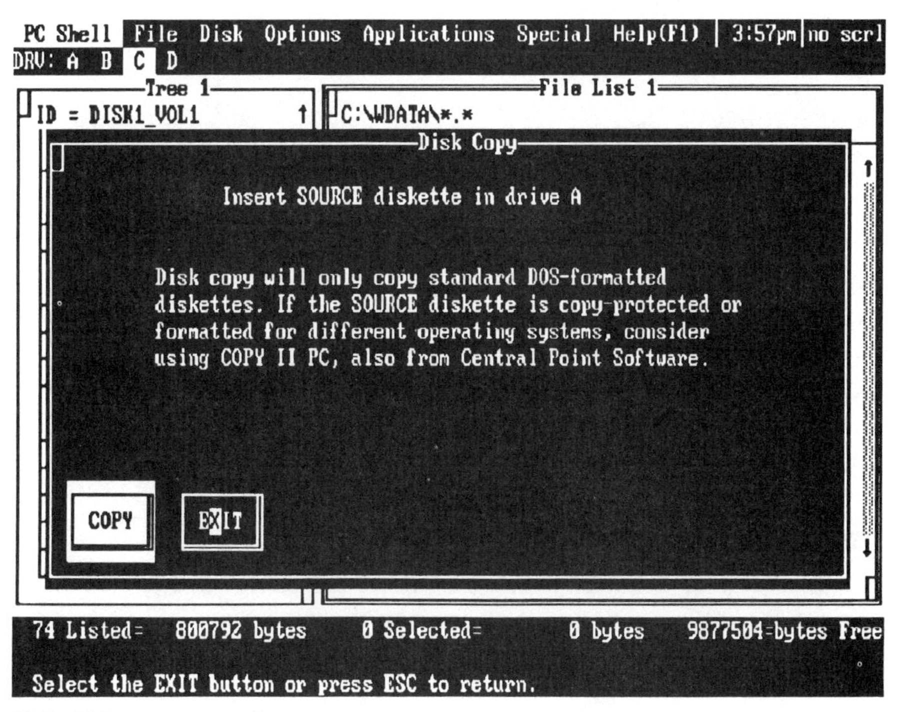

10-3 Disk copy command

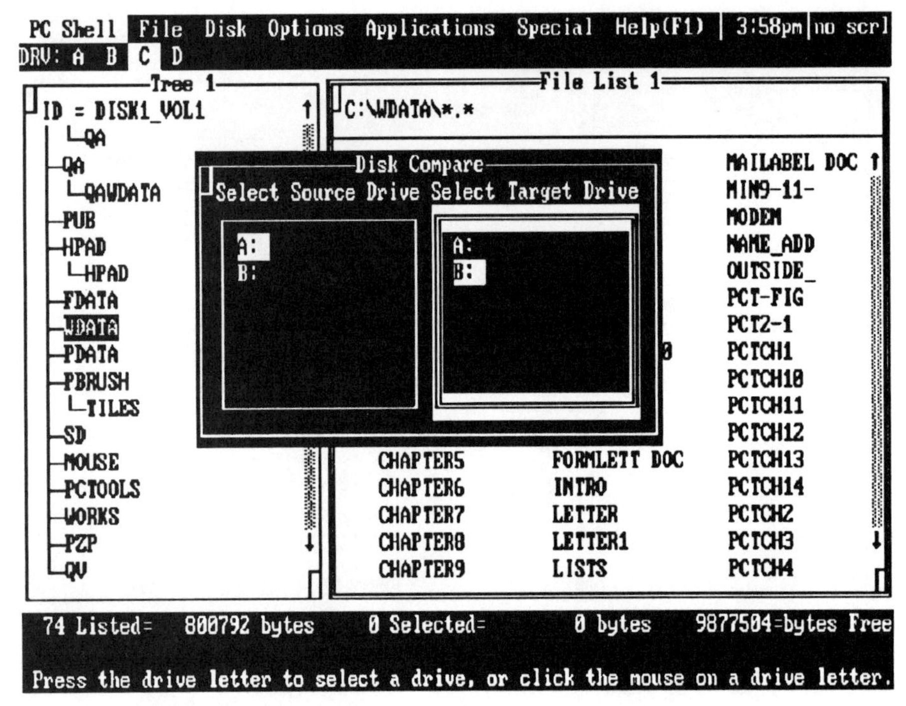

10-4 Disk compare command

target disks, select Continue to start comparing. The screen in Fig. 10-5 appears, showing you the status of the operation. A period appears when each track has been successfully compared, an R if it is reading a track, and a C if it is comparing a track. The Exit command will return you to the PC Shell main screen.

Search

The Search disk screen, shown in Fig. 10-6, lets you inspect the disk for ASCII character strings up to 32 characters in length. If you want to search for hexadecimal values, press the F9 key to change to hex mode, and type in the desired string. Remember that ASCII searches are not case sensitive, while hex searches are case sensitive. When a search is completed, the filename where the string was located will be displayed.

Rename volume

The Rename volume command generates the screen shown in Fig. 10-7. It lets you rename a disk volume label by entering the new name and selecting the Rename box.

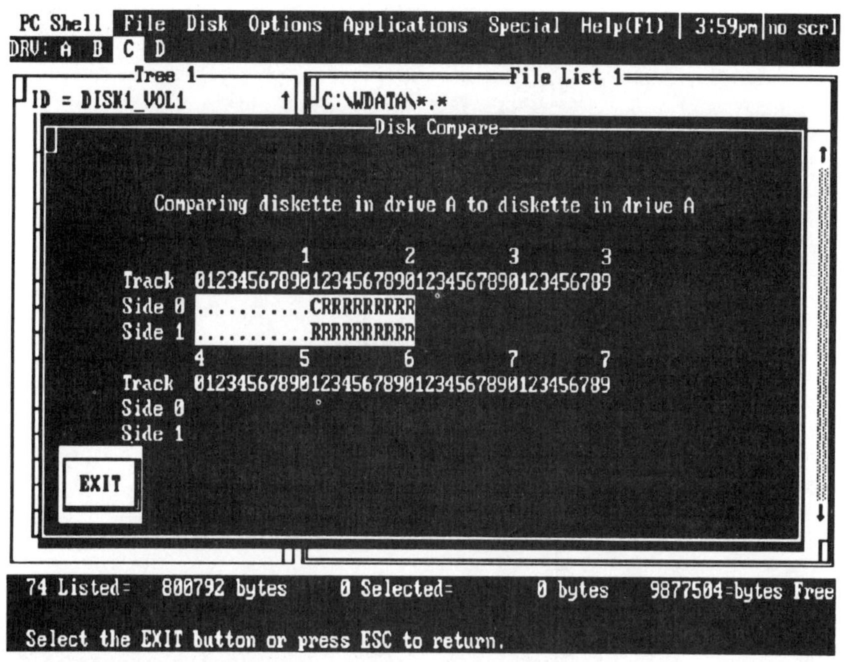

10-5 Status of a disk compare

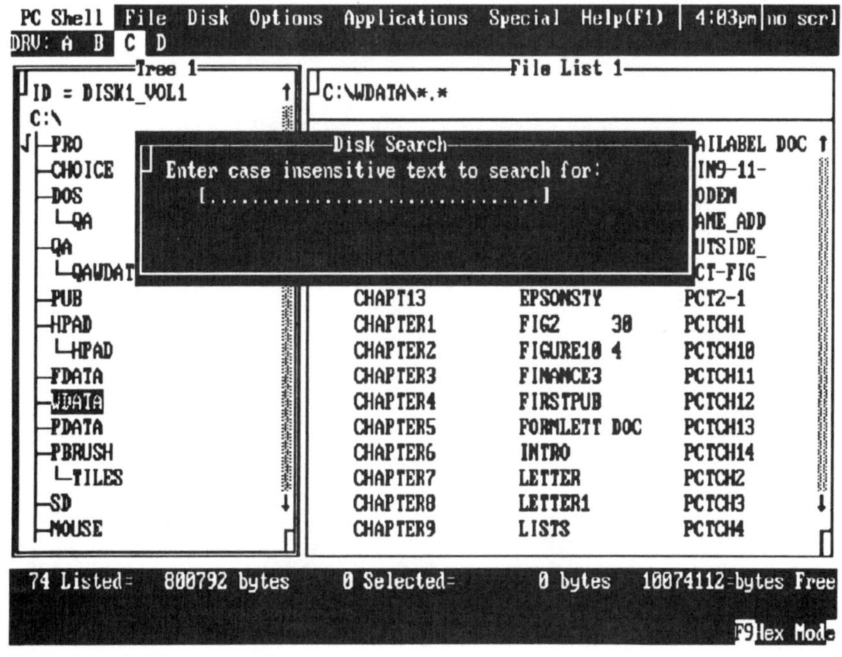

10-6 Search disk command

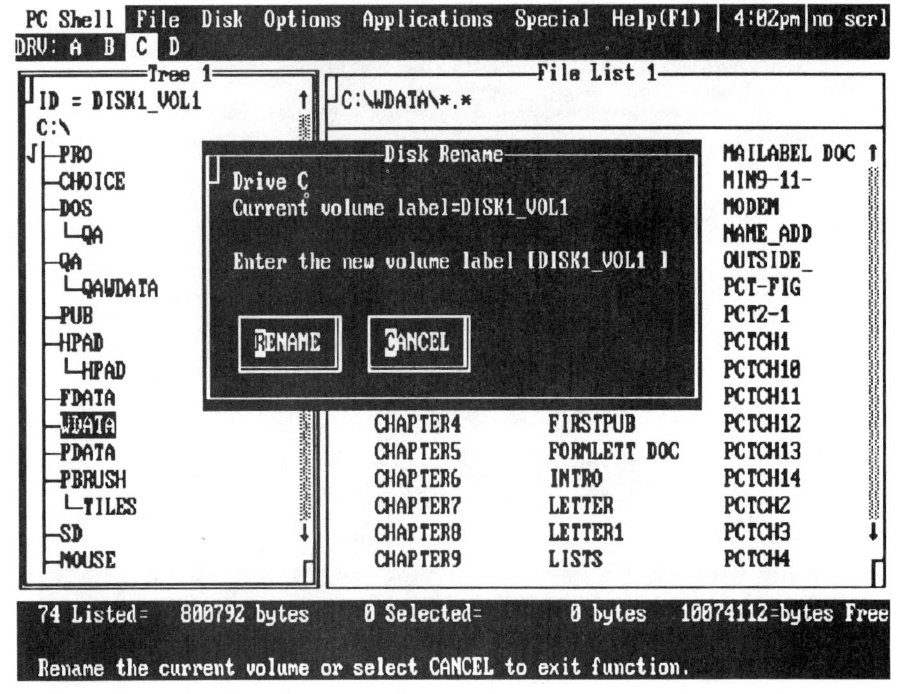

10-7 Rename volume command

Verify

The Verify disk screen, shown in Fig. 10-8, checks the surface of the disk to ensure that all files are readable. If it finds a bad sector, it will display the sector number. It will also tell you the name of the file, if any, that contains the bad sector. If there is no data in the bad sector, the sector is marked so that no future data will be saved on it. If there is data in the bad sector, a message will appear asking you to move the data to a good sector.

View/edit

The View/edit disk screen, shown in Fig. 10-9, lets you see the contents of a disk. It has the same function as the View command of the File menu. The top line of the dialog box shows the filename, relative sector, cluster, and absolute disk sector. The number on the left-hand side of each row is the *displacement* number, a number between zero and 511. The number in parentheses is the displacement number in hexadecimal format. Following those numbers are the 16 bytes of information contained in that sector. Each byte is a hexadecimal number. The characters on the right side of the window are those numbers in ASCII format.

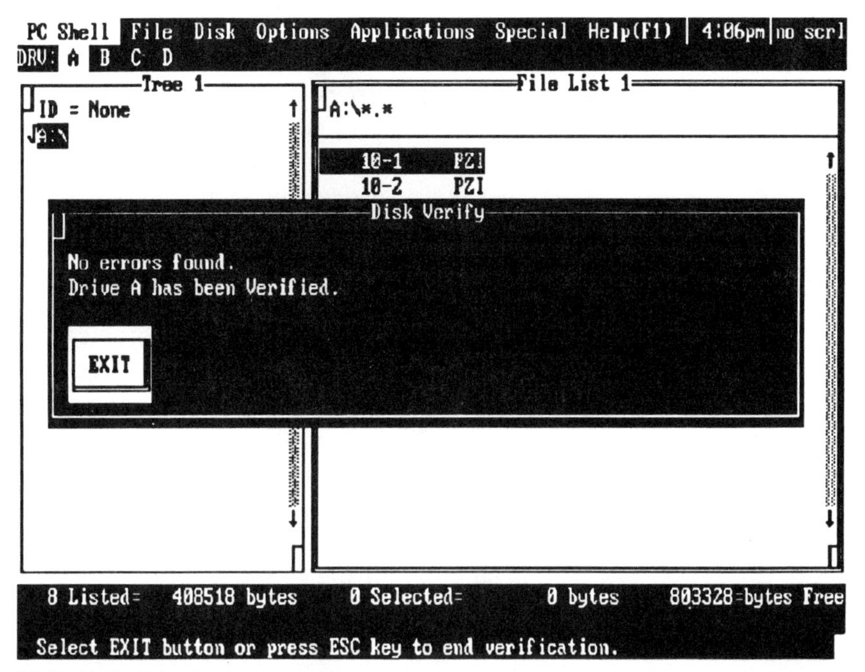

10-8 Verify disk command

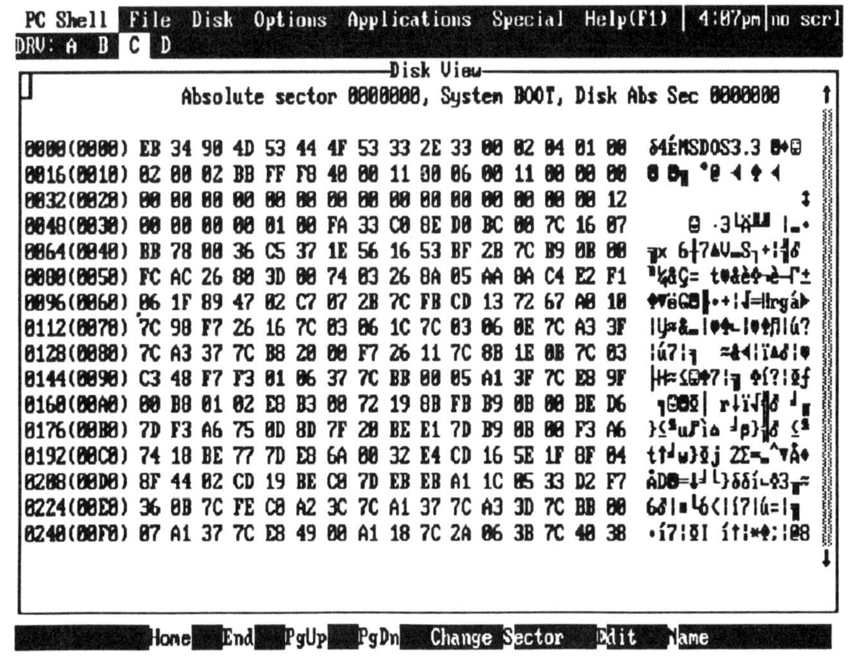

10-9 View/edit disk command

To change any data on the screen, select Edit, place the cursor over the byte you want to change, and type in the corrected changes. You can edit either the hex or ASCII bytes by simply typing over the characters you want to replace. The cursor can move between the hex and ASCII area to do the editing. Changes in hex are automatically reflected in the ASCII area, and vice versa. The ASCII/hex option, or the F8 key, will also toggle the cursor between the two character modes. Selecting Home will move the cursor to the beginning of the file, while selecting End will move the cursor to the end of the file.

All changes to the file are reflected by a change in color if you are using a color monitor, or they are highlighted if you're using a monochrome monitor. Press the F5 key or select Save to write the changes to disk, or hit Esc to exit and cancel the changes.

Selecting Change Sector generates the screen shown in Fig. 10-10, which gives you a menu of six cursor locations. This command is useful for editing when you know where you want to move the cursor. You can move the cursor to the first byte of the *boot sector*, to the first byte of the *FAT sector*, to the first byte of the *root DIR sector* (root directory), and to the first byte of the *data sector*. You can also move to a cluster number with Change cluster #, and to a sector number with Change sector #.

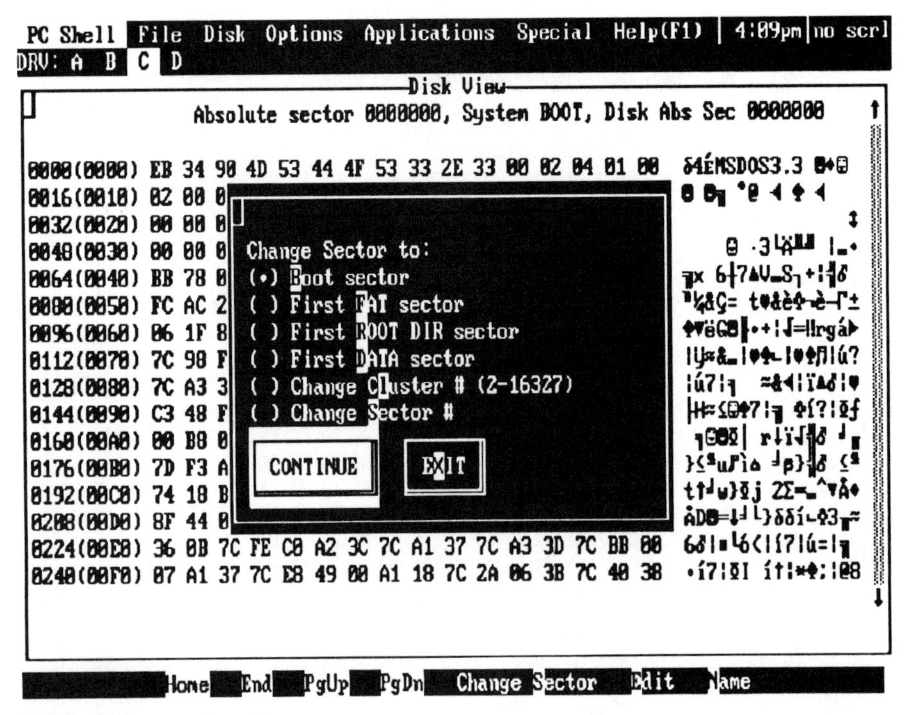

10-10 Change sector option

The *file allocation table*, or FAT for short, is a table that describes each file in a disk. It is essentially the library or index of the disk. It also keeps a record of all unused portions of the disk, so the program knows where to store all future files. The first sector in each disk holds the boot record, which is part of the operating system startup. Disks also contain FAT sectors, directory sectors, and space sectors. The Change sector command lets you view and edit any of these sectors.

Locate file

The Locate file command, shown in Fig. 10-11, lets you find files on your floppy or hard drive. If you are searching for a file on the hard drive, it will search through all your directories to find it. You can search for a certain filename or certain extensions. Figure 10-11 shows the specifications to find all files with the dBASE extension .DBF. The result of this search is shown in Fig. 10-12.

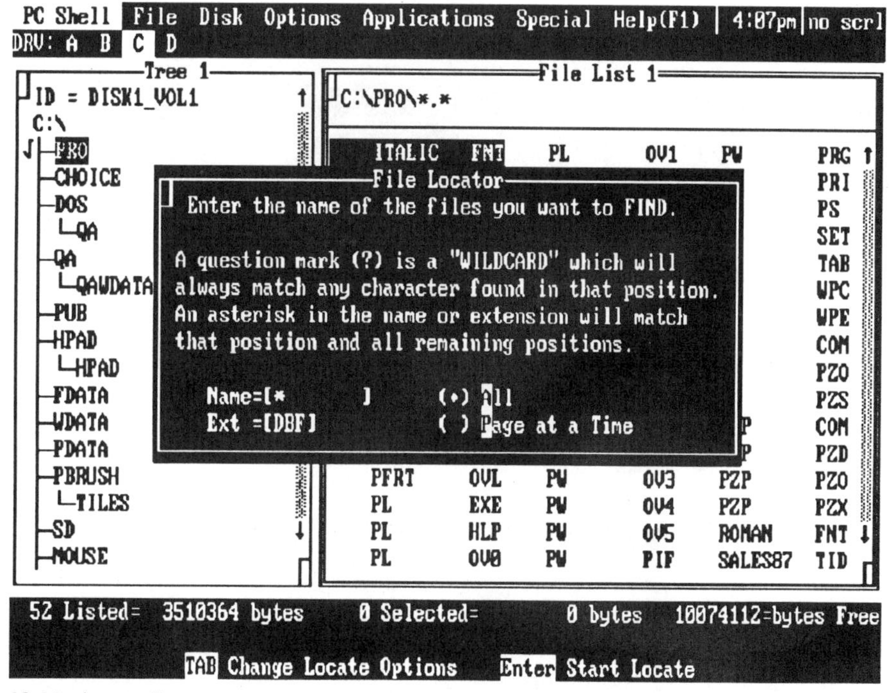

10-11 Locate file command

Wildcards can be used in the search. The asterisk can substitute for an entire name or extension. The question mark can replace a single character. For example, if you search for a file with the following criteria: D??e, you might obtain Duke, Dome, Dale, Dare, or any other four-letter combination that starts with a D and ends with an e.

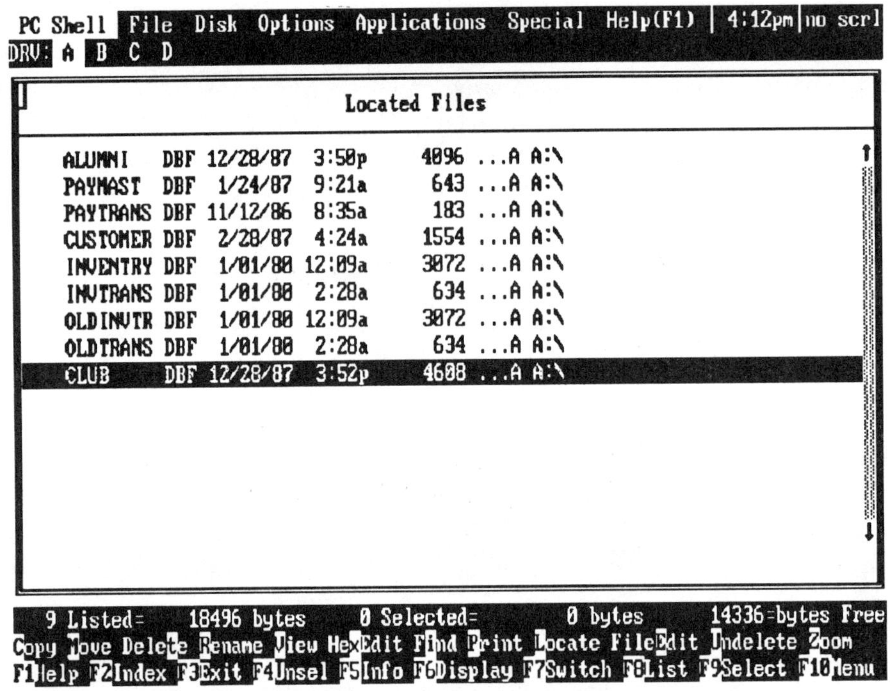

```
PC Shell  File  Disk  Options  Applications  Special  Help(F1) | 4:12pm|no scrl
DRV: A B C D

                              Located Files

   ALUMNI   DBF 12/28/87  3:50p      4096 ...A A:\
   PAYMAST  DBF  1/24/87  9:21a       643 ...A A:\
   PAYTRANS DBF 11/12/86  8:35a       183 ...A A:\
   CUSTOMER DBF  2/20/87  4:24a      1554 ...A A:\
   INVENTRY DBF  1/01/88 12:09a      3072 ...A A:\
   INVTRANS DBF  1/01/88  2:20a       634 ...A A:\
   OLDINVTR DBF  1/01/88 12:09a      3072 ...A A:\
   OLDTRANS DBF  1/01/88  2:20a       634 ...A A:\
   CLUB     DBF 12/28/87  3:52p      4608 ...A A:\

  9 Listed=    18496 bytes     0 Selected=         0 bytes     14336=bytes Free
Copy Move Delete Rename View HexEdit Find Print Locate FileEdit Undelete Zoom
F1Help F2Index F3Exit F4Unsel F5Info F6Display F7Switch F8List F9Select F10Menu
```

10-12 Finding dBASE files

The All option will display all the files in the Locate window, while the Page at a time option lets you perform functions such as copy, move, compare, and delete on these selected files. This is very useful if you want to save or delete files with the same extension. All database, spreadsheet, or word processing files can be moved, copied, or deleted at once even if they are in different directories.

Format data disk

The Format data disk command, shown in Fig. 10-13, lets you format a new disk to store data on. Formatting a disk erases any previous data that might be on the disk.

Select the drive that holds the disk you want to format, and choose Format data disk. All the formats available for that particular drive are displayed in the window. Figure 10-13 shows the options available for a 5¼-inch drive. Select the option you want and choose Format.

Figure 10-14 shows the screen display when formatting is taking place. The period indicates that formatting is accomplished in each track while an F indicates that the track is in the process of being formatted. When formatting is finished, you will be asked to enter the name of the new disk. If you want the disk to be "bootable," Fig. 10-15 appears giving you instructions on what to do. A bootable disk contains the essential DOS files, which let it "boot up" the computer—from

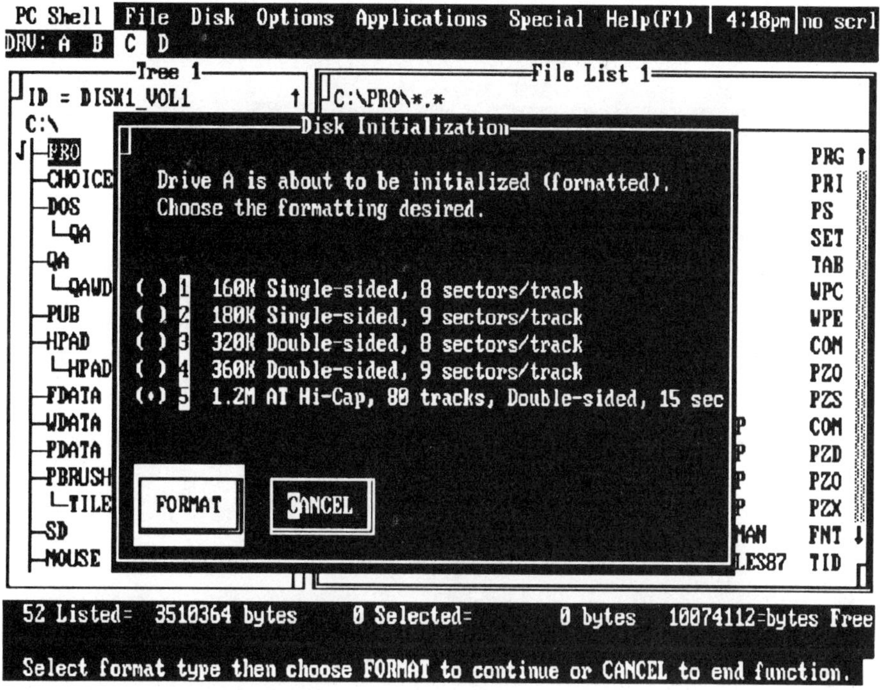

10-13 Format data disk command

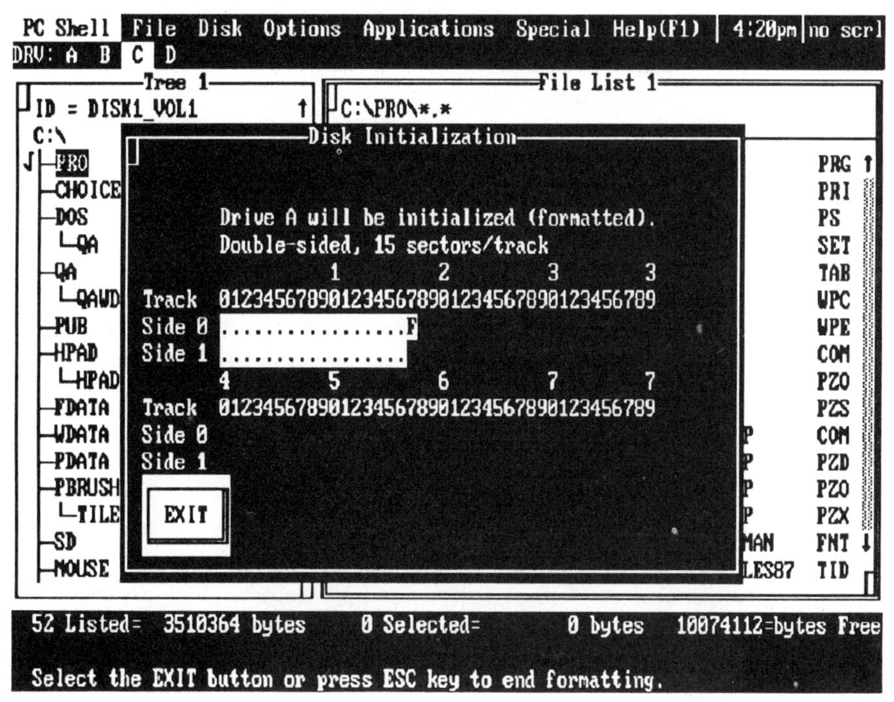

10-14 Status of disk initialization

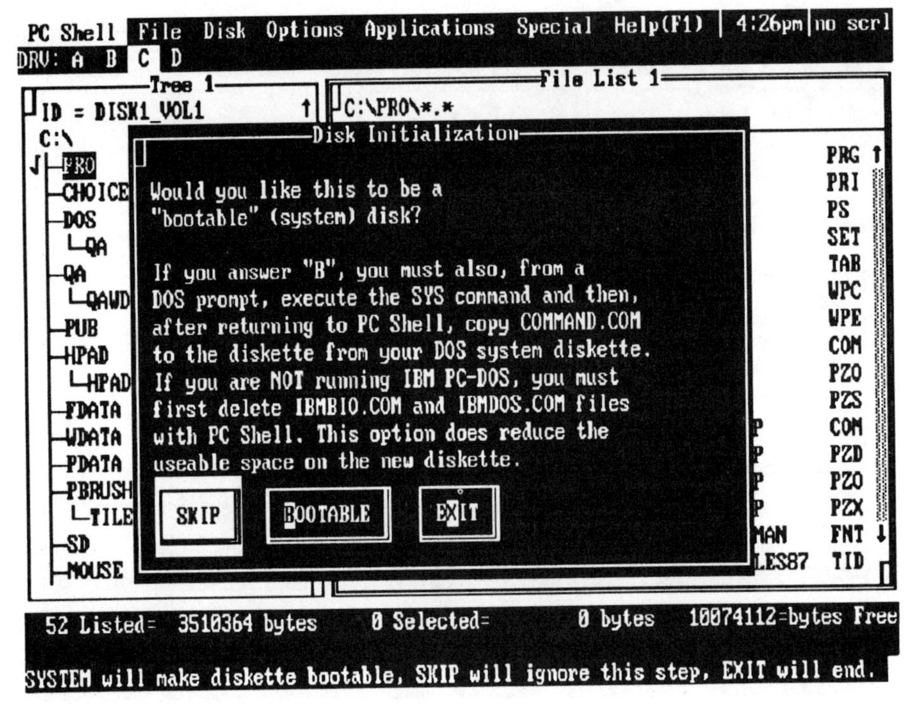

10-15 Making a bootable disk

Off to a DOS ready prompt. Bootable disks are usually programs. You can also make an already formatted disk bootable by making it a system disk, which is shown in the next section.

Make system disk

The Make system disk command is another way to create a bootable disk. When you select this command, Fig. 10-16 appears, asking you for the number of the drive that contains the pre-formatted disk. Figure 10-17 appears, asking you to confirm your decision. Select System to start transferring the DOS system files to make a bootable disk.

Disk info

The Disk info command, which generates the screen shown in Fig. 10-18, provides a lot of useful information about your disk, including total disk space, free space, hidden files, user files, directories, and bad sectors. Disk information can be obtained at any time by pressing F5 key.

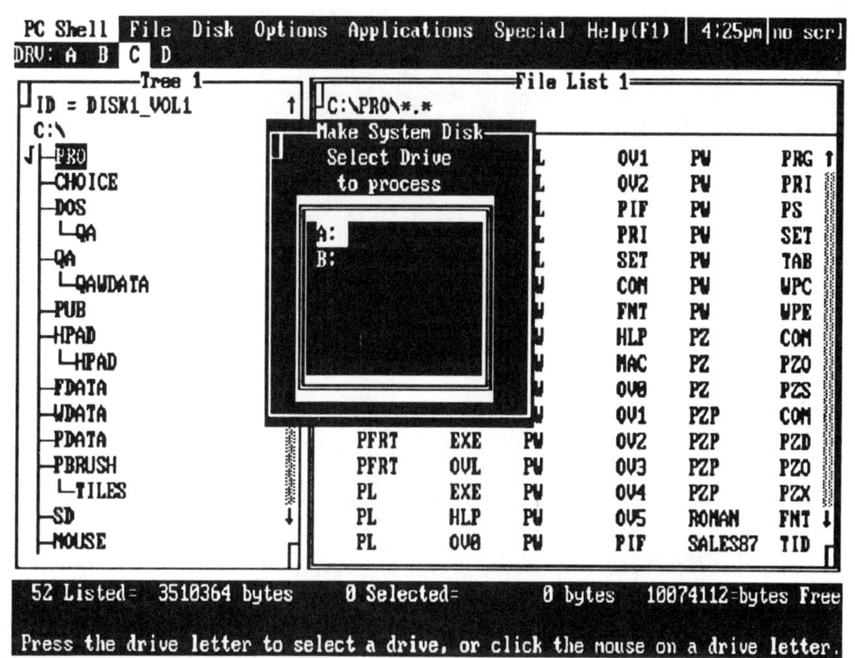

10-16 Make system disk command

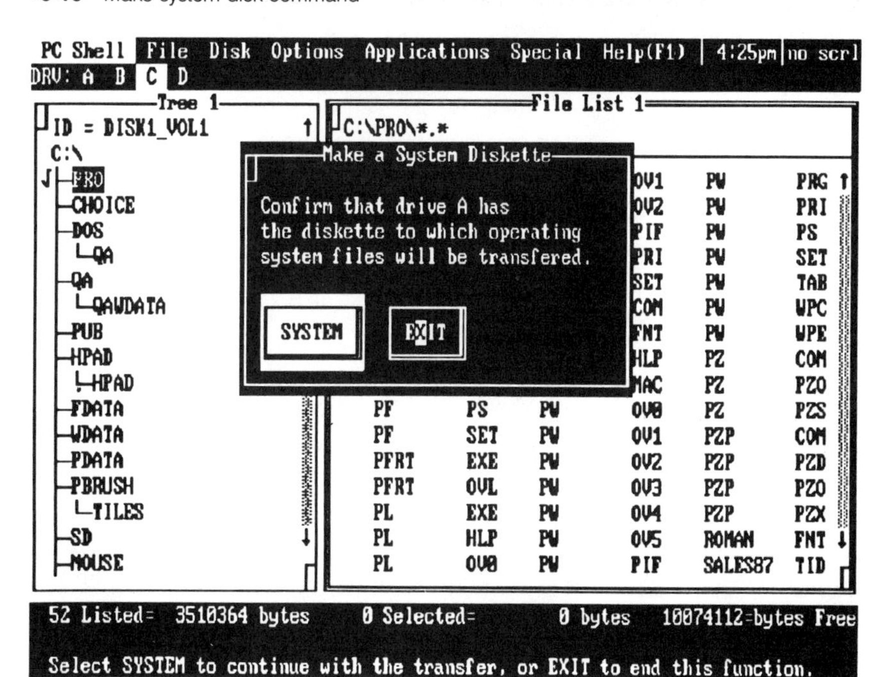

10-17 Confirming the transfer of files

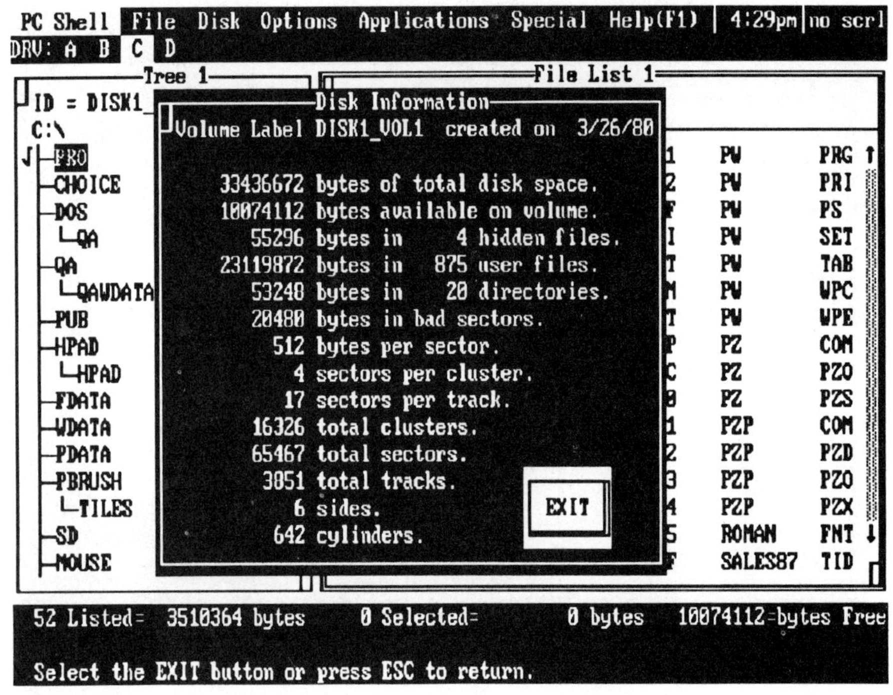

10-18 Disk info command

Park disk

The Park disk command, shown in Fig. 10-19, parks or positions the head of the disk drive over an unused portion of your hard disk. It ensures that there will be no data loss when you transport your computer.

Directory Maintenance menu

The Directory Maintenance menu is shown in Fig. 10-20. The Maintenance menu lets you add, rename, and delete directories (subdirectories). If you are adding a directory, move the cursor to the Tree List, where you want the directory added, choose Add a subdirectory from directory maintenance, and type in a name. The extension is optional. Selecting Continue will add the directory.

To rename a directory, move the cursor in the Tree List to the directory that you want renamed, choose Rename a subdir from the Directory menu and give it a new name. Select Continue to change the name.

When deleting a directory, you must first make sure it is empty. Delete the files with the File menu before you use this command. The root directory cannot be deleted. Move the cursor in the Tree List to the directory you want deleted, and choose Delete a subdir from the Directory menu. Select Continue to delete the directory.

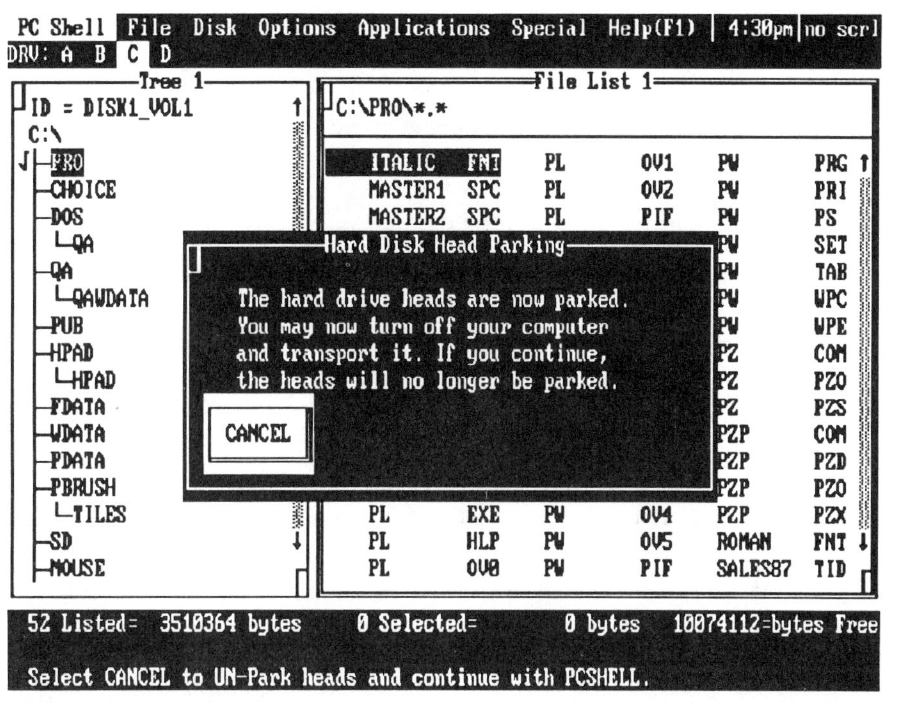

10-19 Park disk command

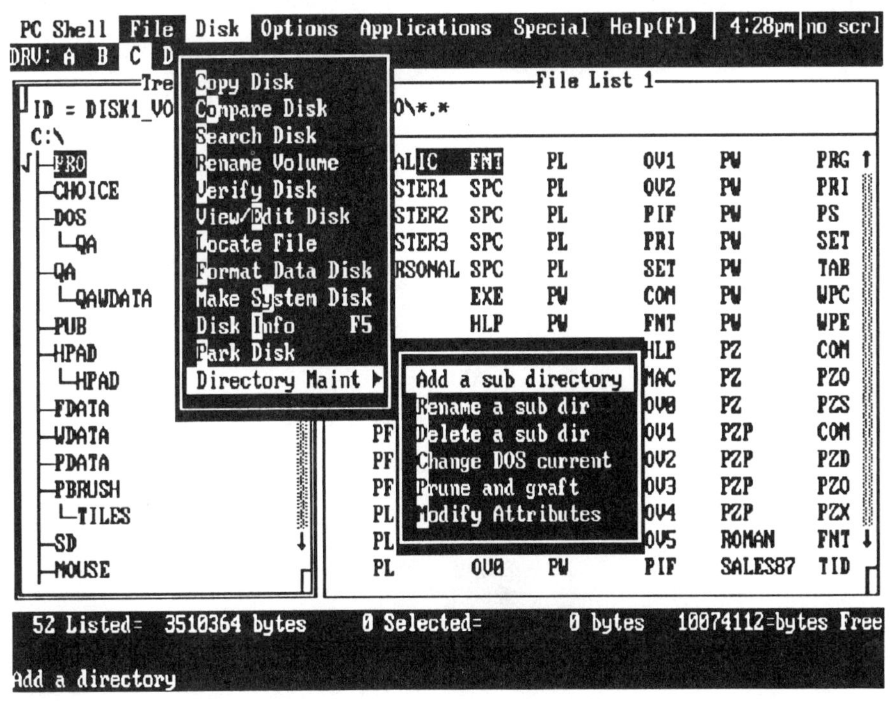

10-20 Directory Maintenance menu

The Change DOS current command lets you change from one directory to another. The highlighted directory will become made, or you can select the directory from the Tree List.

The Prune and graft command lets you move directories from one place to another on your disk. You can "prune," or move, a directory and "graft," or attach, a directory. To move a directory, choose the Prune and graft command. Select the directory to prune (move) by highlighting it with the cursor in the Tree List. Select Continue from the dialog box and the directory will receive a > symbol next to it. Another dialog appears asking you where you want it grafted, or attached. Highlight the directory in the Tree List and select Continue. The directory will now be moved to its new location.

Modify attributes

The Modify attributes command is very similar to the attributes command for a particular file. Select the directory you want to change the attributes of by highlighting it in the Tree List, as shown in Fig. 10-21. Select the Modify attributes command. The directory can be hidden (H), meaning it will not be listed in the DOS directory command but will be in PC Shell; system (S), meaning it will not be listed in the DOS directory command; read only (R), meaning it can be read but not altered or deleted; and archive (A), meaning it will be backed up during the next backup session. Select Update, or Cancel to exit to PC Shell.

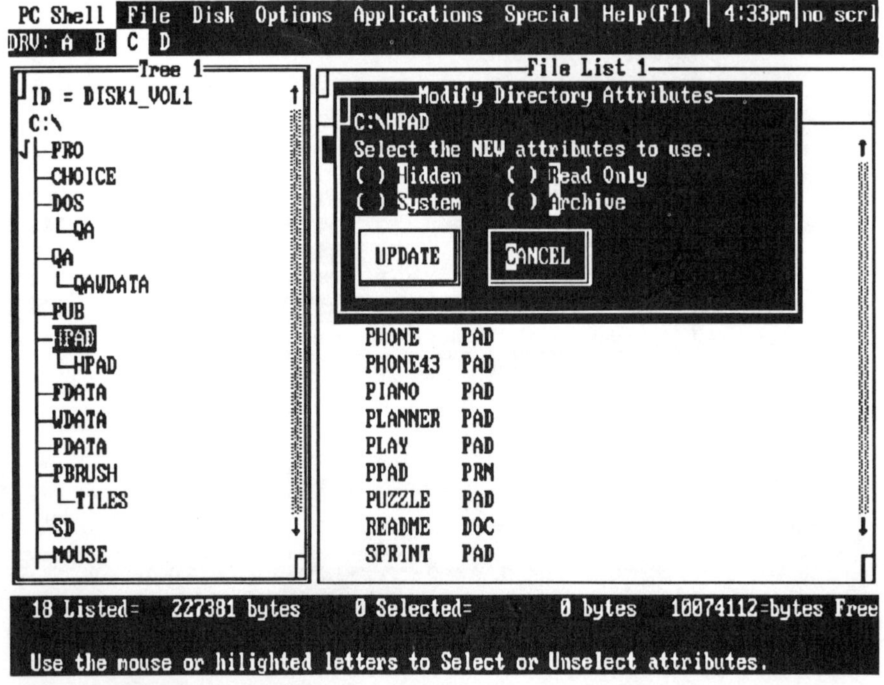

10-21 Modifying directory attributes

11
CHAPTER

Disk options

This chapter discusses the disk options, or settings, that can be used in PC Shell. These settings include advancing from window to window and from file to file; changing the color of backgrounds and windows; selecting files to be displayed and to be marked for copying, moving, or deleting; moving and changing the size of a window; changing the date and time; and adding applications to a pull-down menu.

One- and two-list displays

The Disk Options menu is displayed in Fig. 11-1. A few of the commands, such as the Two list display and One list display, have already been discussed in previous chapters. The Two list display command displays a second window that contains the programs or files in another drive or directory. It lets you copy, compare, or move files from one directory or drive to another. A second window can also be created with the short cut command Ctrl−Alt, plus the drive letter. For example, Ctrl−Alt−A will open a second window, displaying the files contained on drive A. Use the Active list switch command or the F7 key to toggle from one window to the other.

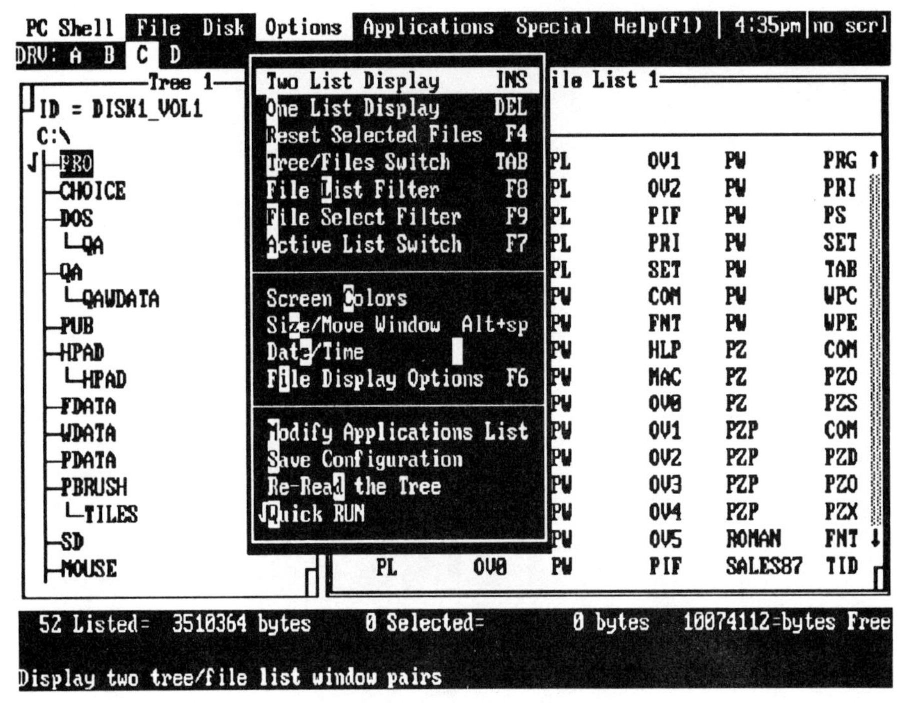

11-1 Options menu

The One list display command removes the second window and returns you to a single-list display.

Reset selected files

The Reset selected files command "unselects" all selected files. The selected files are ones chosen to be copied, moved, or deleted. If you accidentally select the wrong files or change your mind, use the Reset selected files command.

Tree/files switch

The Tree/files switch command lets you move the cursor from the Tree window to the File List window. The active window is the window with the highlighted, double-lined border. You can also press the Tab key to switch from one window to the other.

File list and file select filter

The File list filter command, shown in Fig. 11-2, lets you limit the files in the File List window, while the File select filter command, which generates the screen in Fig. 11-3, lets you select files that have common characters, filenames, or extensions. You can use the wildcard characters, like the asterisk for complete names or

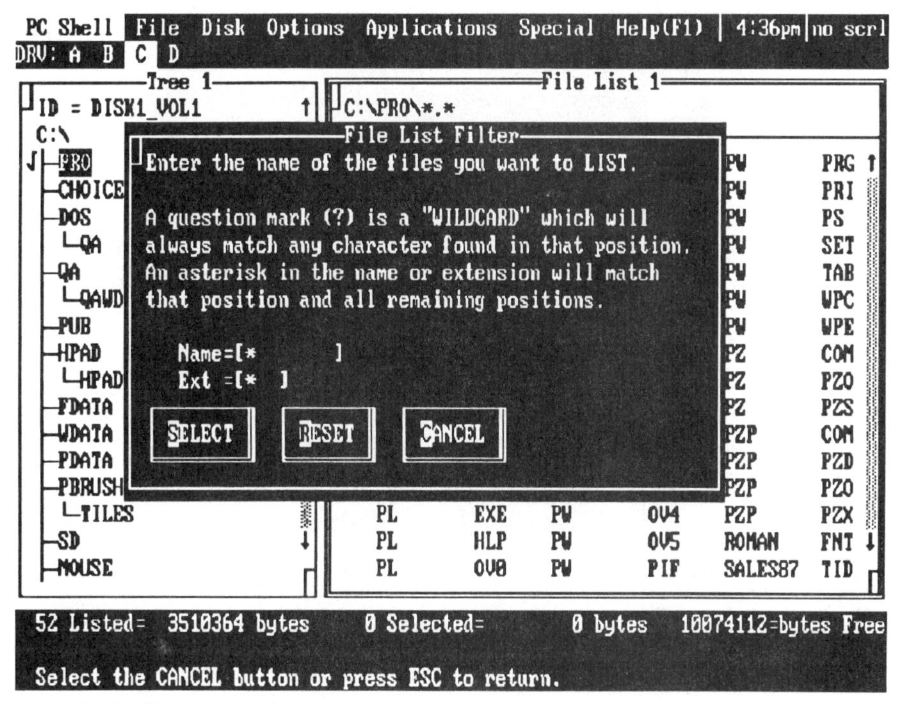

11-2 File list filter command

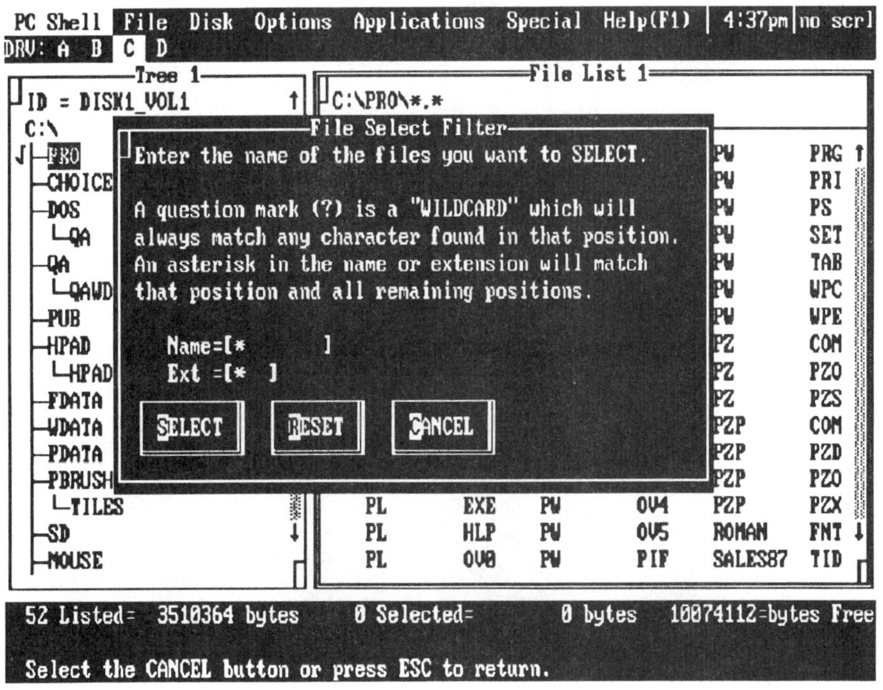

11-3 File select filter command

the question mark for single characters. The File list filter command lists all files that have the same extension, like .DOC or .COM. The File select filter command lets you select files so you can copy, move, or delete them. Press Select to initiate the search, Reset to show all files, and Cancel to exit the operation.

Active list switch

The Active list switch command lets you switch from one window to the next in a two-list display. You can also toggle between windows with the F7 key. Use the Tab key to go from the tree list to the file list.

Screen colors

The Screen colors command lets you change the colors of the menu bar, window, dialog box, and message box. This option is not available on a monochrome monitor.

Size/move window

The Size/move window command, shown in Fig. 11-4, lets you resize or move a window around the screen. To resize a window, move the cursor into the window

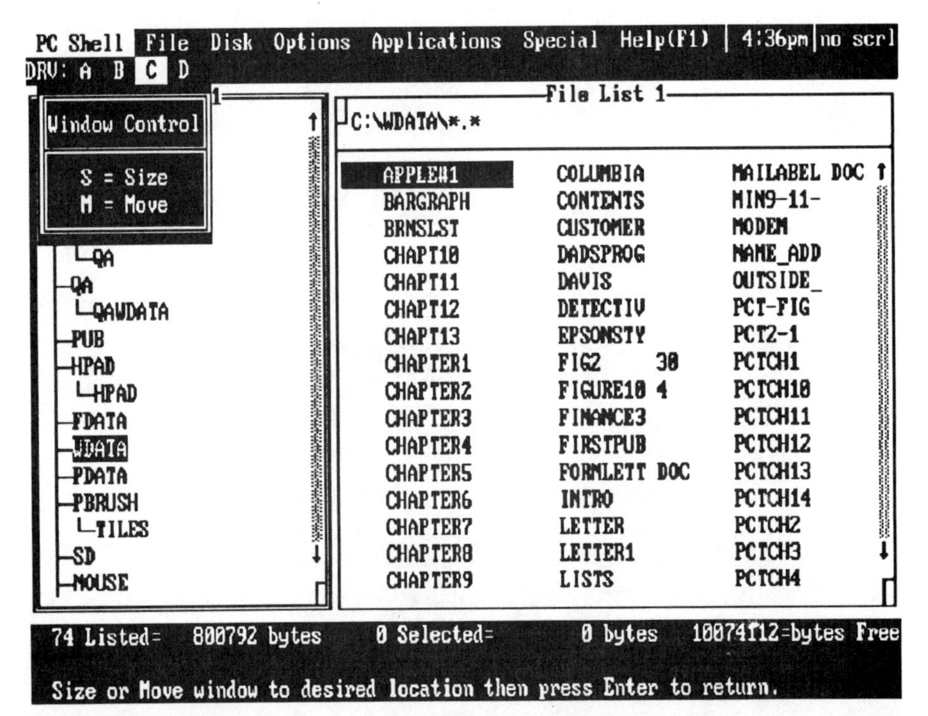

11-4 Size/move window command

you want to resize, making it active. Choose the Size/move window command and select the S = size option. Use the arrow keys to resize the window, or position the mouse on the size box in the lower right-hand corner of the screen. The window will change as you move the mouse. The top left corner of the window will remain fixed. You can resize either the tree or tree list window.

To move a window, position the cursor in the window you want to move, making it active. Choose the Size/move window command and select the M = move option. Use the arrow keys to move the window, or position the mouse on the top window border. The window will move as you either drag the mouse or use the cursor keys. You can also use the Zoom command on the bottom line of the screen to toggle back and forth from a full-screen window to the split screen.

Date/time

The Set date and time command, which displays the screen in Fig. 11-5, lets you change the time and date of the system clock. Type in the date in the form MM-DD-YY, and the time in the form HH:MM. Select Set to implement the changes or Exit to cancel them.

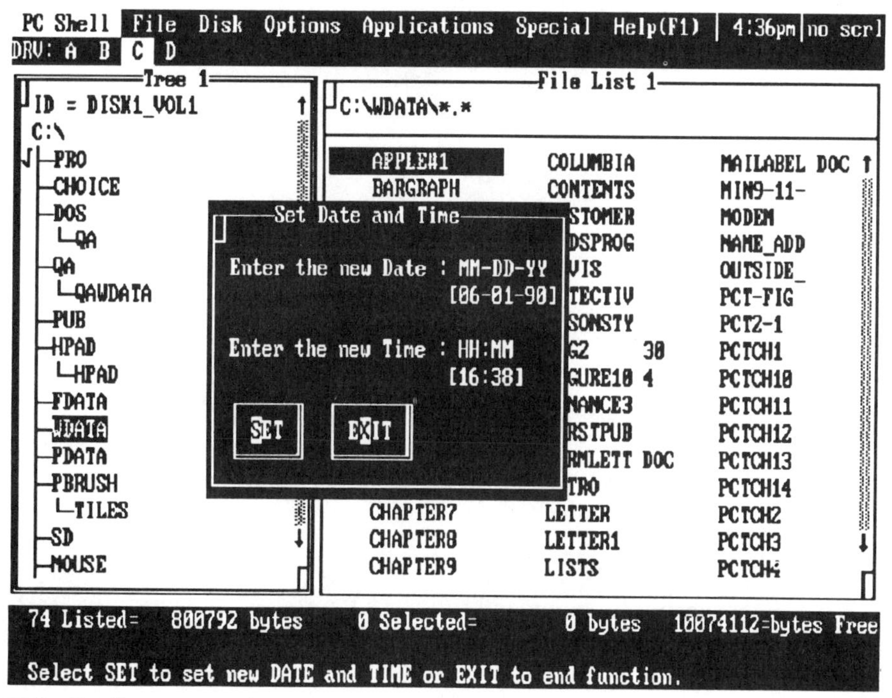

11-5 Date/time command

File display options

The File display options command, shown in Fig. 11-6, lets you determine how files will be displayed in the file list window. The files can be sorted in ascending or descending order. The default setting is a display by filename, unsorted. You can select more than one option with which to display the files.

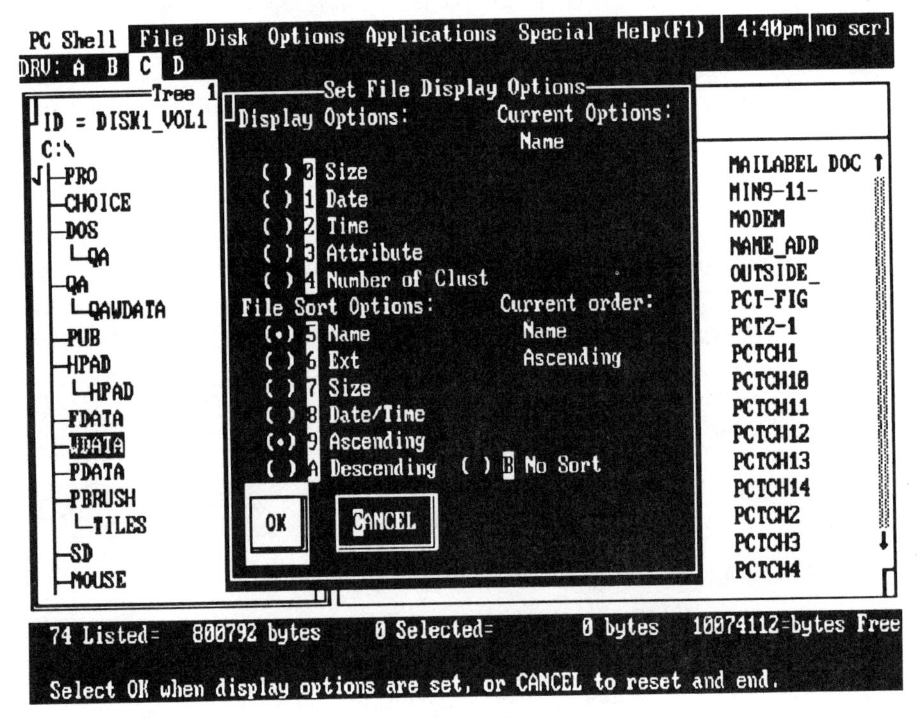

11-6 File display options command

The display options include the file size, the date and time the file was created, file attributes, and the file's starting cluster number. Files can be sorted in ascending or descending order by name, extension, size, date, and time.

Figure 11-6 displays the options for sorting a list in ascending order by filename. The resulting list is shown in Fig. 11-7.

Modify applications list

The Modify applications list command produces the screen shown in Fig. 11-8. It lets you add or delete applications such as dBASE or Lotus to the list of programs found in the Applications pull-down menu. This enables you to run programs from within the PC Shell menu system.

To create or edit an application, select New from the options at the bottom of

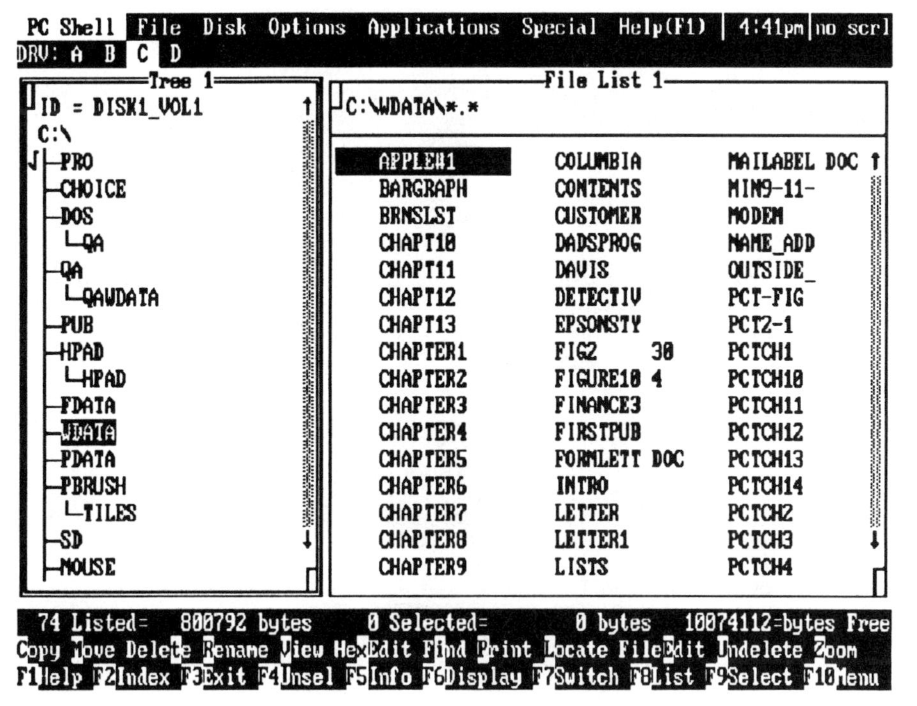

11-7 Sorting files in alphabetical order

the screen. Type in the title of the program, preceded by the ^ symbol. Type the initial directory, including the path where the program is located. Type in the path where the program will be executed. Enter the run filename used to start the application. Check the owner's manual of the program to see if any run parameters are necessary for startup. Type Y or N in answer to the "Wait on last screen?" prompt, which causes the program to pause before returning to PC Shell.

The Run with selected files option lets you select a file to be run with the program. For example, if a dBASE data file was run with the dBASE program, dBASE would access that data file.

The Associated file extensions option lists the file extensions that go with the programs, such as .DBF for dBASE and .WK1 or .WKS for Lotus.

The Pull-down Application menu option displays the pull-down Application menu at startup, if you select Y for yes.

Select OK to edit or add the application.

Save configuration

You can save all the changes you've made to the settings, including color, list filters, file display, and added applications, by selecting Save configuration. The next time you use PC Shell these settings will be used at start-up.

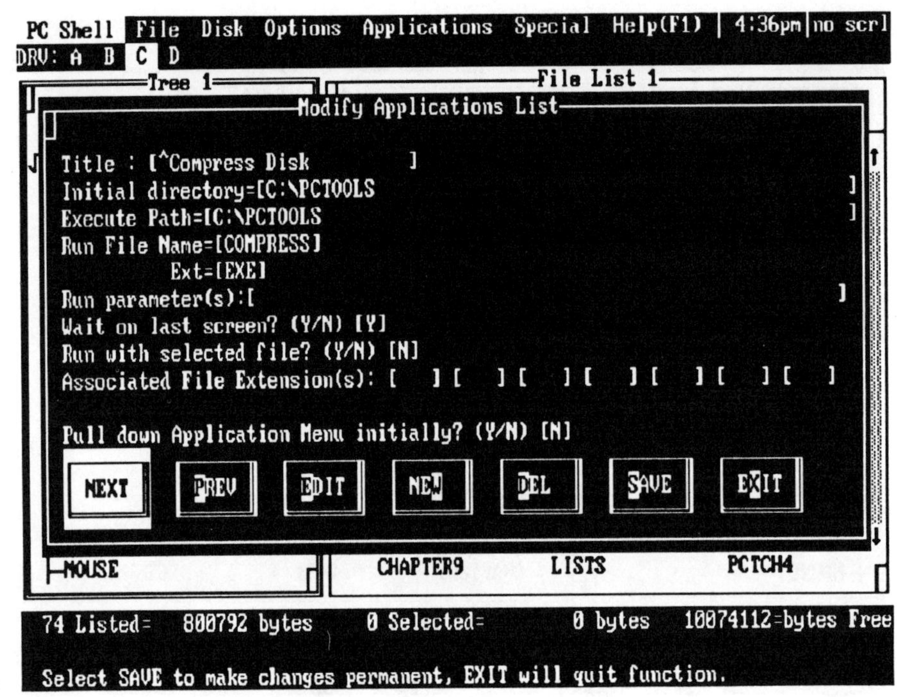

11-8 Modify applications list command

Reread the tree

The Reread the tree command redisplays the directory when the directory structure has changed. This occurs when you switch drives. Any changes to the tree structure do not register in the directory until you force PC Shell to physically read the drive itself. The purpose of this is to make PC Shell operate as fast as possible.

Quick run

The Quick run command appears in the Option menu *only* when PC Shell is nonresident. This selection can be toggled on or off. When the toggle is on, memory is not freed to run an application. If the toggle is off, memory is freed to run an application. Turn the toggle off position when you get an out-of-memory error.

12
CHAPTER

Disk applications

This chapter discusses the PC compress, Mirror, PC secure, and PC format commands that are available in the Disk Applications menu. The Compress command lets you consolidate blank or fragmented space on the disk to improve disk access. The Mirror command lets you create a duplicate of all the files in the directory so they can be recovered if they are accidentally erased or formatted. The PC secure command lets you use passwords to encrypt and decrypt your files to make them more secure. The PC format command replaces the DOS format command, and also allows data to be recovered after accidental reformatting.

Applications menu

The Applications menu, shown in Fig. 12-1, displays the applications available in PC Tools and any applications you might have installed (refer to Chapter 11—the Modify applications list command). The applications list can contain programs that were added, like Lotus, dBASE, or WordPerfect. The commands in the Applications menu help optimize the function of the hard drive.

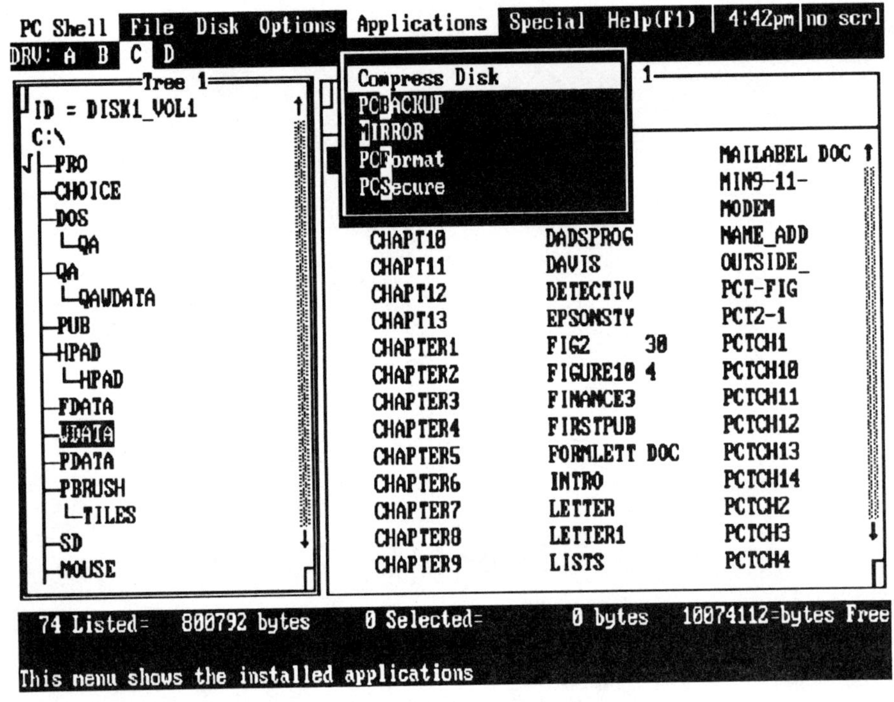

12-1 Applications menu

Compress disk

The Compress disk command lets you consolidate the hard drive by rearranging the files or programs into one continuous area. As programs or files are added or deleted to a hard disk, gaps of space are created which slow down the speed of the drive. This results in a hard drive that has a lot of gaps or is called fragmented.

The purpose of compressing a disk is to eliminate this unnecessary space between files to improve disk access. It is easier to recover a file that is continuous than one that is fragmented over the whole disk. The compress command will also find bad clusters on your disk and block them out.

To unfragment a hard disk, select Compress disk from the Applications menu, and Fig. 12-2 will appear. The drive you want to compress will be highlighted. The message bar at the bottom of the screen gives you messages to begin compressing, disk and surface analysis, and ordering options. A legend is located below the disk graphic picture telling you the status of the compression.

Sort Directory menu

You can specify that directories be sorted before you compress, as shown in Fig. 12-3. Directories can be sorted by date/time, filename, extension, and size. They can be sorted in either ascending or descending order.

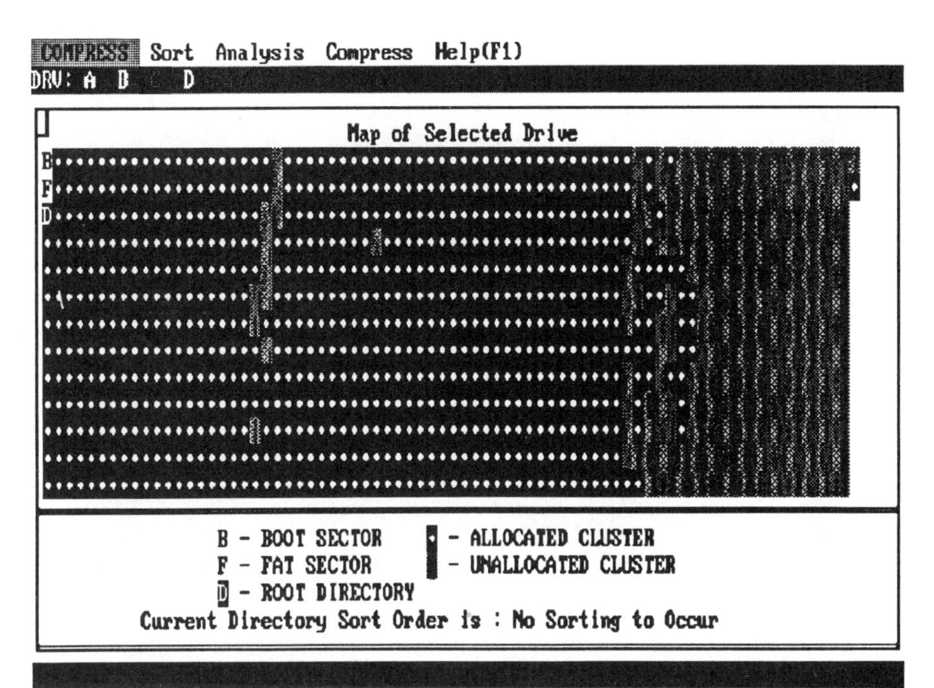

12-2 Compressing a disk

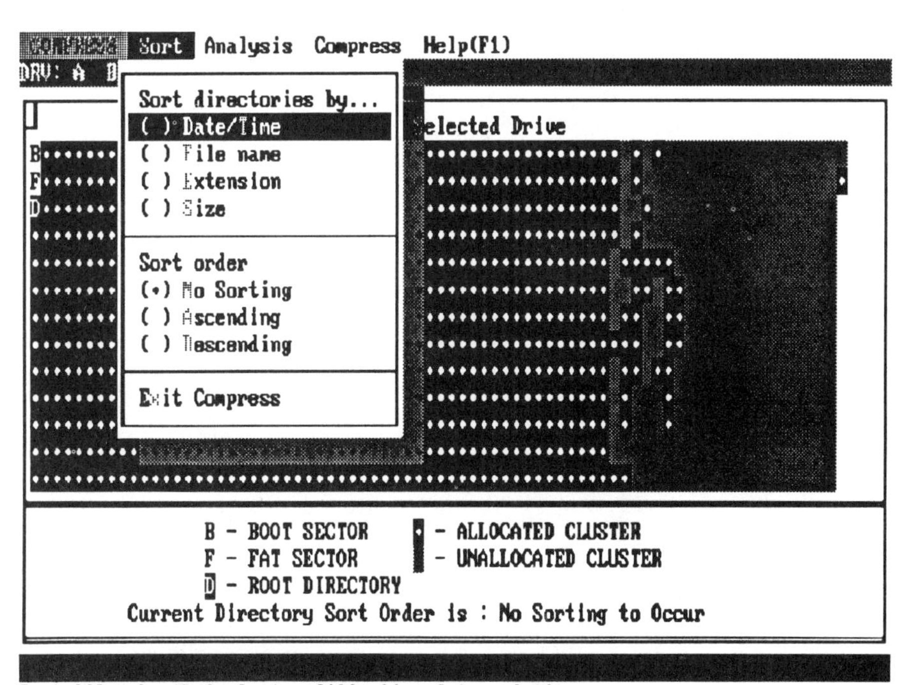

12-3 Sorting directories

Analysis menu

The Analysis menu, shown in Fig. 12-4, determines the condition of your hard drive to see if there are any bad clusters or fragmented files. If there are any bad clusters that are empty, it will mark the clusters. If there is data in the bad clusters, it will attempt to move the data to a good, unused part of the disk.

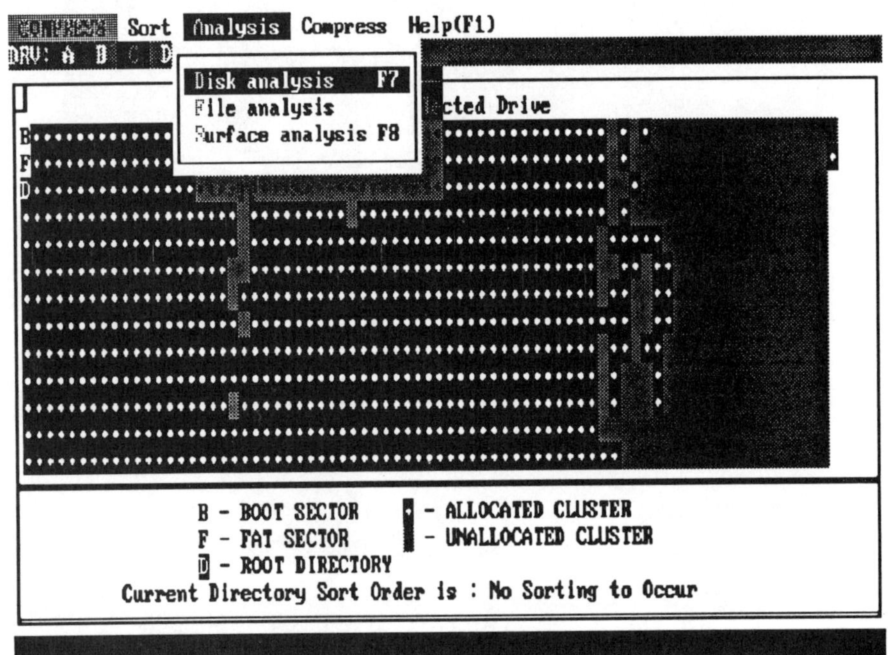

12-4 Analysis menu

If you choose the Data analysis command, which can be accessed either from the Analysis menu or with the F7 key, Fig. 12-5 will appear. It gives you the amount of disk space that is used (allocated clusters), free disk space (unallocated clusters), number of bad clusters, total number of files and directories (file chains), number of fragmented files (fragmented file chains), percentage of file fragmentation, noncontiguous free space areas, number of cross-linked chain files, number of unattached file clusters, and number of bad clusters within file chains.

An unattached file cluster is a cluster that has no data in it but is marked as being used. This cluster is inaccessible and, therefore, wasted space. A cross-linked chain file is a cluster used by two or more different files. Because data usually belongs to only one file, this is an indication that the cluster is probably damaged. These errors are normally caused by the hard disk itself.

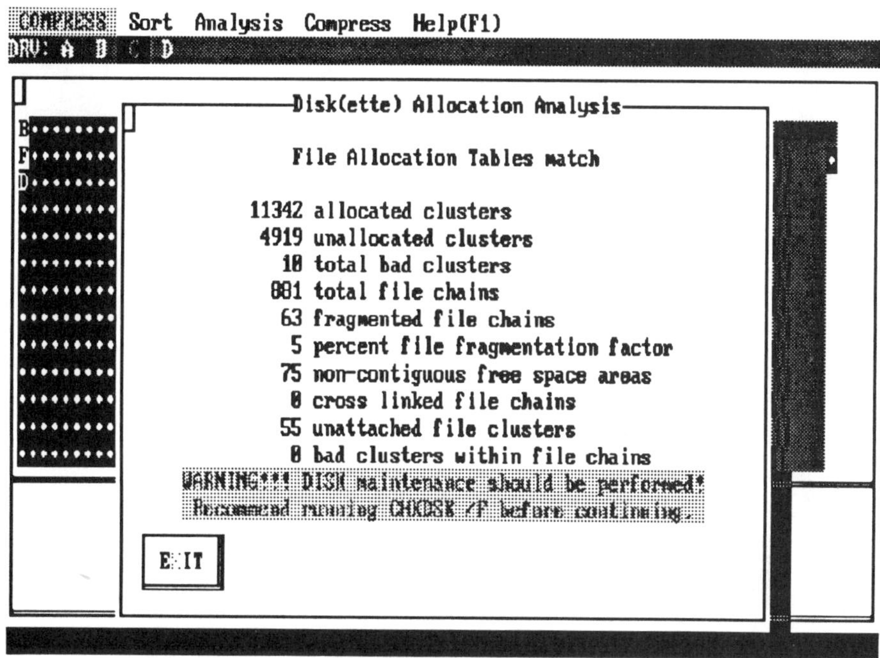

12-5 Disk allocation analysis

If you encounter unattached file clusters or cross-linked clusters, the Data analysis command will give you the following message:

> WARNING! Unattached clusters have been encountered!
> WARNING! Disk Maintenance should be performed!
> Recommend running CHKDSK/F before continuing.

This message means that you will have to run the CHKDSK/F command from DOS before you can compress your files. The CHKDSK/F command releases all unattached clusters so they can be used with other files. The cross-linked cluster has probably damaged one of the files accessing it. The CHKDSK/F command will list all the cross-linked files so you can either delete them or move them to a better storage area.

The File analysis command lets you look at each cluster on the disk, as shown in Fig. 12-6. It lists the filename and number of clusters it occupies. It also shows the number of parts a file is broken into, and the percent of fragmentation. If the number of parts and percent of fragmentation are both high, it is time to compress the disk.

The Surface analysis command checks the surface of a disk to see if it is okay to hold data. If it finds a bad cluster with no data, it will mark it as a bad cluster and prevent data from being saved there. If it finds a bad cluster with data already in it, it will attempt to move the data to a sound area of the disk. This operation

COMPRESS Sort Analysis Compress Help(F1)

DRV: A B C D

File Allocation Analysis

Path=C:\

Name		Clusters	Areas	Pct	Name		Clusters	Areas	Pct
IO.SYS		11	1	0%	QA	<DIR>	1	1	0%
MSDOS.SYS		15	1	0%	PUB	<DIR>	1	1	0%
COMMAND.COM		13	1	0%	HPAD	<DIR>	1	1	0%
DISK1_VOL1	<VOL>				FDATA	<DIR>	1	1	0%
DM.EXE		38	1	0%	WDATA	<DIR>	2	2	50%
DMDRVR.BIN		4	1	0%	PDATA	<DIR>	2	2	50%
CONFIG.SYS		1	1	0%	PBRUSH	<DIR>	2	2	50%
ONLINE.HLP		14	1	0%	SD	<DIR>	1	1	0%
PRO	<DIR>	1	1	0%	MOUSE	<DIR>	2	2	50%
CHOICE	<DIR>	1	1	0%	PCTOOLS	<DIR>	3	3	66%
PRINTER.TST		1	1	0%	AUTOEXEC.SAV		1	1	0%
DOS	<DIR>	1	1	0%	MIRORSAV.FIL		1	1	0%
AUTOEXEC.OLD		1	1	0%	MIRROR.FIL		25	18	96%

PREV DIR | NEXT DIR | FIRST DIR | LAST DIR | EXIT

Press Esc key or F3 key to terminate File Analysis

12-6 File allocation analysis

might take several hours to complete. When you select the Surface analysis command from the Analysis menu, or press the F8 key, Fig. 12-7 appears. It asks you for the number of passes in which to scan the disk, or you can let it run continuously until you press the Esc key. It also asks, as shown in Fig. 12-8, if you want to create a report to send to the printer or disk.

As the computer scans the disk, your screen will look like Fig. 12-9. The screen gives pictorial information on the boot sector, fat sector, bad clusters, root directory, allocated and unallocated clusters, and unreadable clusters. It also displays the elapsed time and the percent of scanning completion.

Compress menu

The Compress menu, shown in Fig. 12-10 lets you select the compression technique that you want to execute. The Unfragment only command compresses only the unfragmented files. It compresses all the free space that remains on the disk so that all new files will be saved in contiguous space. It places all the free space at the back of the disk.

The Full compression command unfragments all existing files and places all free space at the back of the disk.

The Full compression – clear command not only unfragments all existing files, but also clears or erases all old data in unused sectors.

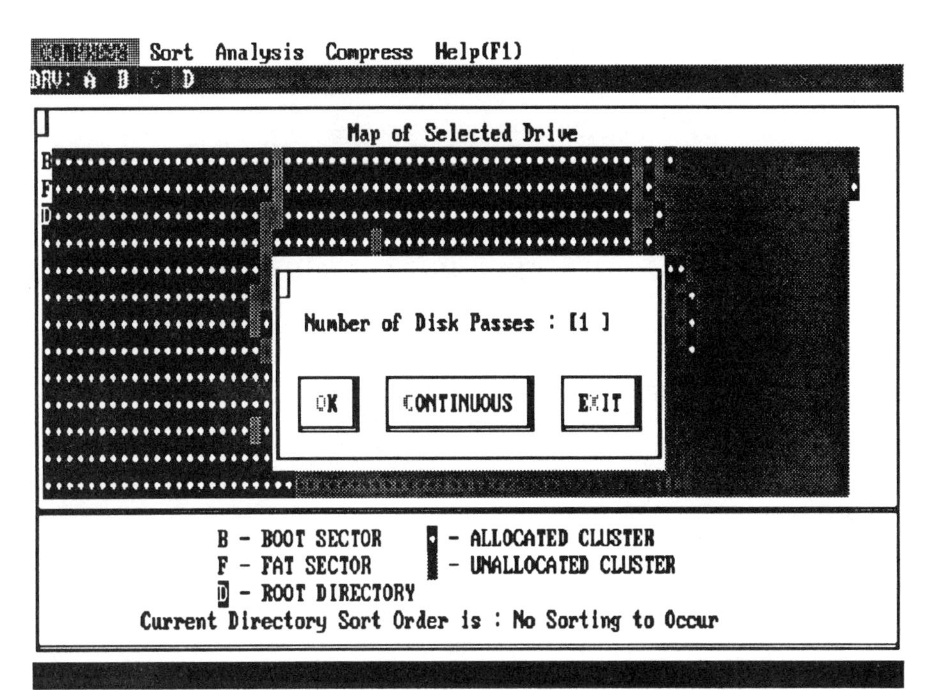

12-7 Surface analysis command

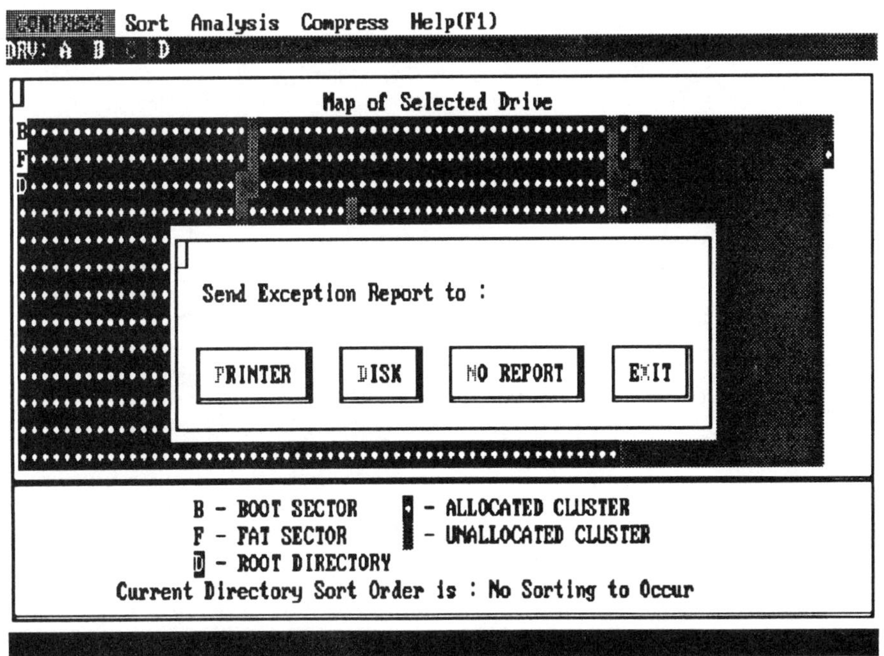

12-8 Creating an exception report

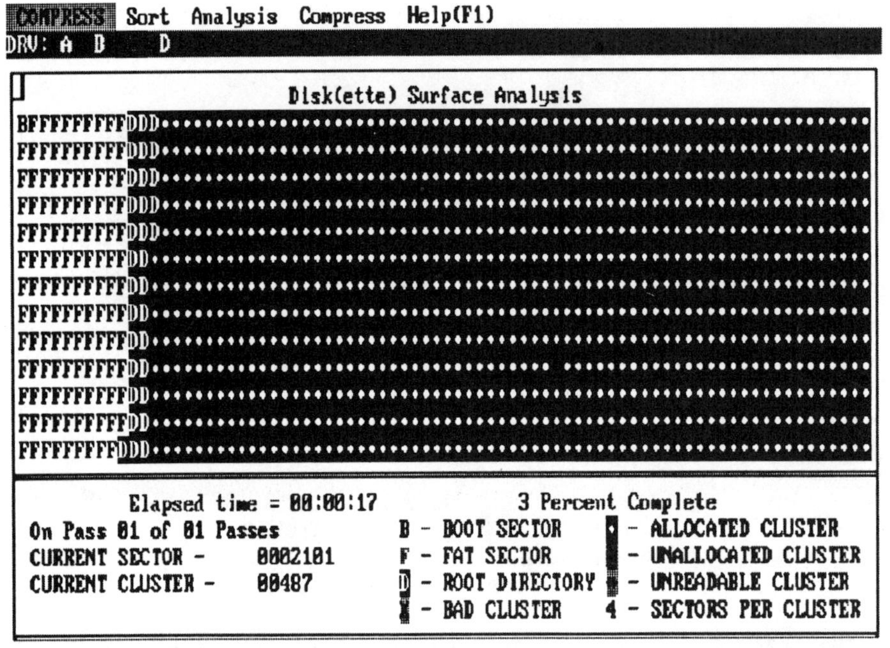

12-9 Disk surface analysis

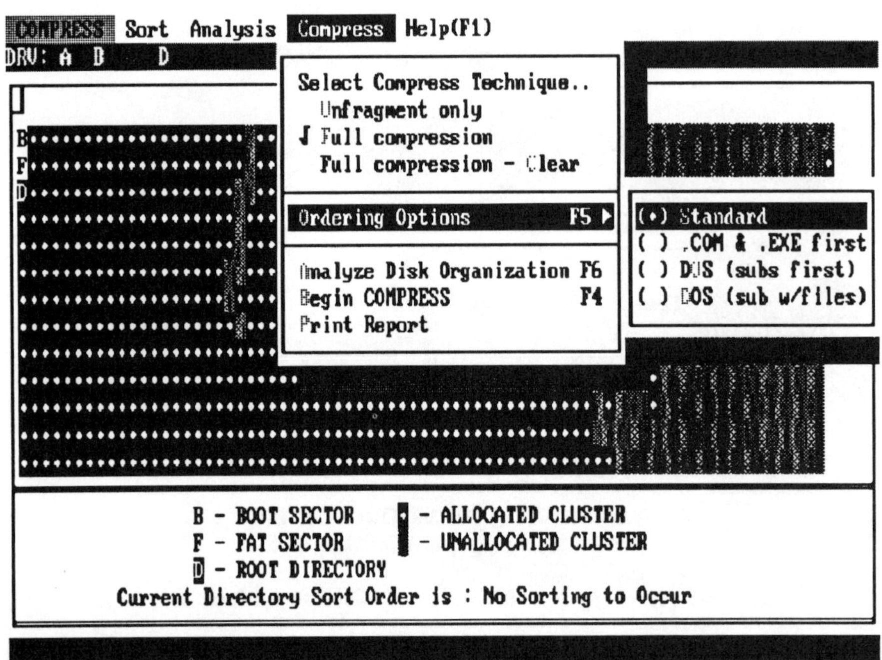

12-10 Compress menu

The Ordering options command generates a menu of its own. It gives you the choices with which to order a hard disk. The Standard option places your files in any physical order the compress function wants. This is the fastest way to compress if you are not placing any restrictions on the program.

The .COM & .EXE first command places the .COM and .EXE files at the beginning of the disk so they may be accessed first.

The DOS (subs first) command places all the directories at the front of the disk, while the DOS (sub w/files) command places both the directories and all their accompanying data files at the front of the disk. This option eliminates a lot of the disk-access time.

The Analyze disk organization command displays the message shown in Fig. 12-11, letting you know if compression is recommended.

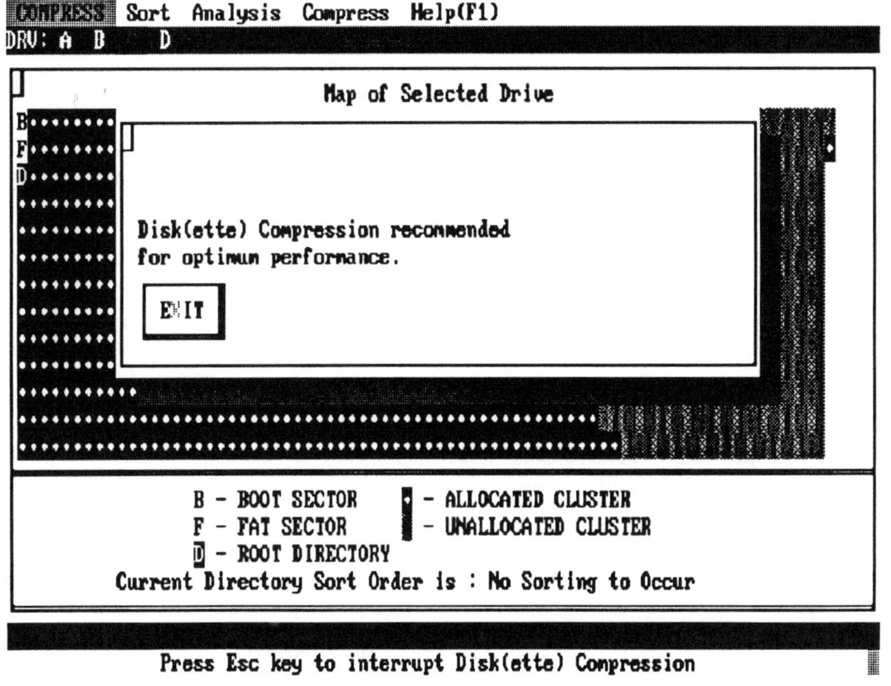

12-11 Analyze disk organization command

Selecting the Begin Compress command displays the warning shown in Fig. 12-12. It tells you to back up the disk before proceeding, and to eliminate all memory-resident programs except PC Tools. Select Continue to start the compress or Exit to cancel it. Compression will now take place. The status of the compression will be shown graphically in a dialog box, giving you the percentage of completion. All the programs and files will be moved to the front of the disk, and all the unused space to the back of the disk.

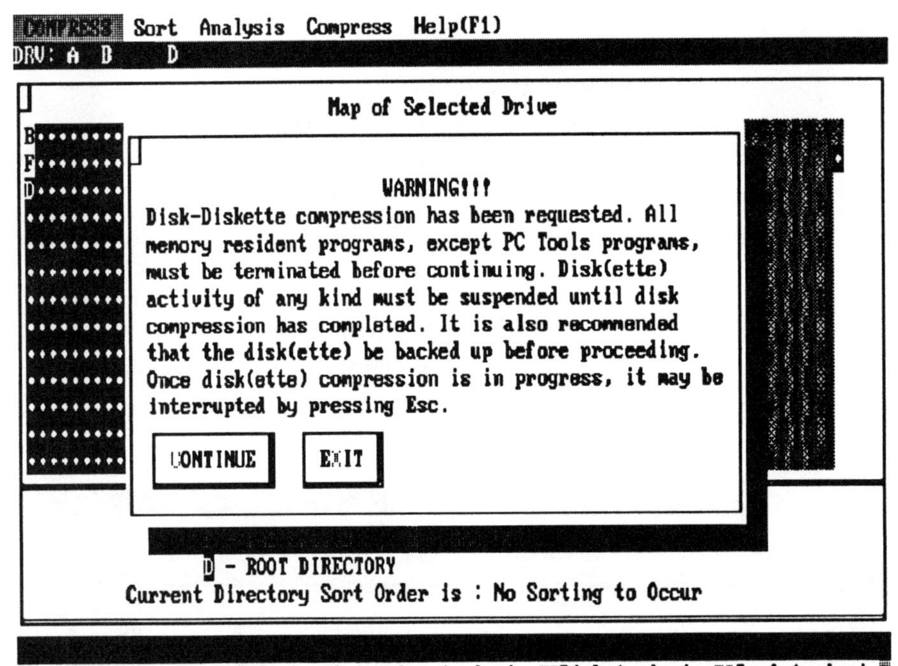

12-12 Beginning compression

Upon completion of the compression, you will be asked if you want a printed report. This report includes the time it took to do the compression, the options selected, and the number of used, unused, and bad clusters on the hard drive.

Files that are copy-protected, or have hidden files will not be moved by the Compress command. This is because there is a chance that files could be damaged or incompletely copied during such a procedure.

Mirror

Mirror is a program that keeps a copy of the file allocation table (FAT) and the root directory in a hidden file. The FAT contains a record that shows where each file is located. It is the index of the files on a disk. The FAT also has an index of all the disk's free space, so it knows where future programs will be stored. The FAT and the root directory take up very little space on a disk. Most of the disk space is used to store directories, files, and their accompanying data.

What Mirror does is makes a copy of the FAT and the root directory so that, if a hard disk is reformatted or accidentally erased, the hard disk can be restored with its original information. Mirror is normally run after compressing the disk, but can be run at any other time as shown in Fig. 12-13. Compressing a disk moves all the files and directories around so Mirror should be run after each disk compression.

```
PC Tools Deluxe - MIRROR  R5.5
(C) Copyright 1987-1989  Central Point Software, Inc.
Unauthorized duplication prohibited.

Creates an image of the SYSTEM area.

Drive C being processed.

MIRROR successful.







Press any key or a mouse button to re-enter PC Shell
```

12-13 Creating a mirror image

The Mirror program should be included, during setup, in the AUTOEXEC .BAT file. This way, any programs you add or delete will be automatically updated every time you turn on the computer.

Mirror saves all the deleted file information in a delete tracking file. This file is automatically created during setup.

PC Format

Disks are divided into units of storage called *sectors*. Each sector contains 512 bytes of information. Disk sectors are created by the series of concentric circles on a disk called *tracks*. A floppy disk will normally have about 40 tracks, with nine sectors on each track. A 20-megabyte hard disk has approximately 1200 tracks and 17 sectors on each track. A 360K disk has 720 sectors, while a 20-megabyte drive has 41,000 sectors.

Formatting a disk creates the tracks and sectors. A disk cannot be used before it is formatted. A format command (DOS or PC Tools) divides a disk into tracks and sectors, and also checks for defective sectors on the disk.

PC Format will format any floppy or hard disk, as shown in Fig. 12-14. It differs from the DOS format command in that the data or files saved on a PC Tools-formatted disk can be recovered using the Rebuild command. PC Format cannot recover data from floppy disks that have been formatted using the regular DOS format command.

```
PC Tools Disk Formatter 5.5
Copyright 1987-89 by Central Point Software, Inc.  All rights reserved.

Will format drive A:  (physical # 00h, type= 1.2M 5.25-inch)

Formatting 15 sectors, 80 cylinders, 2 sides.
Press Enter when ready...
```

12-14 Formatting a floppy disk

When formatting a hard disk, PC Format works the same way the DOS Format does in that it changes only the filenames in the root directory and the position of files in the file allocation table. It does not physically overwrite the files.

All the options that can be used for a floppy disk with PC Format are shown in Fig. 12-15. The list of options can be obtained at any time by typing PCFORMAT at the DOS prompt. The general form of the format command is:

FORMAT d: [/S] [/1] [/8] [/V] [/4] [/N:S/T:t] [/T:xx] [/F:ddd]

with the d: being the letter of the drive you want to format. The /V option places a volume label on the disk. The label is physically placed on the disk, and helps you keep track of what is stored on the disk. It can be up to 11 characters in length. Whenever you generate a directory of a disk, the volume label is also displayed.

```
C:\PRO>pcformat
PC Tools Disk Formatter 5.5
Copyright 1987-89 by Central Point Software, Inc.  All rights reserved.

The PCFormat options are:
/V        Add volume label.
/S        Include DOS; make it bootable.
/Q        Quick erase of a previously formatted disk.
/P        Echo all messages to printer LPT1.
/TEST     Disk isn't really written.
/1        Single-sided.
/4        360k disk in 1.2 Meg drive.
/8        8 sectors per track.
/N:s/T:t  Sectors/track (s=8,9,15,18), Tracks (t=40,80).
/F:ddd    Diskette size (160,180,320,360,720,1200,1440).
/R        Read each track, reformat, and rewrite data.
/F        Like /R, but also delete all files.
/DESTROY  Format & erase all data.
```

12-15 PC Format options

The /S option copies DOS on the disk to make it bootable. If you normally boot from a hard disk, you can omit the /S option, because DOS will take up unnecessary room on the floppy disk. If you format a disk without DOS, you cannot later add DOS to it unless you use the /S option. Unfortunately, this will destroy all data saved on it.

The /Q option erases the directory and file allocation table from the disk, but leaves its data.

The /P option prints the formatting information out on the printer.

The /TEST option simulates a format without actually accomplishing it.

The /1 option formats a single side of the disk.

The /4 option formats a 360K disk in a 1.2-megabyte drive. This lets a high-density drive read a low-density disk.

The /8 option formats a disk with 8 sectors instead of the normal 9. This creates disks that are compatible with earlier versions of DOS.

The /N:s/T:t option tells the formatting command how many sectors are in each track.

The /F:ddd option specifies a full format, whereby each track is read, formatted, and rewritten with data.

The /R option reformats and writes every track.

The /F option is the same as the /R option but deletes all files.

The /DESTROY option formats the disk, then erases the formatting.

The /T option, not listed in Fig. 12-15, specifies the number of tracks to format.

The options that can be used with a hard disk are: d:, /S, /V, /P, and /TEST.

PC Secure

The PC Secure selection from the Applications menu lets you encrypt important data so that it is secure from anyone viewing it. It uses a widely used encryption and decryption technique called DES, which is almost impossible to decipher without a code. It uses keys or passwords to encrypt and decrypt files. PC Secure can encrypt all files except those that have copy protection. Encrypted files can be sent via modem, as long as both the sending and receiving computers have access to the same password or key.

Start PC Secure from the DOS prompt by typing PCSECURE and hitting a Return. There are no spaces between any of the letters. You can also specify the black and white option (PCSECURE/BW) or the government option (PCSECURE/G). The /G option causes the original file to be erased, leaving only the encrypted file on the disk.

When you enter PC Secure, Fig. 12-16 appears, asking you to enter a password. This password, or key, enables you to encrypt or decrypt any file. The key can contain anywhere from 5 to 32 numeric or alphanumeric characters, and is case sensitive. This means that the passwords AGLOWBOOK, AglowBook, and aglowbook are all different.

The password should be a series of keys that will not be easily guessed, but that you should not forget. Write it down if you have to, and hide it in a secure place. When alphanumeric characters are typed in, an asterisk is displayed on the screen. You can also enter a key in hexadecimal by pressing the F9 key. After you enter the key, PC Secure will ask you to verify the key by typing it again. After

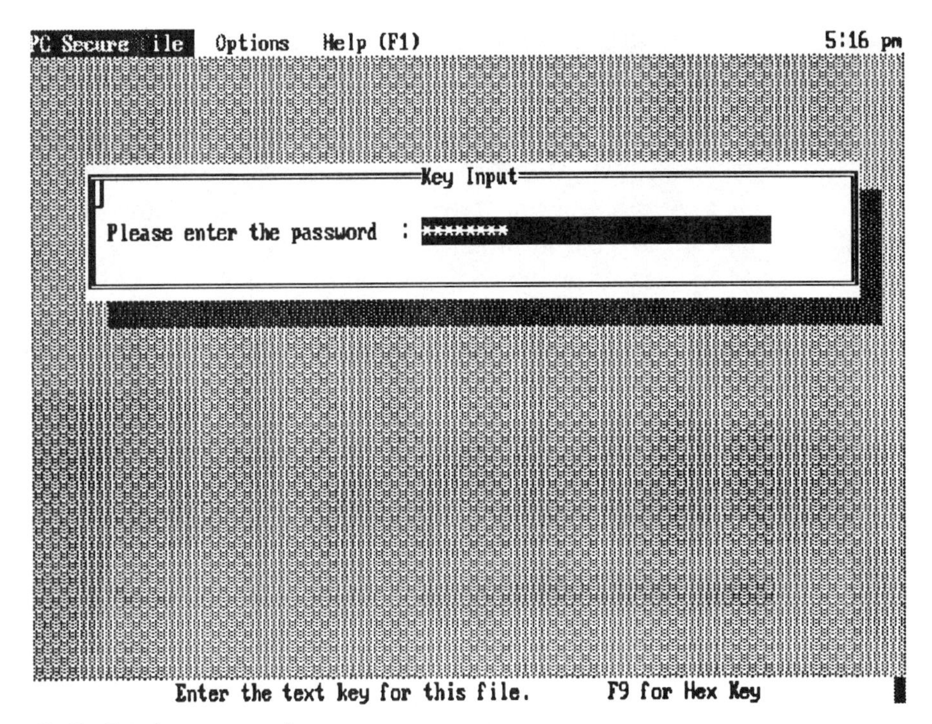

12-16 Entering a password

you do that, a message will appear in a dialog box telling you the master key has been installed.

The PC Secure screen, with its pull-down File menu, is shown in Fig. 12-17. The File menu lets you encrypt and decrypt a file, provides you with information about the compressed encrypted file, and lets you exit the program.

The Encrypt file command lets you encrypt a file by randomly scrambling its data. You can encrypt either a single file or all files in a directory. When you select command from the menu, or press the F4 key, Fig. 12-18 appears asking to select a file. You can select a file from the list, or type one in the text box. Use either the cursor keys or the mouse to scroll through the file list box to select the file. After the file is selected or typed in, press the Enter key, type Alt−E, or select the Encrypt button to encrypt the file.

Directories can be encrypted the same way by selecting the Directory box in Fig. 12-18. You will be given a message saying that more than one file will be affected, and asking you if you want to include all subdirectories associated with that particular directory.

You will be asked to enter a password to encrypt the file, and to repeat it again for verification. Make sure to always use different passwords for different files. The master key password can encrypt any file, while the password entered now can be used only for this specific file. If you forget the password, you can use the master key password to encrypt a file.

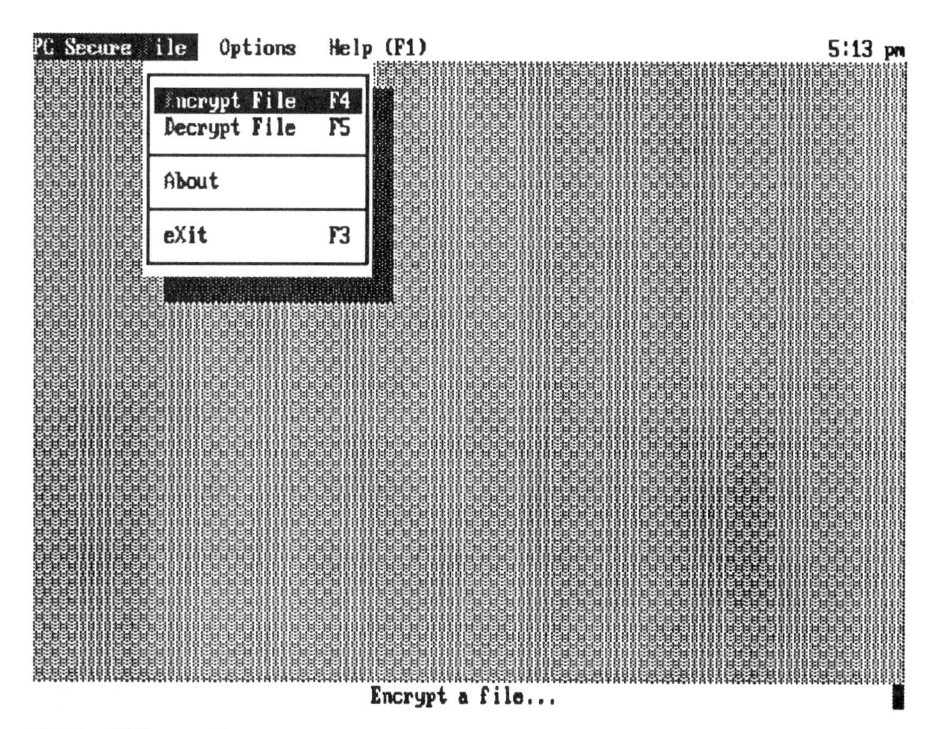

12-17 PC Secure File menu

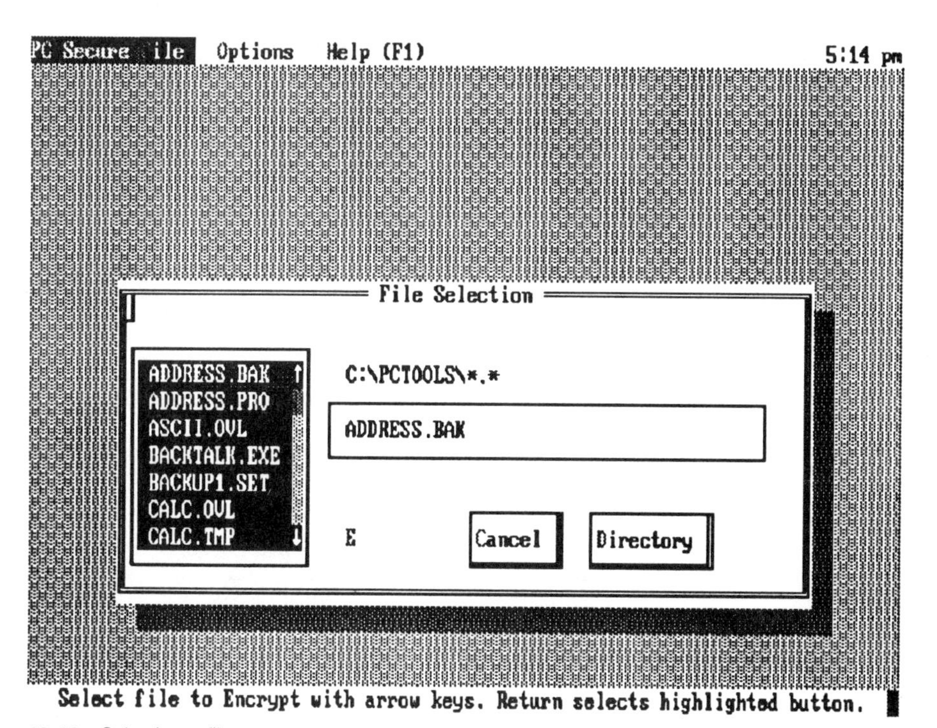

12-18 Selecting a file to encrypt

After you type the password a second time to verify it, Fig. 12-19 will appear, showing the progress of the encryption. If you are encrypting a directory, a message will tell you how many files are processed and the number of bytes read and written.

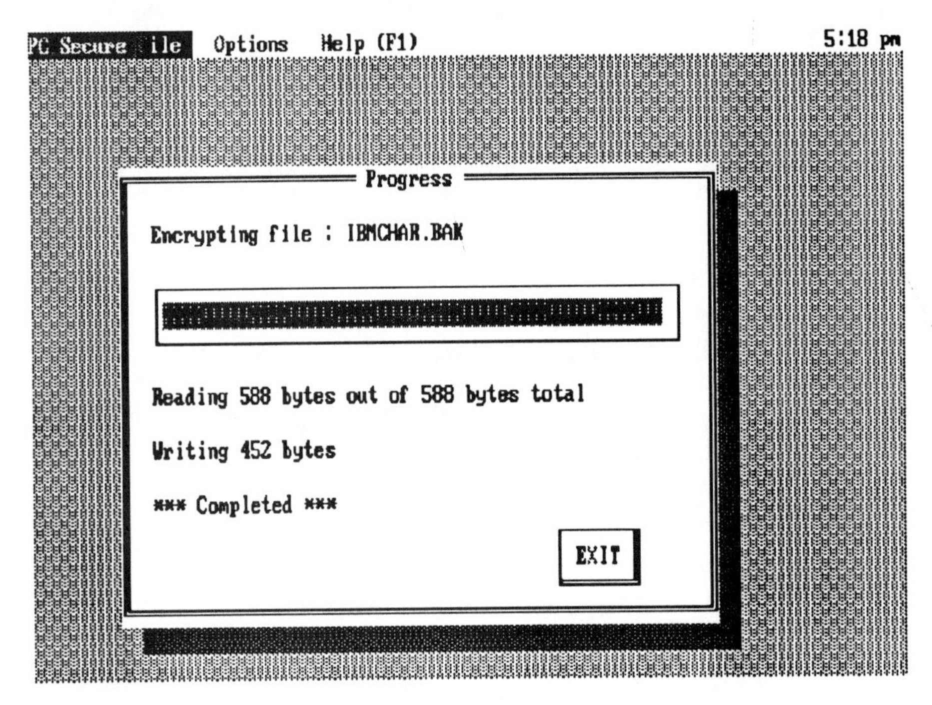

12-19 Encrypting a file

Decrypting a file is the opposite of encrypting a file; it lets you return an encrypted file to its original format. Select Decrypt file from the Applications menu, or press the F5 key. The file selection dialog box will appear, as shown in Fig. 12-20. Select the file or directory you want to decrypt. Choose Decrypt from the dialog box.

You will be asked for the password to decrypt the file, twice for verification. After you type in the password and select OK, the decryption takes place, as shown in Fig. 12-21. If you are decrypting a directory, you will see a message indicating that more than one file will be affected.

The About command, shown in Fig. 12-22, gives you information on the initial size of a file, the final size after compression, and the percent of reduction. Notice that, in some cases, instead of a reduction there is actually an increase in the size of the encrypted file. This might occur if you are encrypting a word processing spelling checker, a thesaurus, or files that are already compressed.

The Exit command lets you quit PC Secure.

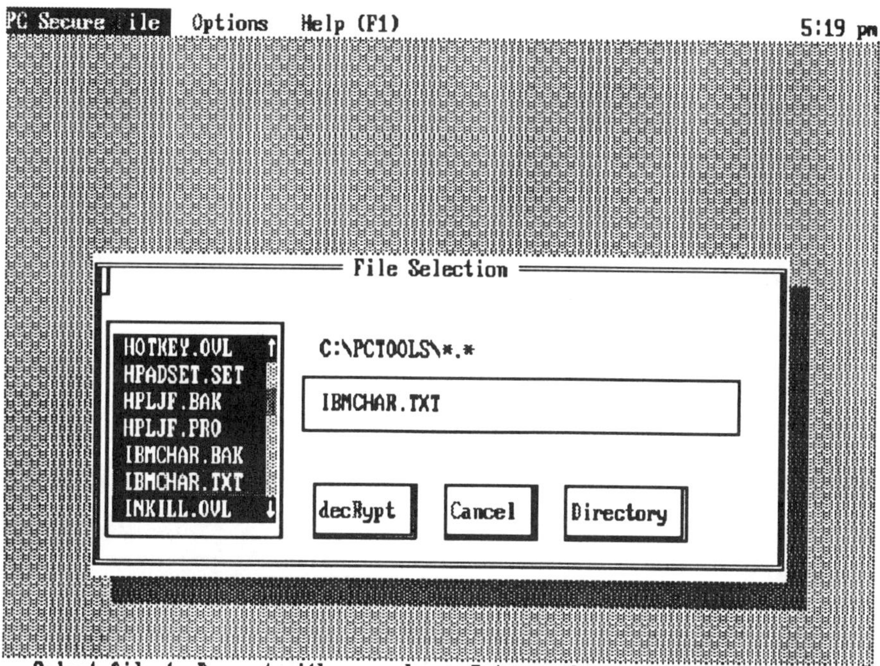

12-20 Selecting a file to decrypt

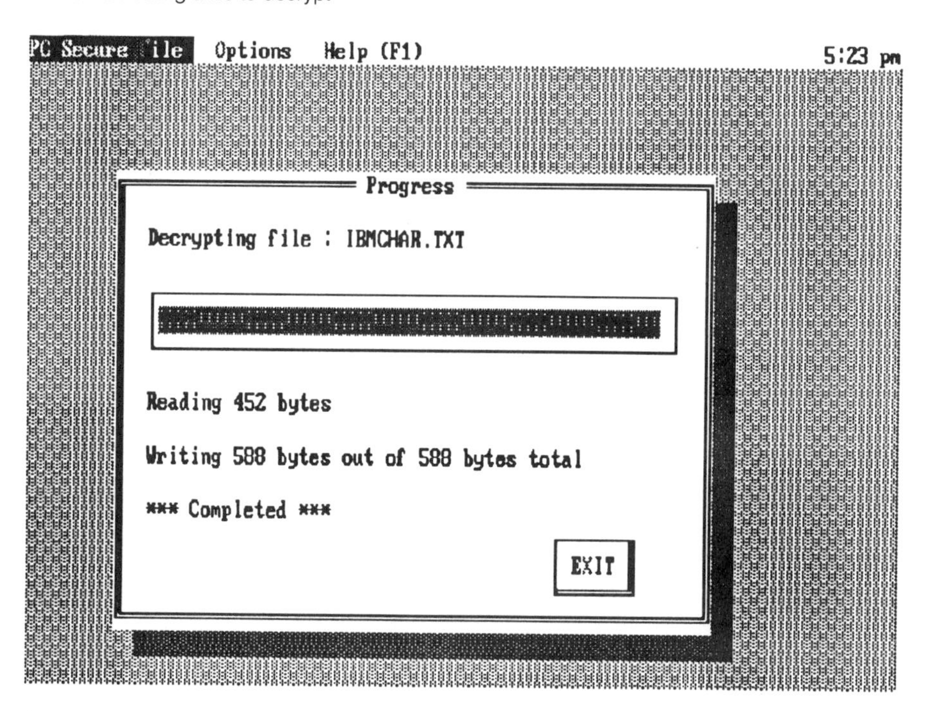

12-21 Decrypting a file

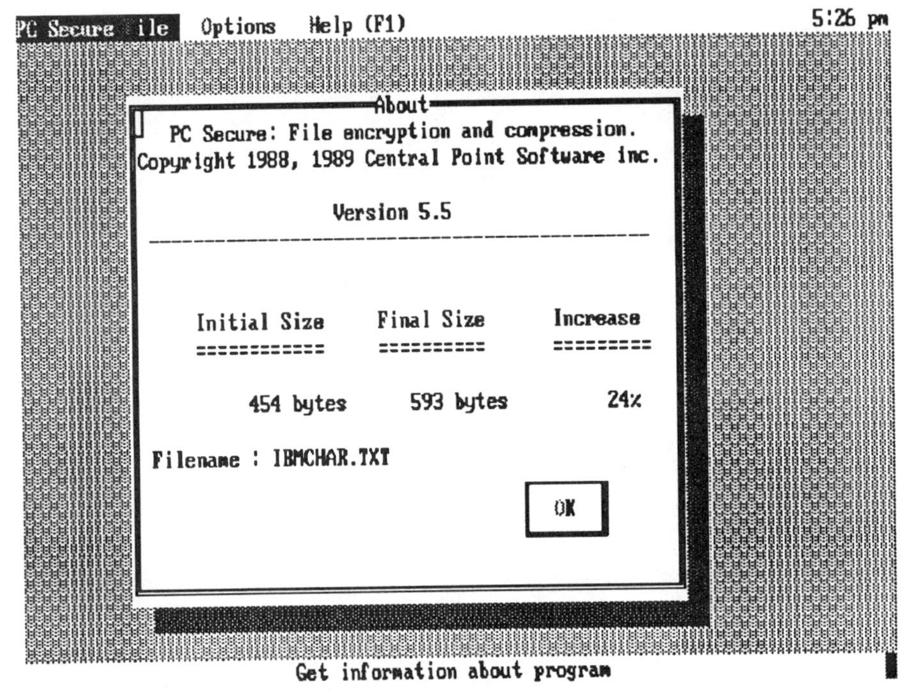

12-22 The About command

Options menu

The Options menu, shown in Fig. 12-23, lets you toggle options used in the encryption, decryption, and compression of files. The Full DES encryption command runs half as fast as the Quick encryption command, but is more reliable because it does eight times the number of encryptions. The Quick encryption command can be toggled on or off. If both of these commands are off, the program will do only compression.

The Compression command is a toggle command that reduces the size of the file as it is being encrypted. If the file has already been compressed, you can turn off the toggle switch for compression.

The One key command, when turned on, lets you use the same key or password for every encryption or decryption. When turned off, it will ask you for a password each time you encrypt or decrypt a file.

The Hidden command will keep the encrypted file from being listed during a normal DOS directory. It will be listed in the directory using PC Tools.

The Read-only command prevents a file from being accidentally deleted.

The Delete original file command is a toggle that, when on, encrypts a file and then deletes the original. If it is off, you will have two files, the original file with the extension .SEC, and the encrypted file. This option must be turned on if you are using the /G government option discussed in the last section.

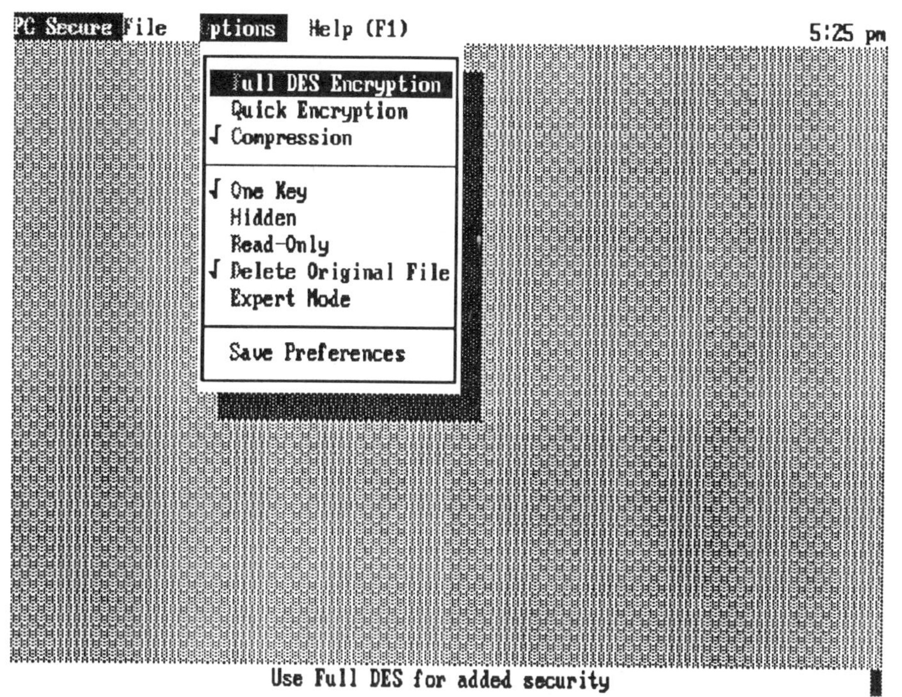

12-23 PC Secure Options menu

The Expert mode command lets you encrypt a file with only the file password or key. If you forget the file password, you cannot use the master key to decrypt it. This option should be reserved for high priority data.

The Save preferences command lets you save the options chosen in the Options menu.

Options help

Just as the other modules of PC Tools have a help index and help screen so does PC Secure, as shown in Fig. 12-24. If you forget a command's name or function, you can access the help screen by pressing the F1 key.

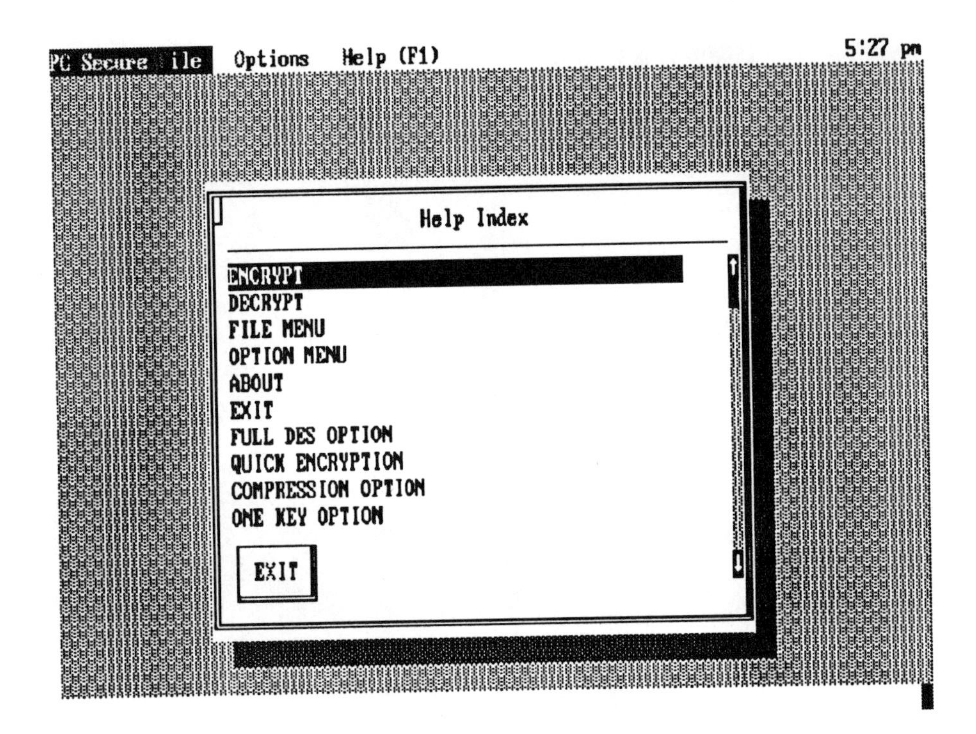

12-24 PC Secure Help index

13
CHAPTER

PC Secure special menu and PC Cache

This chapter discusses techniques on how to undelete a file or directory; how to produce a file, disk, or memory map; how to sort a directory; how to remove PC Shell from memory; and how to use PC Cache.

System info

The System info command is the first option in PC Secure's Special menu, shown in Fig. 13-1. When that option is selected, Fig. 13-2 appears. The system information screen provides some very useful information about your computer. It displays the type of computer, if it can recognize it. If it cannot recognize the computer, it will omit this line.

It tells you when the basic input/output system (BIOS) was changed. This is useful when comparing BIOS versions, or if you want to know what version of BIOS you have. It displays the current version of DOS you are using, the number

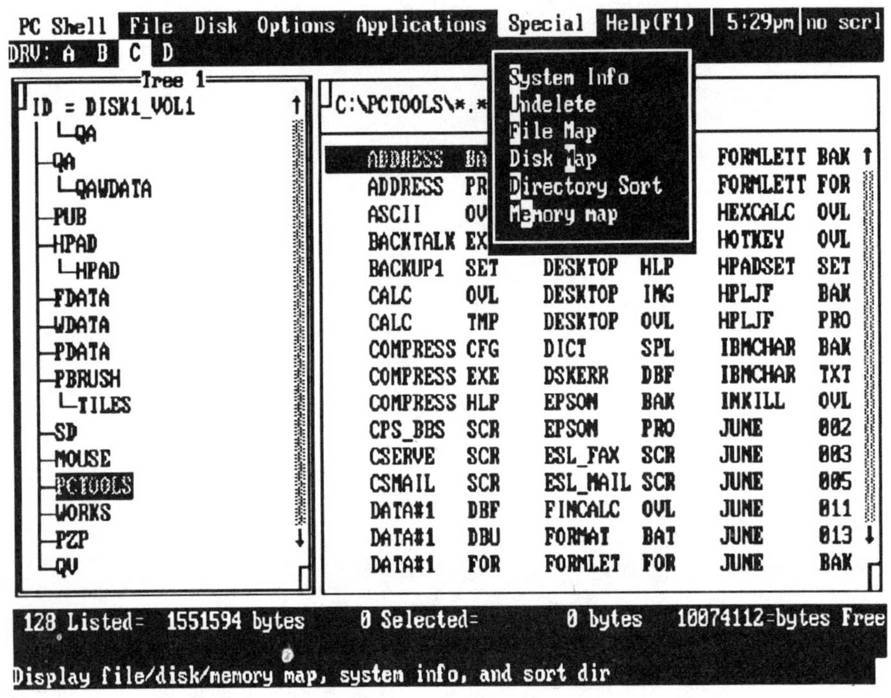

13-1 PC Shell Special menu

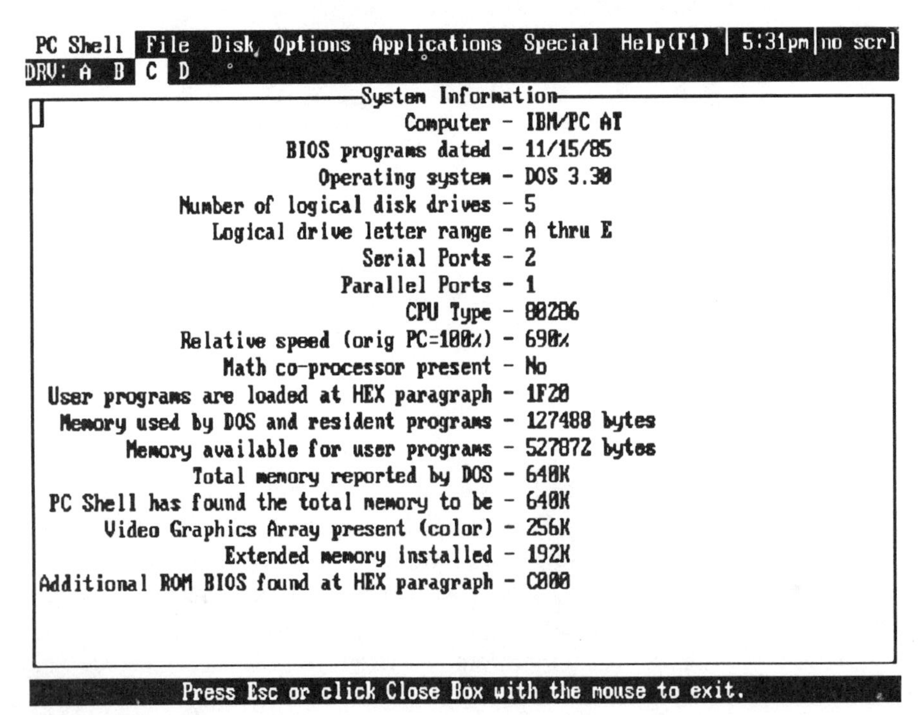

13-2 Special information command

of logical disk drives (the default is five, even though you won't have that many), and the drive letter names used with these drives.

It displays the number of serial and parallel ports in your computer, the CPU type, and the relative speed. The speed is relative to the 4.77 MHz of the first IBM PC. It tells you if a math coprocessor is present. It displays the hex address for user programs. It shows the amount of memory used by DOS and resident programs, and the amount of free memory.

It displays the total memory reported to DOS and the total memory found by PC Shell. It shows the type of display monitor used and whether or not extended memory is installed. Extended memory is added following the first 1,048,576 bytes of memory in the standard PC. While this memory cannot be used by applications such as spreadsheets, it can be used to create performance enhancements like RAM drives, disk caching, or print buffers.

Lastly, it displays additional ROM BIOSs, if they are used in your machine, and tells you if any deluxe option boards are installed.

Undelete

The Undelete command lets you recover lost files or directories that have been accidentally deleted. Use this command right after you accidentally erase a file, so the data in the file is not completely overwritten. I recommend that you always back up your disks, so if there is a problem with the Undelete command, you can use another disk.

Delete tracking method

If PC Shell has been installed on your hard disk drive, all the deleted file addresses are saved in a special file. The file is called PCTRACKR.DEL and is saved in the root directory of the first partition of the disk. This method is called the *delete tracking method*.

To "undelete" a file or directory, select the file or directory and choose the Undelete command from the Special menu, and Fig. 13-3 appears. Choose Continue to confirm that you want to delete the file.

The Undelete dialog box appears, asking you if you want to delete a file or directory, shown in Fig. 13-4. If you select either File or Subdir, Fig. 13-5 appears. Select Deltrack to indicate that you want to select the delete tracking method of file undeletion. Once selected, Fig. 13-6 appears, showing a list of deleted files. Notice that several files have a character to the right of the extension. The @ symbol indicates that the file can be restored with no problem. The * symbol indicates that some of the file clusters are not available. The files that have no symbol to the right of the extension means that the file is impossible to recover.

Selecting Go will undelete the highlighted file. It will continue to list each individual file until you are done with the recover. Each file that is recovered should be checked to make sure it runs properly.

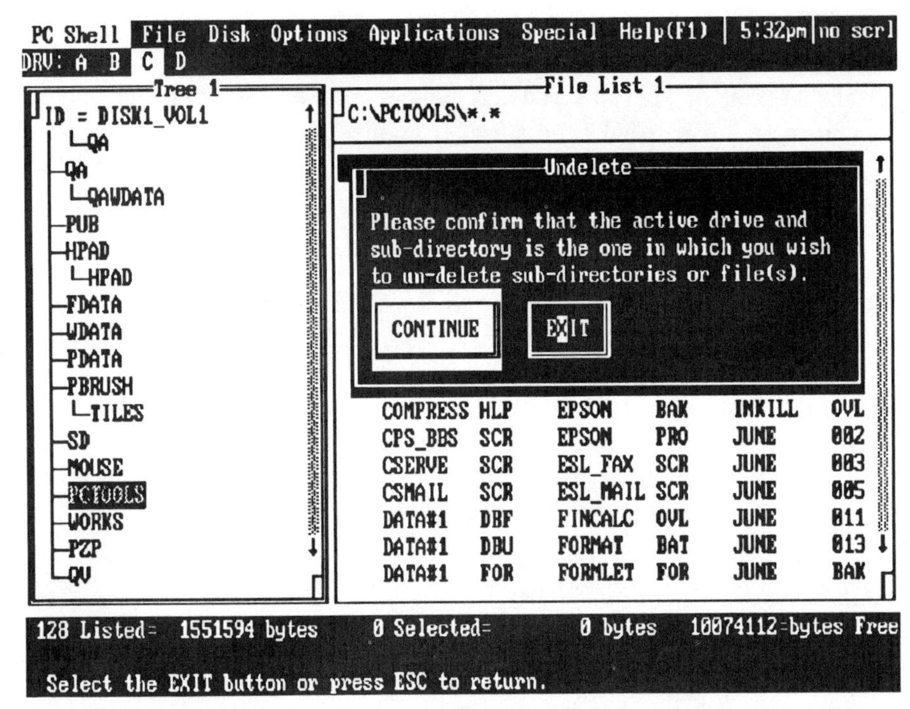

13-3 Undelete command

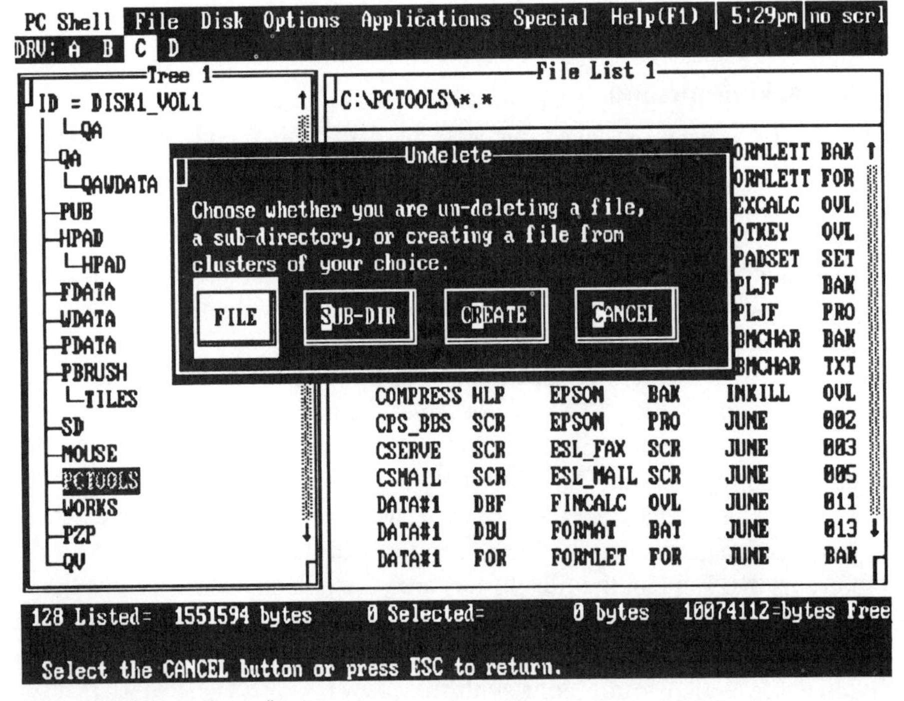

13-4 Undeleting a file or directory

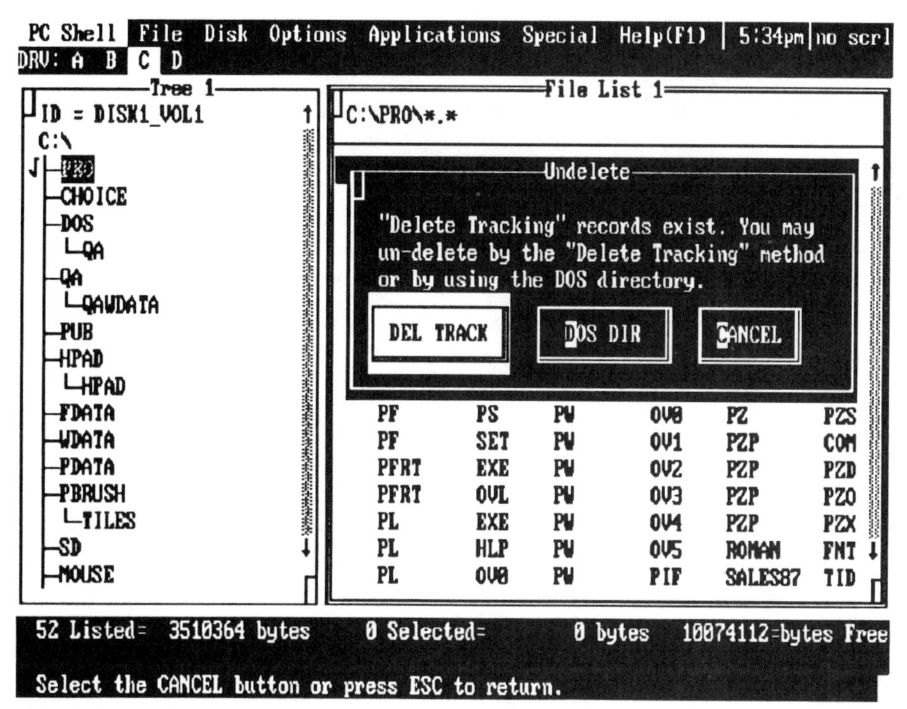

13-5 Delete tracking method

PC Shell File Disk Options Applications Special Help(F1) | 5:36pm | no scrl
DRV: A B C D

Undelete

Name	Ext	Size	Date	Time	Attr	Del Date	Del Time
HPTUTOR2	PAD*	135512	5/05/89	1:00a		12/29/89	11:19a
HPTUTOR	PAD*	55772	8/02/89	6:30p		12/29/89	11:19a
HPAD2	OVL@	66480	5/05/89	1:00a		12/29/89	11:19a
HPAD	PRN	4414	5/05/89	1:00a		12/29/89	11:19a
HPAD	INI@	712	8/02/89	9:13p		12/29/89	11:19a
HOME	PAD@	19853	8/02/89	9:13p		12/29/89	11:19a
HELP	PAD@	49897	8/02/89	9:12p		12/29/89	11:19a
GX2	EXE	8706	5/05/89	1:00a		12/29/89	11:19a
EXPENSES	PAD@	14584	5/05/89	1:00a		12/29/89	11:19a
DOS	PAD@	31149	8/02/89	6:39p		12/29/89	11:19a
DIALER	PAD@	16830	5/05/89	1:00a		12/29/89	11:19a
DAILY	PAD*	24455	5/05/89	1:00a		12/29/89	11:19a
CAP	EXE@	4633	5/05/89	1:00a		12/29/89	11:19a
CALC	PAD@	31811	8/02/89	9:03p		12/29/89	11:19a

GO EXIT

Select file to undelete. Press "G" to proceed.

13-6 File list for the delete tracking method

Standard DOS method of file recovery

You have the option of recovering with two methods. The delete tracking method is the preferred method of file recovery, but can be used only when the Delete Tracker was installed prior to installation. If it was not installed, you can use the standard DOS method.

This method is almost the same as the previous method. Choose the file or directory you want to undelete. Select Undelete from the Special menu. Select Continue to confirm the drive and files to be undeleted. Select whether you will be undeleting a file or directory. Select DOS DIR for the standard DOS selection method. A screen appears, as shown in Fig. 13-7, listing the files you can select to undelete. Notice that, just like the delete tracking method, several files have a character to the right of the extension. The symbols are the same as they are with the delete tracking method.

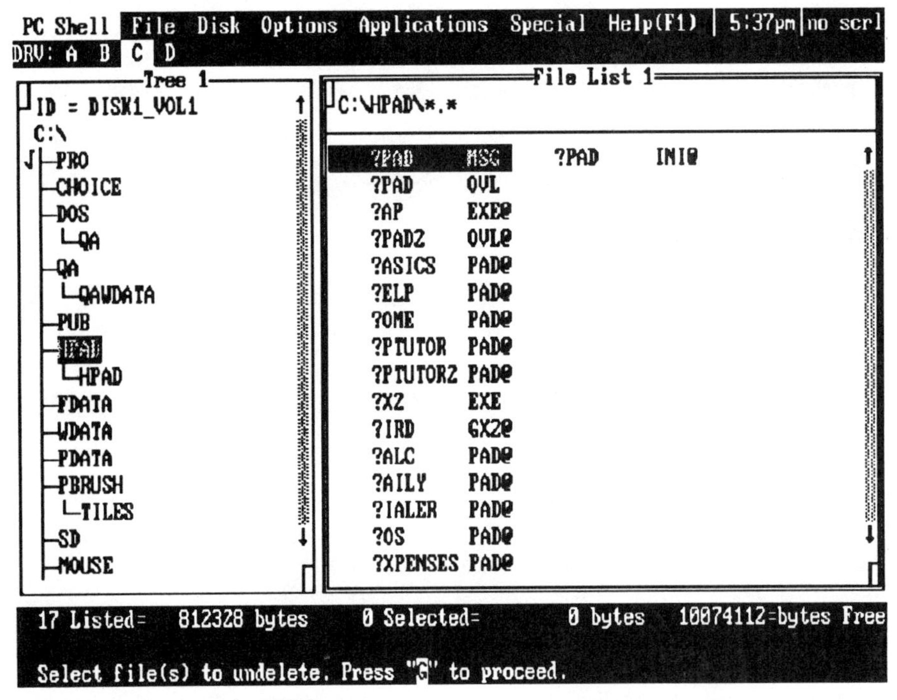

13-7 File list for a standard DOS recovery

Select the file or files you want to delete from the File List window. Select G to delete the files. Each of the files to be undeleted will be listed individually. The first letter of the filename will be missing and you will be asked to type it in, as shown in Fig. 13-8. Type in the first letter of the filename and select Undelete to recover it.

If the file has an @ symbol next to its name, select the Automatic command to retrieve it, as shown in Fig. 13-9. If it does not have this symbol, select Manual to recover it.

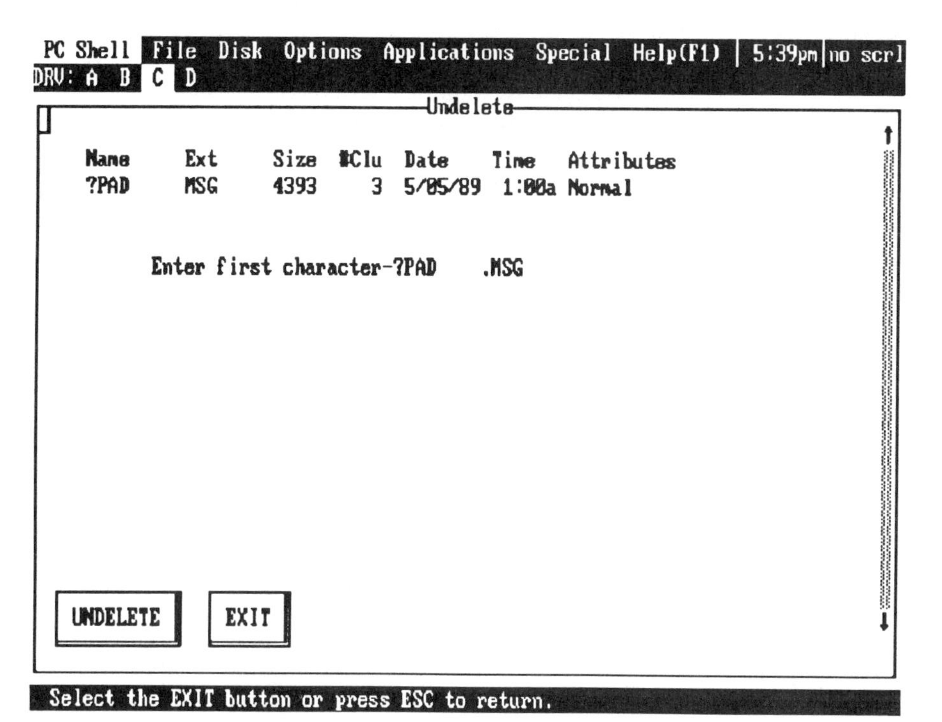

13-8 Undeleting a file with the delete tracking method

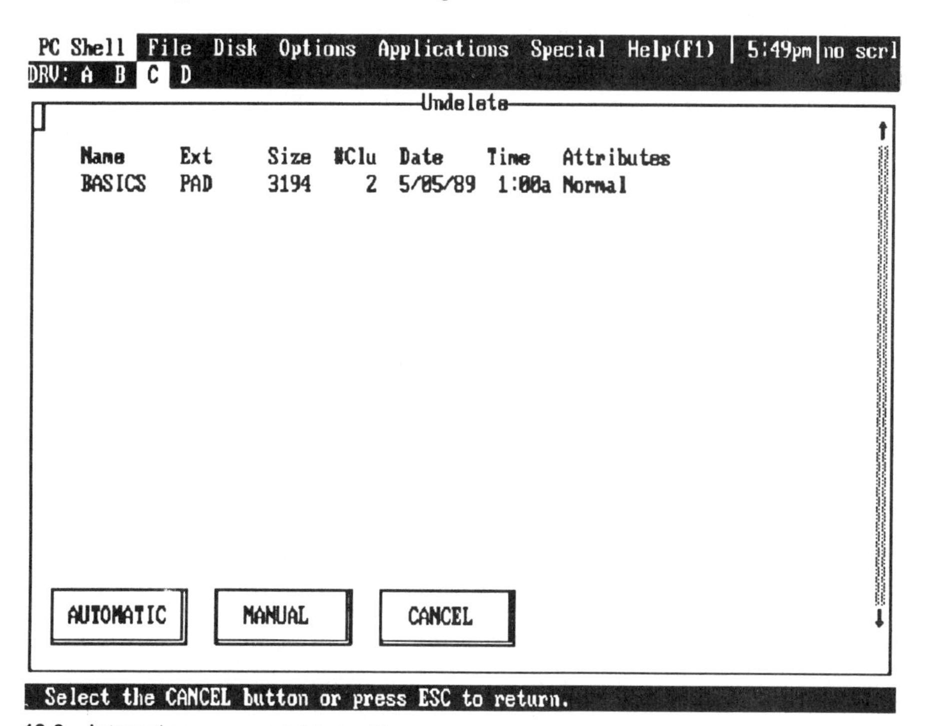

13-9 Automatic or manual delete tracking

If the file can be recovered with automatic recovery, a message will be displayed stating that it has been successfully undeleted, as shown in Fig. 13-10. If it cannot be recovered, the message in Fig. 13-11 will be displayed. It tells you that you must use manual recovery.

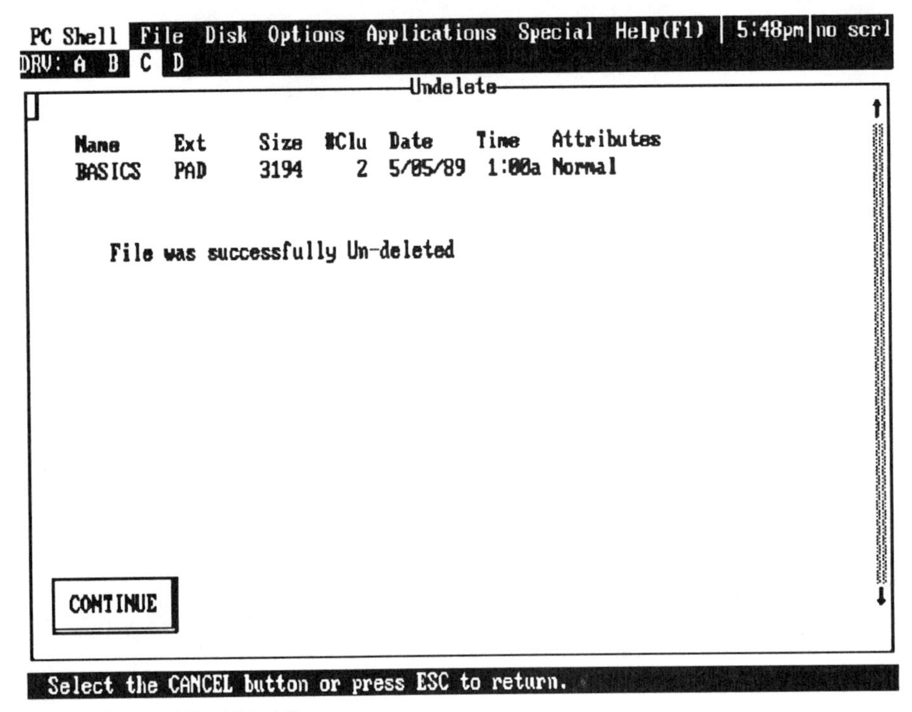

13-10 Successfully deleted file

When you use manual recovery to undelete the file, Fig. 13-12, which depicts the sectors that belong to the deleted file, is displayed. It shows a quarter of a sector at a time. You can use the PgUp key to go to the previous quarter sector and PgDn key to advance to the next quarter of a sector. The Home key moves the cursor to the beginning of the cluster, while the End key moves it to the end of the cluster.

The commands at the bottom of Fig. 13-12 list the options that are available to you. You can add the cluster to the file, skip to the next cluster and not add the present one, save the cluster to the undeleted file, search for information in other clusters to add to the file, select a new cluster, edit clusters in the file by reordering them, and exit from the menu.

Clusters can be moved or removed with the Edit command. After you select Edit, highlight the cluster you want to move or remove, and select Move or Remove, as the case may be. Choosing Exit will inform you if the file was undeleted. You should always try accessing the file to see if it was successfully undeleted.

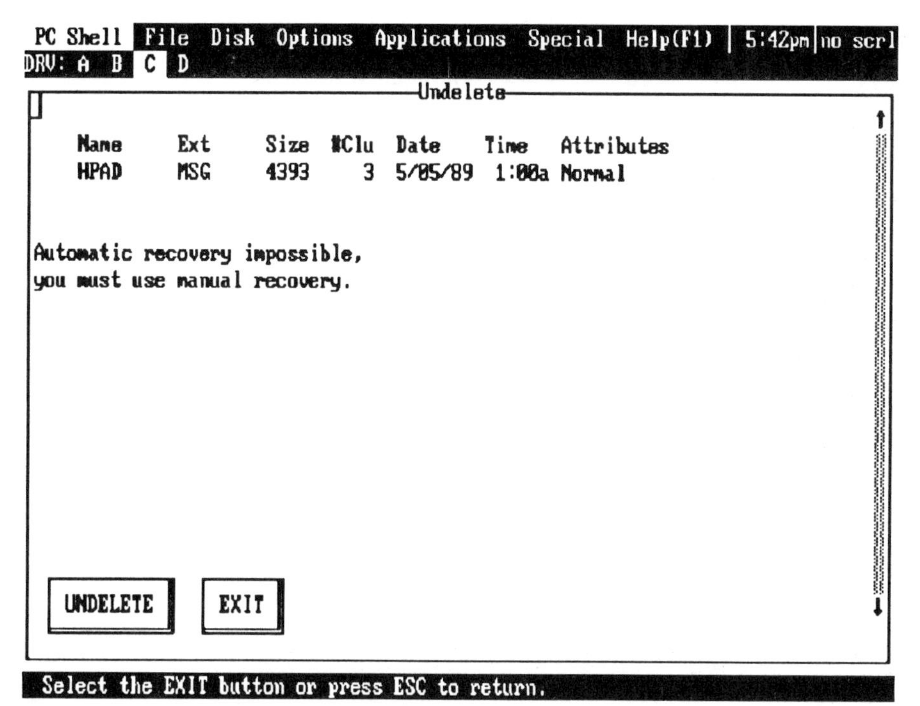

13-11 Automatic recovery failure

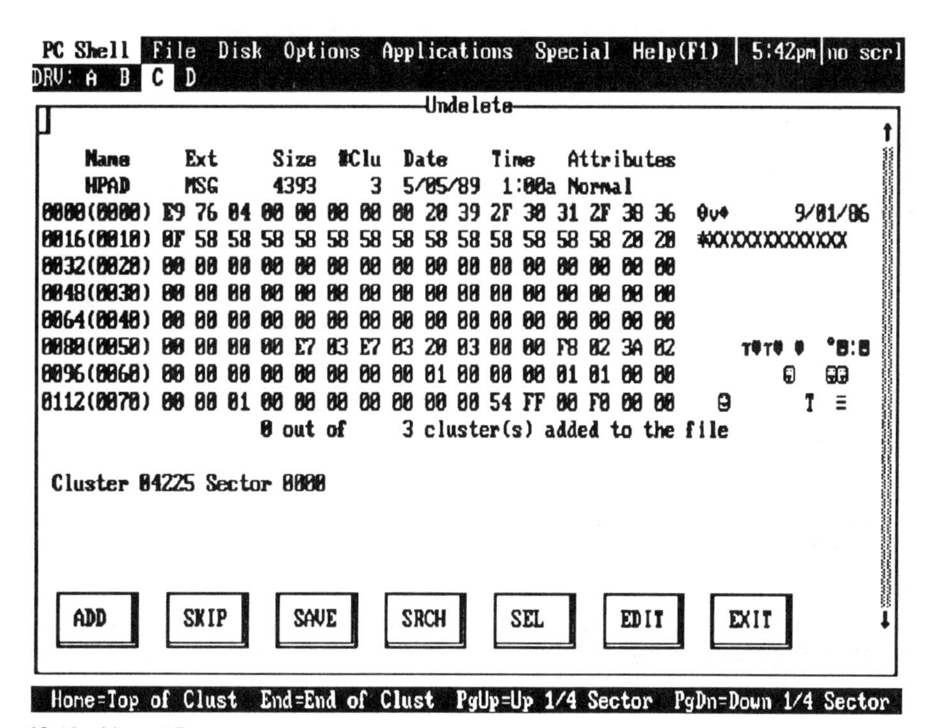

13-12 Manual Recovery screen

File creation method of recovery

The Create file recovery command lets you create a filename and then lets you add clusters to it, as shown in Fig. 13-13. To undelete a file or directory, select the file or directory and select the Undelete command from the Special menu. Choose Continue to confirm that you want to delete a file. Select whether you want to delete a file or directory.

The Undelete dialog box appears, asking you whether or not you want to delete a file or directory. Select the Create command and Fig. 13-13 appears. Type in a filename and extension. Use the manual method of recovery to reconstruct the file.

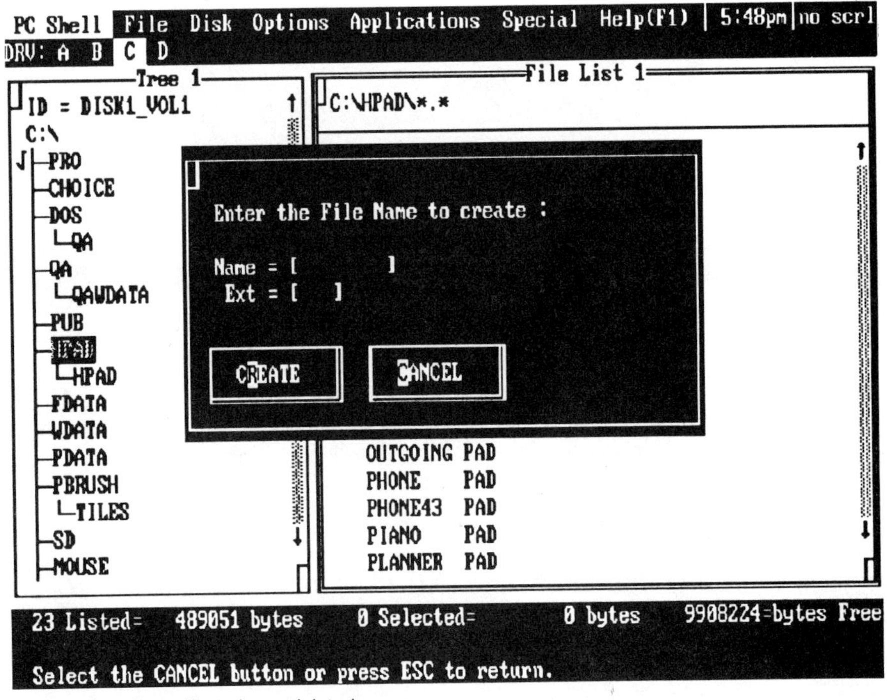

13-13 Creating a file to be undeleted

File map

The File map command, displayed in Fig. 13-14, lets you see the clusters that are in a file. Select the drive and directory you want to view. Choose the files you want to select to be mapped, and pick File map from the Special menu. The map will display the files that were selected. Each dot on the screen represents one cluster. A cluster can be two sectors on a double-sided disk, or four or more sectors on a hard drive. The available clusters, boot record, file allocation table (FAT), directory, allocated and hidden clusters, read-only clusters, and bad clusters are marked on the map.

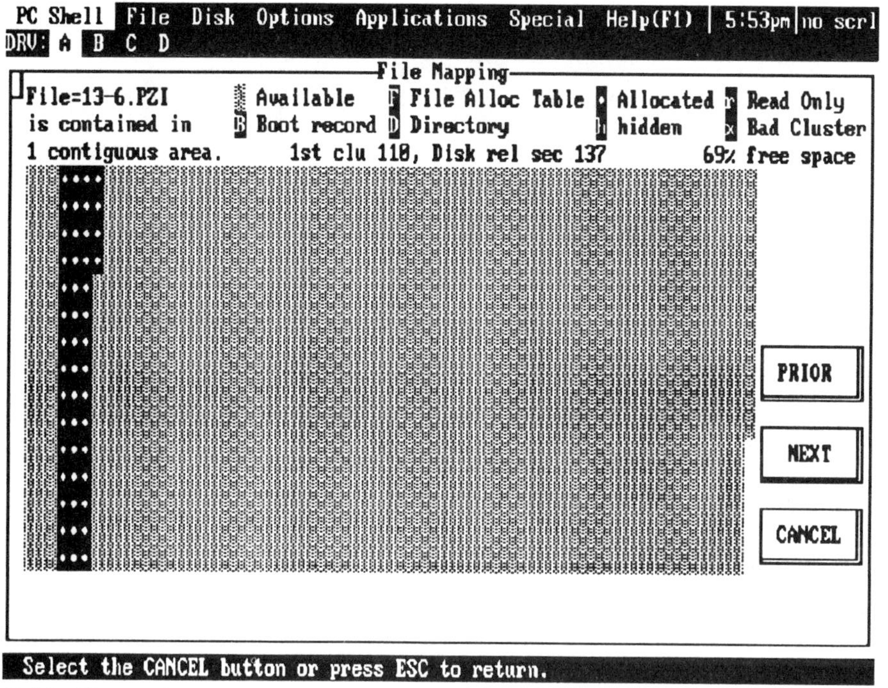

13-14 File mapping

Disk map

The Disk map, shown in Fig. 13-15, displays the sectors or clusters that are free on the disk, and which ones are occupied by files. It shows the same information as a file map, but at the disk level. A pictorial graph of a disk can be very useful because it can tell you if a disk should be compressed or not.

Directory sort

The Directory sort command, shown in Fig. 13-16, lets you sort the files in a directory. You can sort in ascending or descending order, by name, extension, size, date/time, or by numbers associated with the selected files. Select Sort to proceed. You will be given another dialog box which gives you four options. You can view the sorted files, update the directory, re-sort or cancel the operation.

Memory map

The Memory map command, shown in Fig. 13-17, displays the DOS memory blocks with the applications that belong to them. It shows where the memory-resident programs are located. It can display all the blocks occupied by programs, all the blocks occupied by programs with system pointers, all memory blocks, and all memory blocks with system pointers.

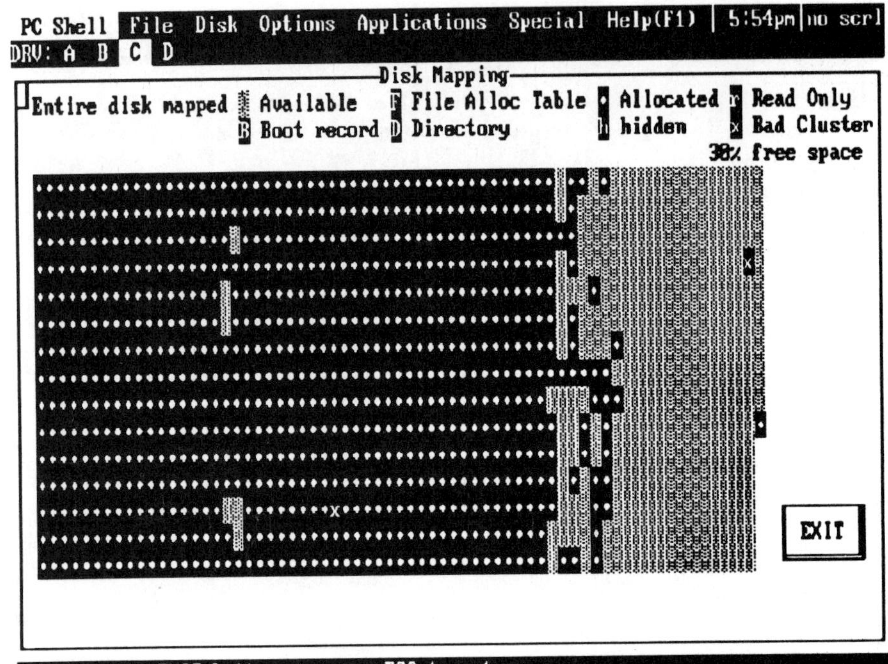

13-15 Disk mapping

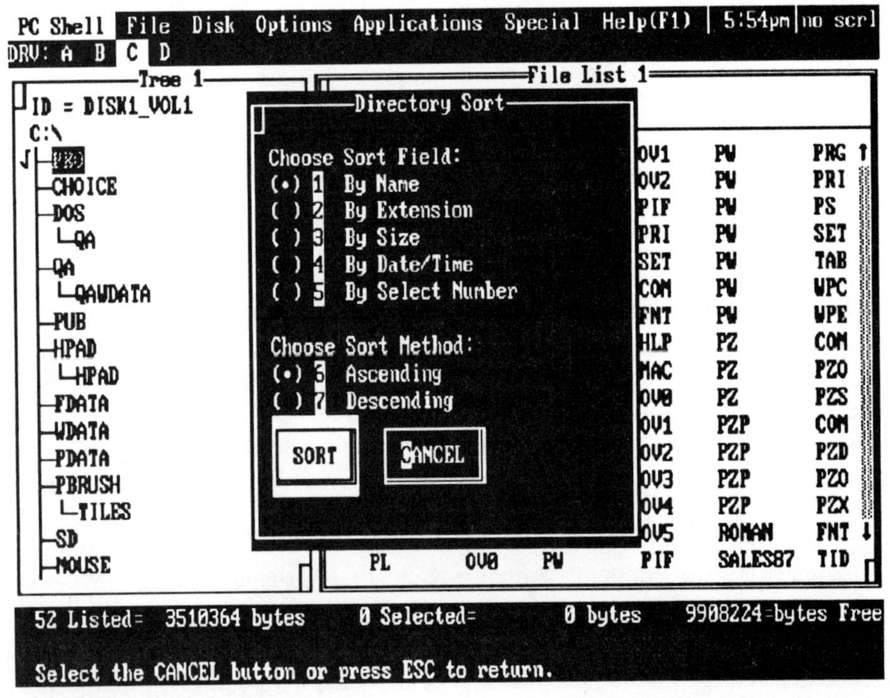

13-16 Using the directory sort

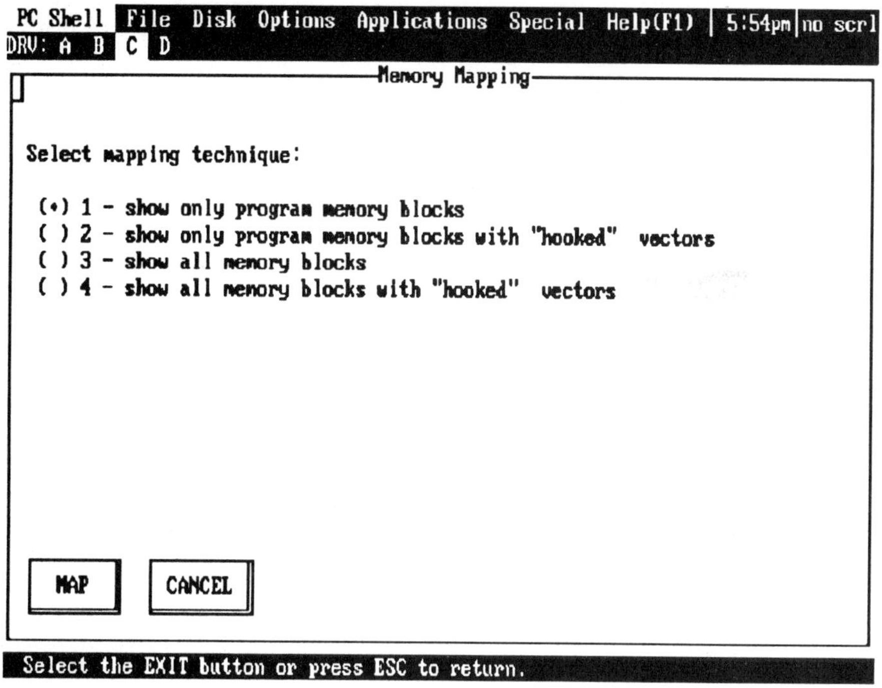

```
PC Shell File Disk Options Applications Special Help(F1) | 5:54pm|no scrl
DRV: A B C D
                         Memory Mapping

 Select mapping technique:

  (•) 1 - show only program memory blocks
  ( ) 2 - show only program memory blocks with "hooked" vectors
  ( ) 3 - show all memory blocks
  ( ) 4 - show all memory blocks with "hooked" vectors

  [ MAP ]  [ CANCEL ]

 Select the EXIT button or press ESC to return.
```

13-17 Memory mapping options

When you select MAP, Fig. 13-18 is displayed. It shows the application program (Prog), system program (Sys), DOS environment (Env), and unallocated memory (Free). It gives the memory area occupied by the program, the number of bytes used by the program, and the name of the program.

Remove PC Shell

When it is a memory-resident program, PC Shell can be removed by selecting the Remove PC Shell command, or typing KILL at the DOS prompt. If you select Remove PC Shell from the Special menu, Fig. 13-19 appears, asking you to verify.

PC Cache

PC Cache lets you use part of the computer's memory to store frequently used commands. It saves time by reducing the amount of access to the floppy or hard disk. PC Cache supports both extended and expanded memory. Any memory over 640K is called *extended* memory. Extended memory cannot be used by DOS directly, but certain programs, such as Lotus, has developed a way to use it. This is called *expanded* memory, or *expanded memory specification* (EMS).

Extended memory is simply extra memory, either on the main system board or on a memory adaptor board. You can get expanded memory by buying special memory boards or by using a program that turns extended memory into expanded

```
PC Shell  File  Disk  Options  Applications  Special  Help(F1) | 5:58pm|no scrl
DRV: A  B  C  D
                          Memory Mapping
Conventional memory.  Total:  640K
Largest executable program:   515K

Type  Paragraphs  Bytes    Owner
Prog  0CCE-0DA0H   3376   0CCEH <itself>
Prog  0DB5-1065H  11024   0DB5H <itself>
Prog  106E-120FH   6688   106EH <itself> C:\PCTOOLS\MIRROR.COM
Prog  1210-15DFH  15408   1210H <itself> C:\PCTOOLS\PC-CACHE.COM
Prog  15E8-1F17H  37632   15E8H <itself> C:\PRO\PZP.COM
Prog  1F20-9FFFH   515K   1F20H <itself> C:\PCTOOLS\PCSHELL.EXE

 MAP    EXIT

Select the EXIT button or press ESC to return.
```

13-18 Using the memory map

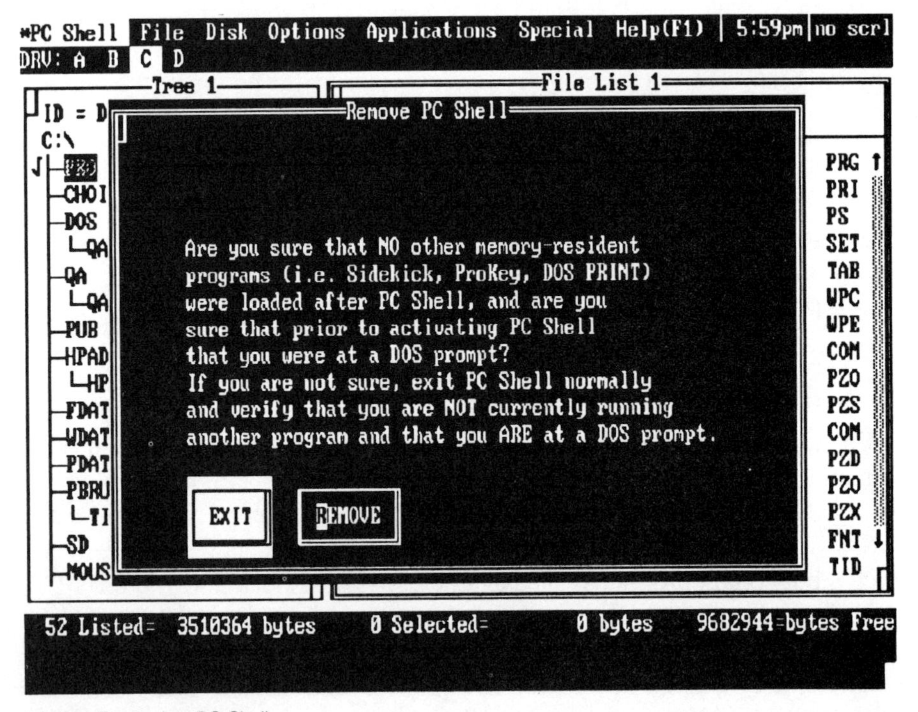

13-19 Removing PC Shell

memory. Expanded memory can be used by programs that are specifically written for it, like Lotus. These are programs, like spreadsheets or databases, that usually require a lot of memory.

PC Cache must be installed after the Mirror program. The commands for PC Cache, which can be obtained by typing PC-Cache/? are shown in Fig. 13-20. If you used PC Setup to install all the PC Tools utility programs, PC Cache is automatically installed in the AUTOEXEC.BAT file.

```
                    PC-CACHE, Version 5.5.
          Licensed exclusively to Central Point Software, Inc.
   Copyright 1986, 1989 Multisoft Corporation, All Rights Reserved.
                      Summary of Parameters

/EXTSTART=xxxxK  Don't use extended memory below xxxxK.
/FLUSH           Flush cache -- set the cache to empty.
/Ix              Do not cache drive x.
/MAX=xx          Read no more than xx sectors ahead.
/MEASURES        Display measurements.
/NOBATCH         Don't batch copy to/from the cache.
/PARAM           Display parameters in effect.
/SIZE=xxxK       Set up xxxK cache in conventional memory.
/SIZEXP=xxxxK    Set up xxxxK cache in expanded memory.
/SIZEXT=xxxxK    Set up xxxxK cache in extended memory.
/UNLOAD          Un-install the cache.
/?               Display this information.
```

13-20 PC Cache options

The /EXTSTART=xxxxK option specifies the start location in extended memory. It must be greater than 1024K, or one megabyte. It is used when other programs are installed in extended memory.

The /FLUSH option empties the cache of information.

The /Ix option says to ignore caching drive x. All drives are cached if you don't specify this option.

The /MAX=xx option limits the number of sectors to no more than xx.

The /MEASURES option displays the number of data transfers between the cache and applications, the data transfers between the disk and applications, the number of physical transfers, and the percentage of transfers saved by PC Cache.

The /NOBATCH option reduce the number of sector transfers from four to one when using extended memory.

The /PARAM option displays the current parameters.

The /SIZE=xxxK option gives the amount of memory allocated to DOS. The default is 64K, and the maximum is 512K.

The /SIZEP=xxxxK option gives the amount of expanded memory given to PC Cache.

The /SIZET=xxxxK option displays the amount of extended memory given to PC Cache.

The /UNLOAD option un-installs PC Cache.

14
CHAPTER

Disk backup

In this chapter, you will learn how to back up an entire hard disk or selected files and directories from a hard disk to floppy disks. You will also learn how to restore the files back to a hard disk after a disk crash.

Purpose of a backup

If you have a hard disk system, you are sure to start loading it up with lots of files and data. The problem is, "How do you protect your data from being accidentally lost?" A periodic backup onto floppy disks is the answer. This can be accomplished with the PC Backup program. It is designed to copy data from a hard disk to as many floppy disks necessary to hold them. It can copy from any type of disk to any other type of disk, and even back up your disks onto a hard drive. If you have more than one partition, or hard disk, you can back up one hard disk to another. The main purpose of PC Backup, however, is to place copies of your hard disk data onto floppy disks for safekeeping.

You can back up the contents of one directory or subdirectory. PC Backup is smart enough to tell how many floppy disks you will need for the backup, and gives you an estimate of how long the backup will take. It supports all types of disks drives: $5^1/_4$ and $3^1/_2$ inch, both high and low density.

Hardware requirements

PC Backup can back up hard drives onto any IBM PC, PCjr, XT, AT, PS/2 series, and most IBM compatibles. It requires at least 512K of memory, a hard drive, and a floppy disk drive or back-up tape. It requires DOS 3.2 or higher. Like the other utilities of PC Tools, it works more efficiently with a mouse.

Starting PC Backup

You must use PC Setup to install PC Backup on the hard drive. During setup, PC Tools will ask you if you want to install PC Backup on your hard drive of a PC or the hard drive of the file server of a network. Look at the installation procedure in Chapter 1 to see how this can be accomplished. The PC Backup parameters to be used are shown in Fig. 14-1.

```
C:\PRO>pcbackup/?
PCBACKUP (c) 1988,1989 Central Point Software 7/07/89
Options are:
 /BW    — Suppress colors
 /LCD   — LCD display screen
 /DOB   — Use Deluxe Option Board for Format
 /NO    — Don't Use Overlapped I/O
 /LE    — Left Handed Mouse (swap left and right)
 /PS2   — Use this option if mouse disappears on PS/2s
 /R     — Automatically start restore mode
 /?     — Help (this display)
 fname  — Automatically start backup with this setup
 d:     — specify a start up drive letter to backup
```

14-1 PC Backup options

The /BW option will start up PC Backup, using black and white colors.

The /LCD option uses the LCD screen display.

The /DOB option lets you use the deluxe option board for backup, which improves the speed of formatting.

Use the /NO option if your computer hangs up or freezes during backup. It stands for no overlap.

The /LE option is for left-handed people who want to use the mouse. It exchanges the left and right mouse buttons.

Use the /PS/2 option for the mouse on IBM PS/2 computers.

The /R option places you in the restore mode, asking you to insert the last backup disk.

To start PC Backup, type PCBACKUP. You can use any of the parameters that were just listed. Do not insert any spaces between PC and BACKUP. The first time you use PC Backup, you will see a Welcome screen, shown in Fig. 14-2, telling you that you will have to configure your computer system before PC Backup can be used. Press Continue to proceed with the configuration setup.

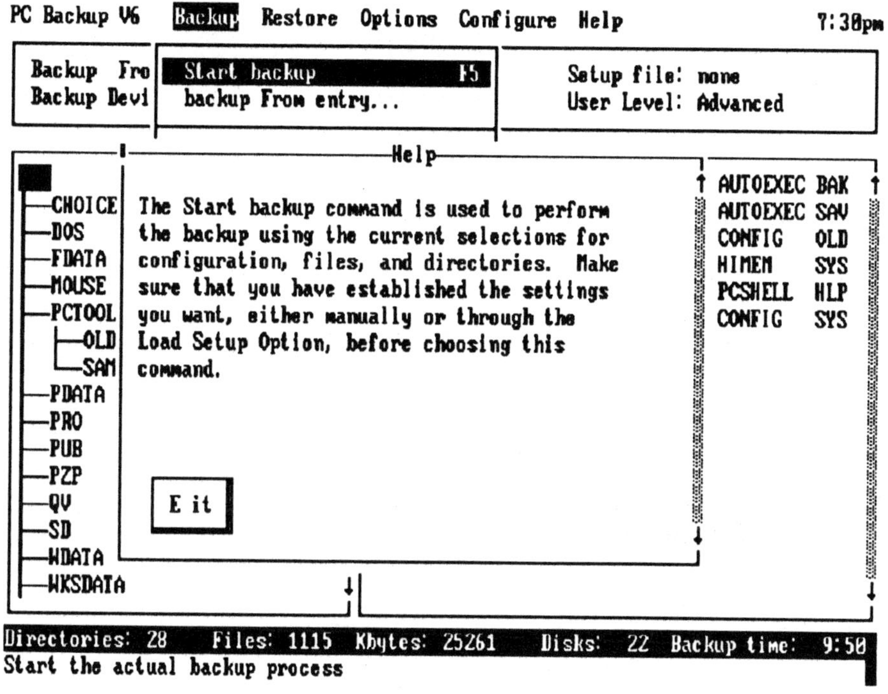

14-2 Configuring PC Backup

The Drive and Backup Type dialog box appears, as shown in Fig. 14-3. PC Backup will try to determine what type of drives are in your machine. They can be either $5^{1}/_{4}$- or $3^{1}/_{2}$-inch drives. It will also ask for the type of disk media, high or low density. If the program cannot determine these specifications on its own, you will have to enter the settings.

PC Backup asks you for the method of drive backup—high-speed DMA or DOS compatible. *DMA* (direct memory access) uses the DMA controller, which enable a computer to read from your hard disk drive and write to a floppy disk simultaneously. These disks are readable by DOS, and can also be read by the PCBDIR command, discussed at the end of this chapter. The DMA backup saves a lot of time in the backup process. If you select the DMA method and the DMA controller is not available in your computer, you will be told that the method is not available.

Use the DOS-compatible method if the DMA backup is not available for your computer, or if you are backing up from a hard drive to a tape device.

If you select the high-speed DMA option, Fig. 14-4 appears asking you to select the drive and media type for the backup. If your two floppy-disk drives are

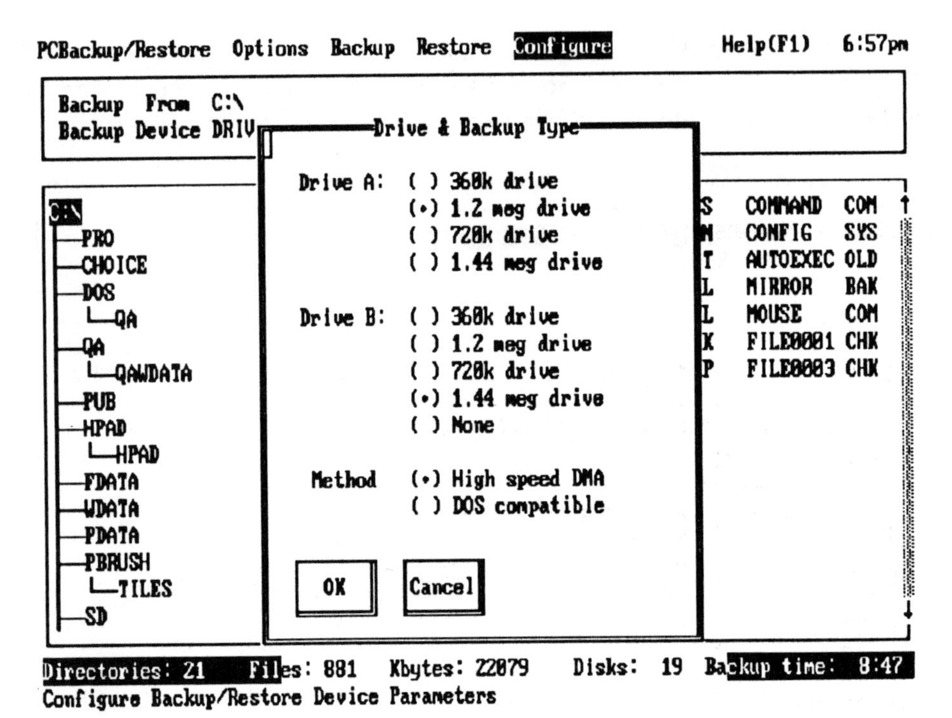

14-3 Selecting drive and backup tape

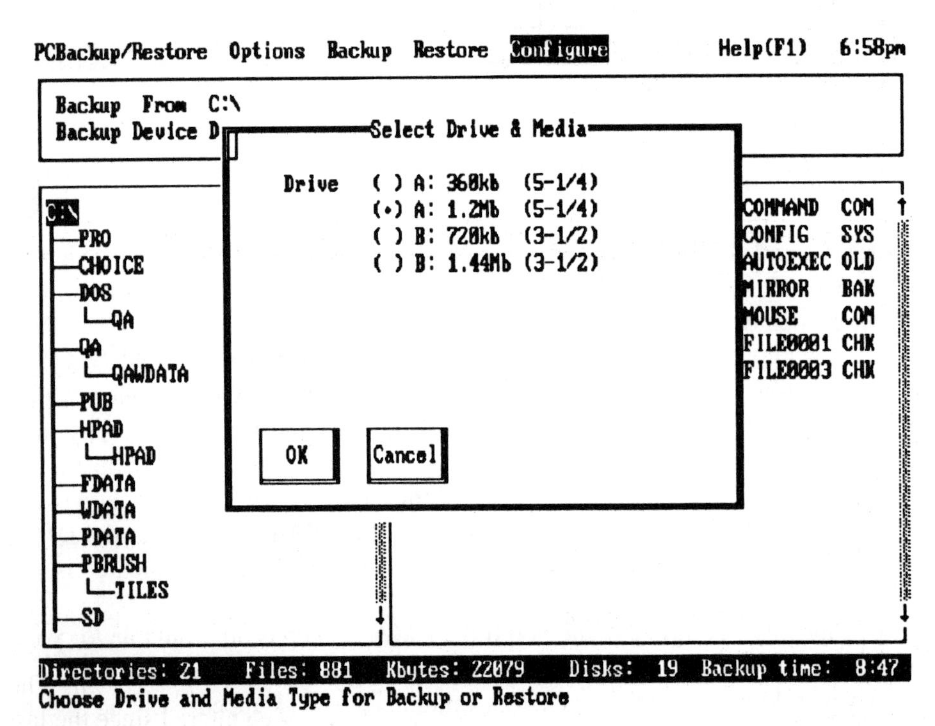

14-4 Selecting drive and media

of the same type, an additional line appears asking you if you want to use one or two drives for the backup. Selecting a two-drive backup will be faster, because once it is finished backing up to the first drive it will continue to back up to the second drive.

Options menu

The Options menu, shown in Fig. 14-5, lets you choose additional settings to enhance a backup.

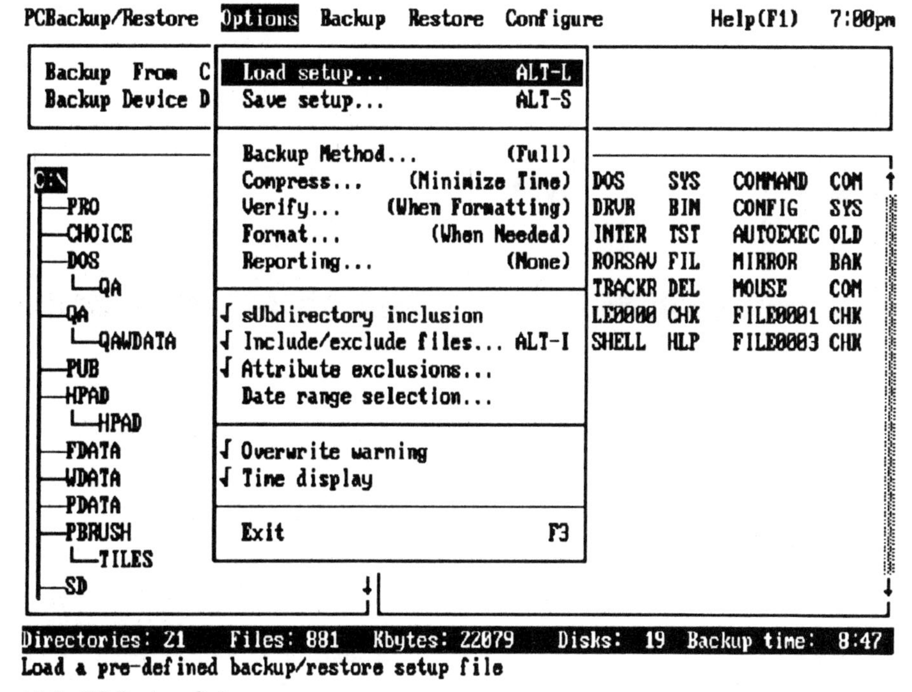

14-5 PC Backup Options menu

The Load setup command, which generates the screen in Fig. 14-6, lets you load previously saved backup settings. This is very useful if you want to perform backups of certain files at certain times. Just retrieve the backup setting and perform the backup.

The Save setup command's screen, shown in Fig. 14-7, lets you save particular backup settings, including the files you want backed up. Once the files or directories are selected, choose Save setup and give the settings a name.

The Backup method option gives you a menu containing four options, as shown in Fig. 14-8. The default setting is the full backup method. This means that all selected files are backed up, regardless of the setting of the *archive bit.* The archive bit tells the program whether or not the file has been altered since the last

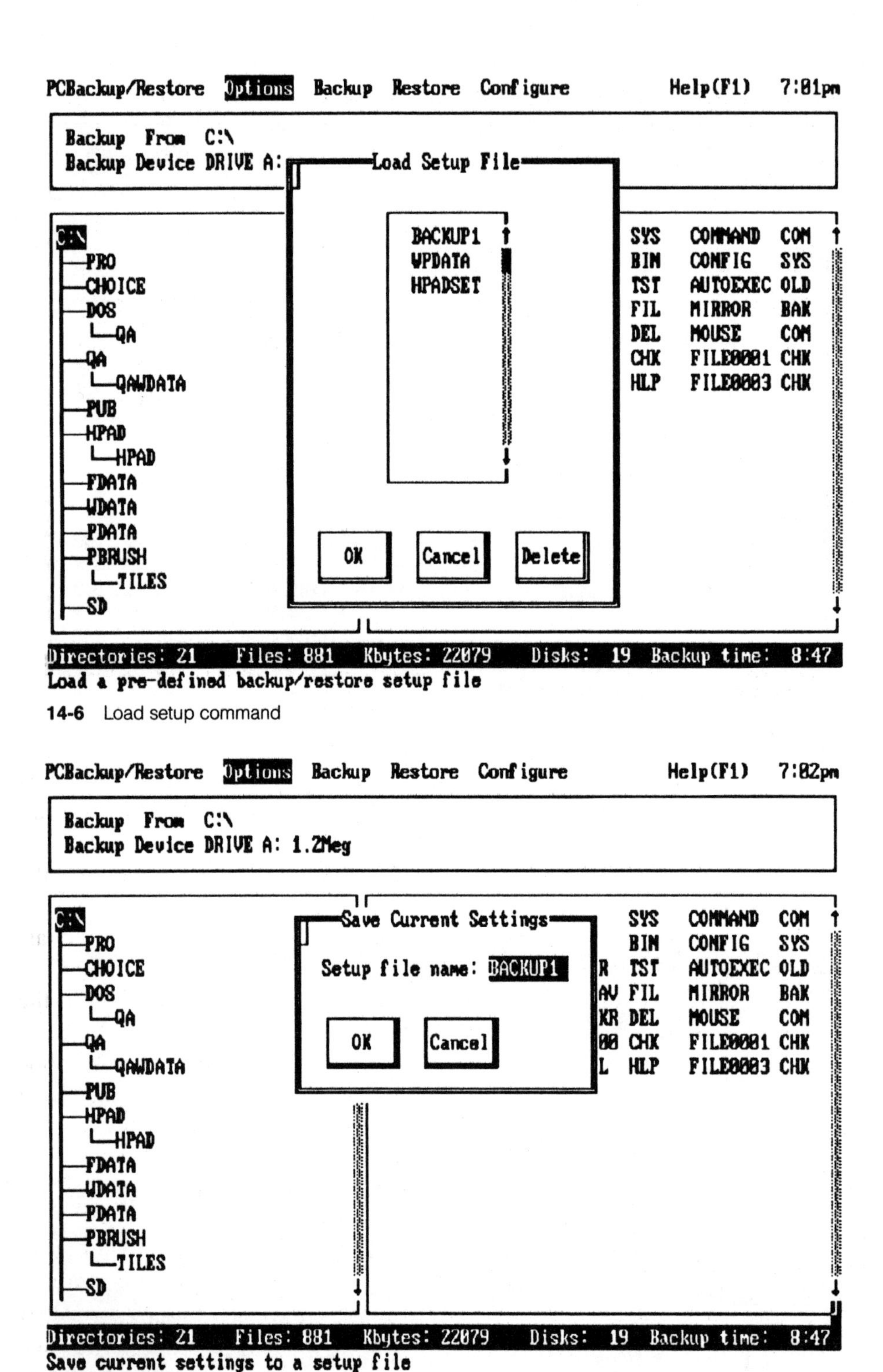

14-6 Load setup command

14-7 Save setup command

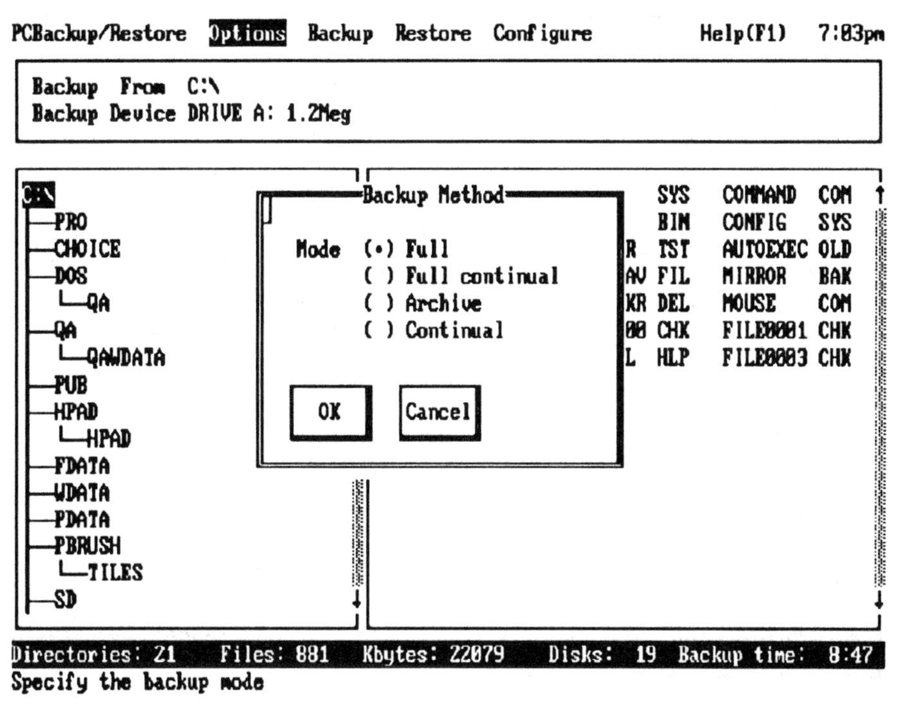

14-8 Backup method command

time PC Backup was run. If the file has been changed, the backup is performed. If the file has not been changed, no backup is performed. A full backup backs up all selected files, despite what the archive bit signifies.

The Full continual option backs up all selected files, but does not reset the archive bit.

The Archive option backs up all the files that have been updated since the last backup, and resets the archive bits.

The Continual option backs up all the files that have been updated since the last backup, but does not reset the archive bits.

The Compress command, shown in Fig. 14-9, has three options. The Minimize disks command minimizes the number of disks during backup compressing the data to reduce space. The Minimize time command, which is the default, reduces the amount of time during the backup, while compressing the data. This option works only on 286- or higher-level computers, with a high-speed DMA setting. The None option specifies no compression.

The Verify command checks the data on the backup disks to make sure it is good. The three verification options are shown in Fig. 14-10. The When formatting option, which is the default setting, verifies the surface of the newly formatted disks to make sure data can be safely saved. The Always option, which results in a slower backup time, verifies the data after it is written. The None option performs no verification whatsoever on the backup disks.

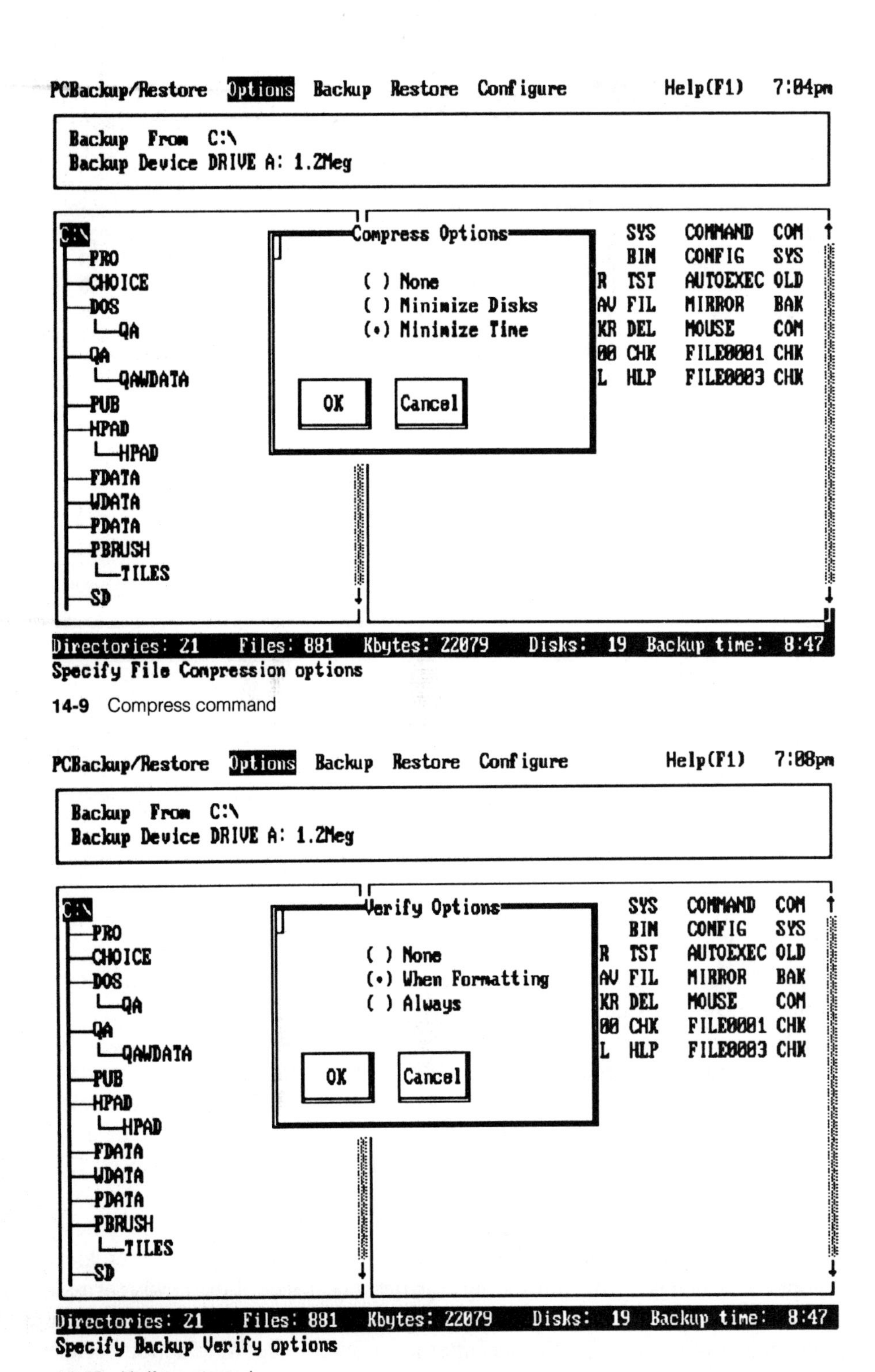

14-9 Compress command

14-10 Verify command

The Format command displays two options, as shown in Fig. 14-11. The When needed option, the default setting, is only used in DMA backups. Disks are formatted only the first time they are used. After that, they are rewritten instead of being formatted again. The Always command will format all disks, even if they have been formatted before.

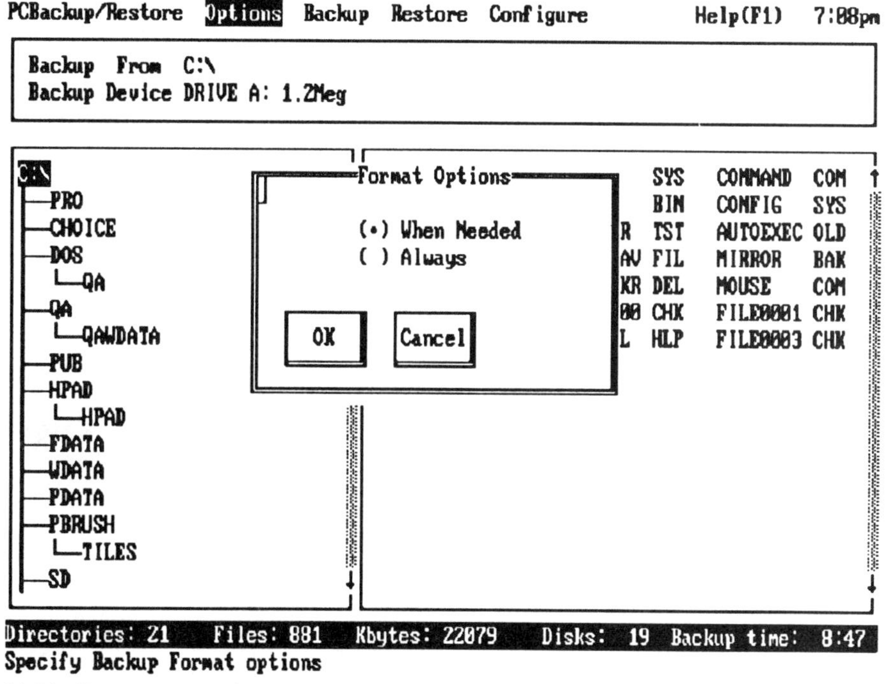

14-11 Format command

The Reporting command gives three options, as shown in Fig. 14-12. The Printer option will print a report after the backup to the printer. The Disk option, which will require you to give a report name, will print a report after the backup to the disk. The None option will not generate any report.

A sample report, generated from the Printer option, is shown in Fig. 14-13. It indicates all the settings that have been used during the backup, including the names and locations of the files.

The Subdirectory inclusion command is a toggle that, when on, includes all the subdirectories contained in a particular directory during backup. When turned off, the subdirectories are unaffected.

The Include/exclude files command, shown in Fig.14-14, lets you select specific files for backup. The default setting is *.*, which means that all files are to be backed up. You can specify any file or group of files. For example, PCTCH*.* will back up all files beginning with the letters PCTCH and having any extension. C:\DBASE*.DBF will back up all dBASE files. You can list several files on consecutive lines in this command window.

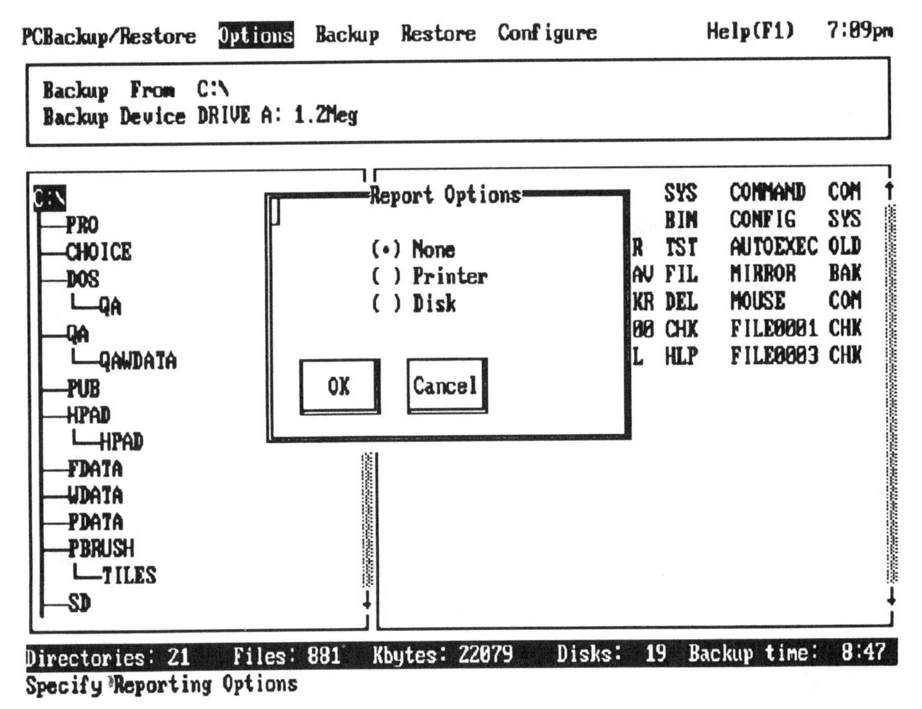

14-12 Reporting command

The Attributes exclusions option lists three options, as shown in Fig. 14-15. The Exclude hidden files option indicates whether or not copy-protected files should be left alone. Normally, when hidden files containing copy-protected programs are moved, they will not work.

The Exclude system files option refers to the DOS files, which also are position sensitive. It doesn't pay to move them as this will also create problems.

The Exclude read-only files option omits the program files that are read-only from being backed up.

The Date range selection command, when selected from the Options menu, generates the screen shown in Fig. 14-16. This command lets you select files before, between, or after certain dates. Change the Date selection option to On and type the Range, in the form MM/DD/YY.

The Overwrite warning command, again selected from the Options menu, is a toggle that signals that there is data on your backup disks. It warns you so you can change the disks, if necessary.

The Time display command toggles the backup time display with the system clock display.

The Exit command lets you leave the Options menu.

```
PC Backup directory report 5.5
Copyright 1989 Central Point Software, Inc.  All rights reserved.
Backup performed on 12-27-89   2:38pm

Backup method: High-speed DMA
Drive type: 1.2M          Media type: 1.2M
Backup type: Full.     Compression of data: Minimize time
Backup from C:\
Backup to A:
Date range: None
Include/Exclude files:
    PCTCH*.*

Total directories:     1
Total files:          12
Disks used:            1

       Name            Size       Date       Time      Attr    Disk#

Directory  C:\

Directory  C:\WDATA
   PCTCH1             10690   12-15-89    8:24am              1
   PCTCH2             28188   11-28-89   10:14pm              1
   PCTCH3              7021   11-29-89    5:47pm              1
   PCTCH4             24241   12-01-89    3:06pm              1
   PCTCH5             16811   12-06-89    3:28pm              1
   PCTCH6             26064   12-10-89   10:57am              1
   PCTCH7             16426   12-12-89   10:05am              1
   PCTCH8             17275   12-17-89    4:57pm              1   Compressed
   PCTCH9             24770   12-18-89    7:30pm              1
   PCTCH11             9337   12-20-89    8:03pm              1   Compressed
   PCTCH12             8911   12-26-89    2:45pm              1
   PCTCH10            13987   12-27-89    1:16pm              1

Total bytes:    199K
```

14-13 Printing out a report after backup

Backup menu

The Backup menu is displayed in Fig. 14-17. The Start backup command prompts you to insert a disk in the floppy drive for backup. The light on the disk drive will stay on continually until the backup process is completed. The bottom of the screen during backup, as shown in Fig. 14-17, indicates the number of disks being backed up, the percent of completion, the elapsed time, and the track number of the floppy disk.

When the backup is completed, as shown in Fig. 14-18, a dialog box appears, listing the total directories, files, number of kilobytes and number of disks backed up, backup time, and transfer rate.

The Backup From entry command, shown in Fig. 14-19, lets you back up a selected hard drive, if you have several. Enter the drive and/or path specification and press the Enter key. The bottom of the screen indicates the directories, number of files, kilobytes to be backed up, number of disks needed for backup, and estimated backup time.

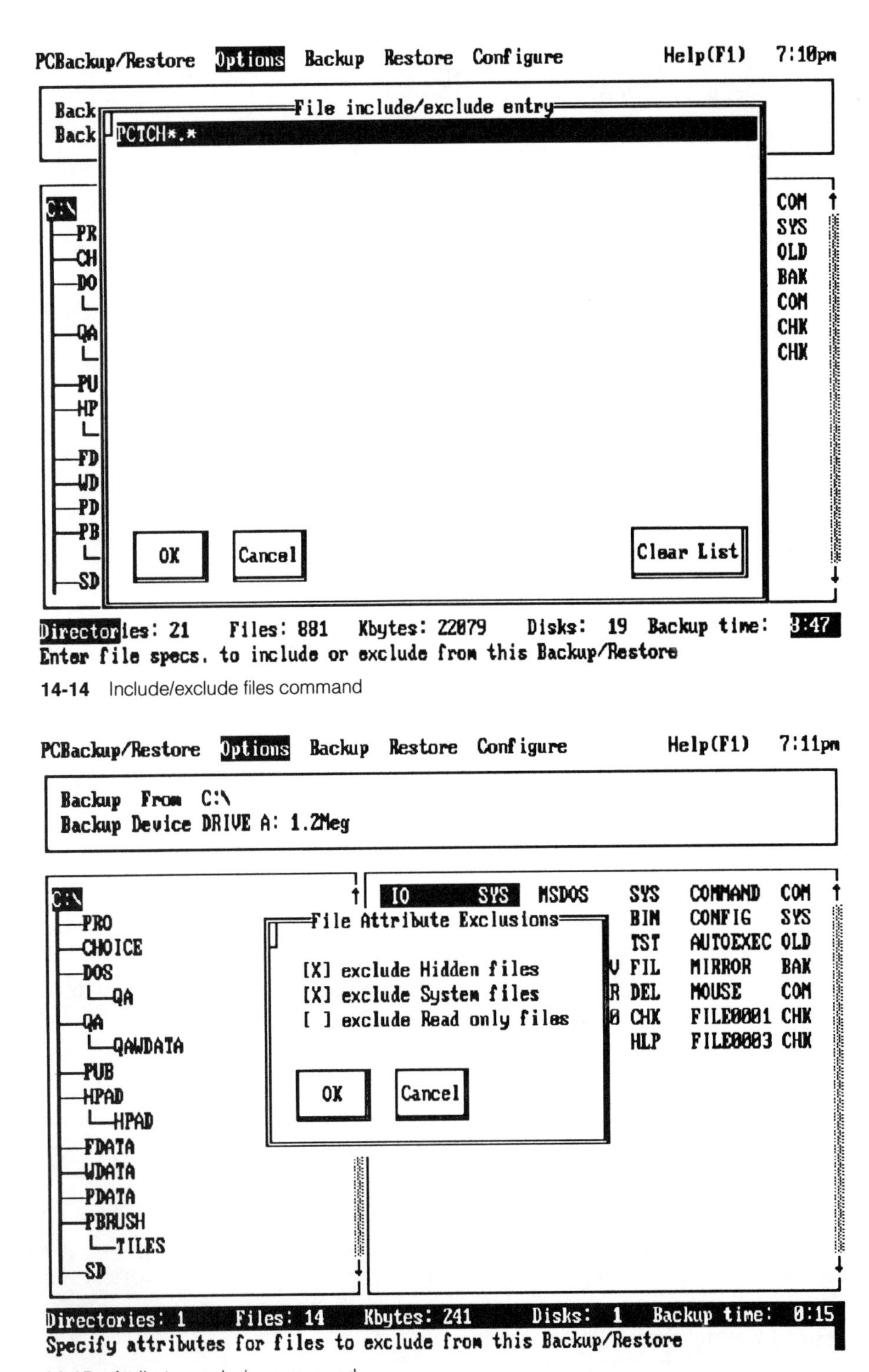

14-14 Include/exclude files command

14-15 Attributes exclusion command

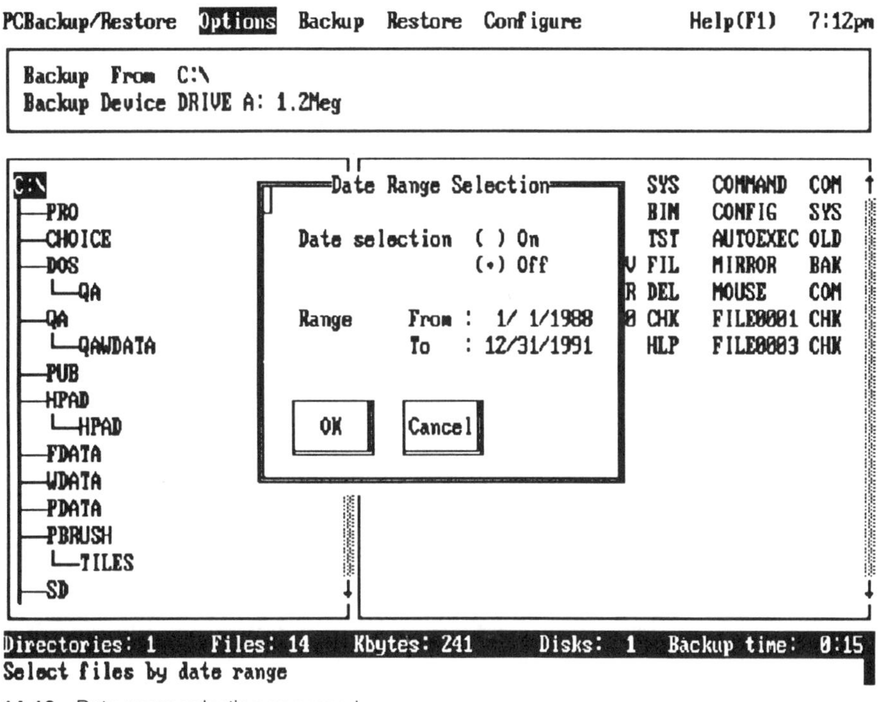

14-16 Date range selection command

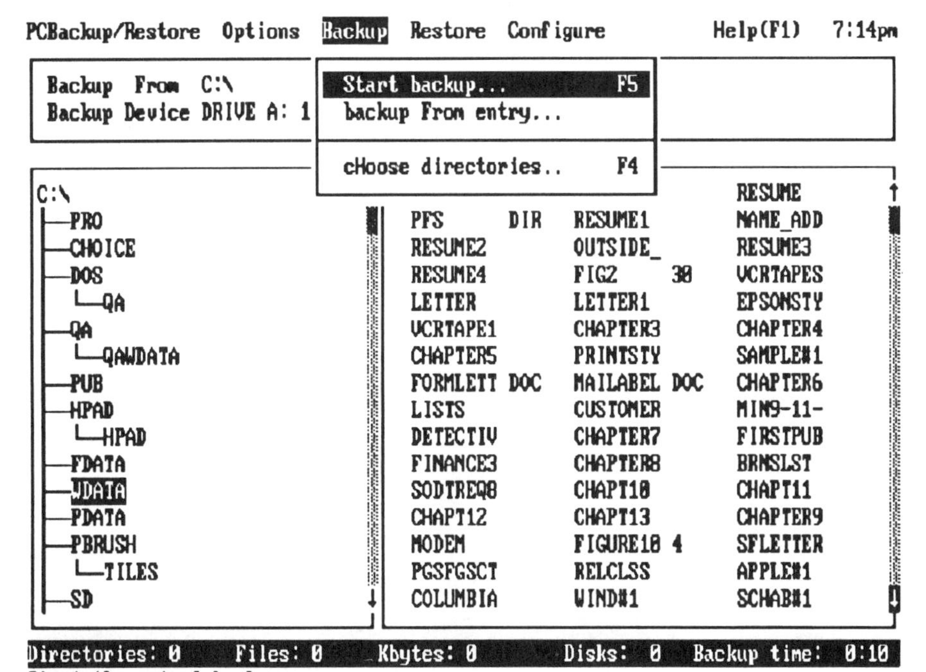

14-17 Backup menu

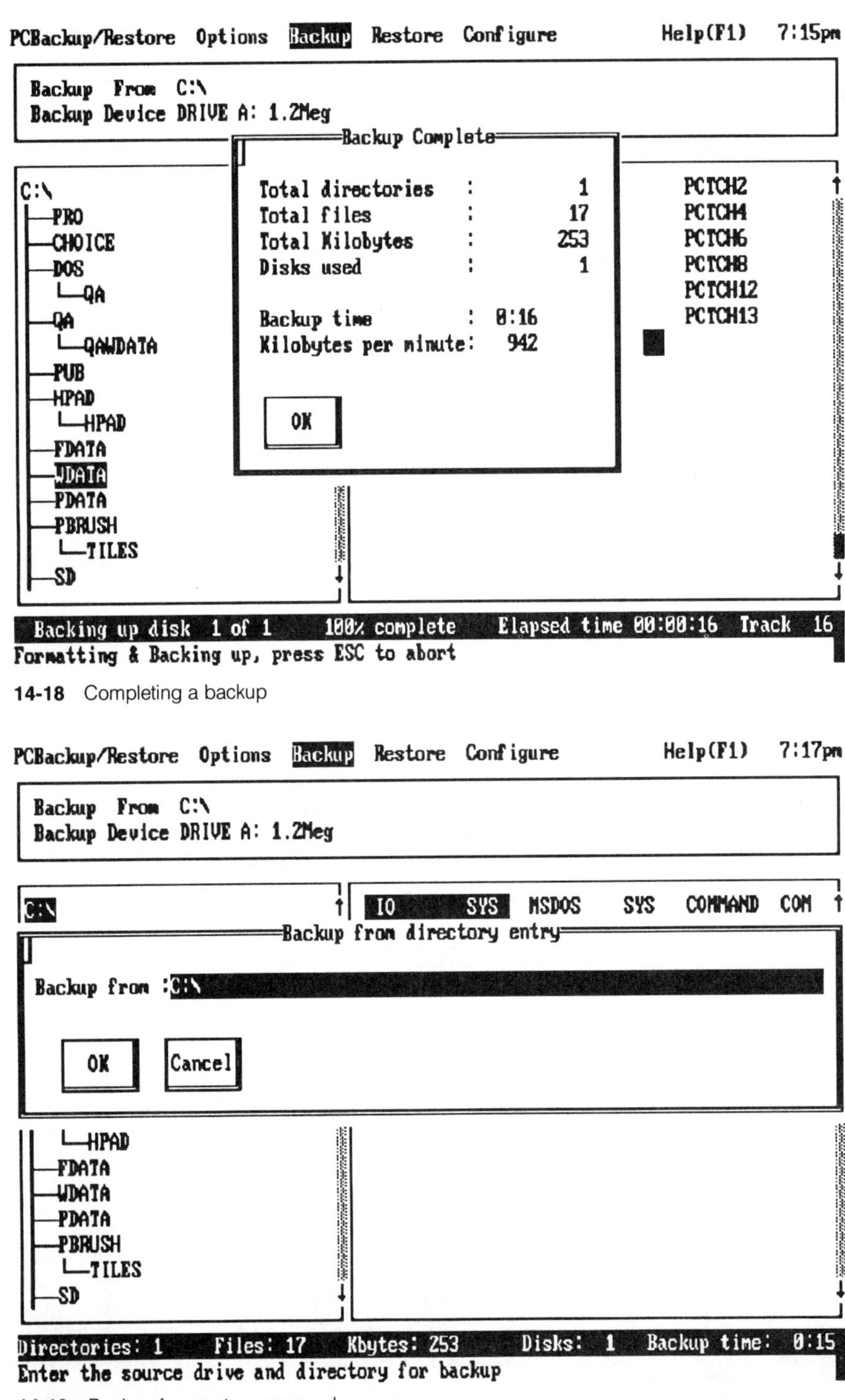

14-18 Completing a backup

14-19 Backup from entry command

The Choose directories command lets you select the directory and the files you want to back up. You can use the cursor keys or the mouse cursor to select the appropriate directories or files. The Tab key will move you to the Tree list window at any time.

Restore menu

The Restore menu is shown in Fig. 14-20. The restore function is used to recover files that might be lost or damaged after a disk crash. Notice that the menu is very similar to the Backup menu. You cannot restore any files unless DOS and PC Backup are installed on the disk.

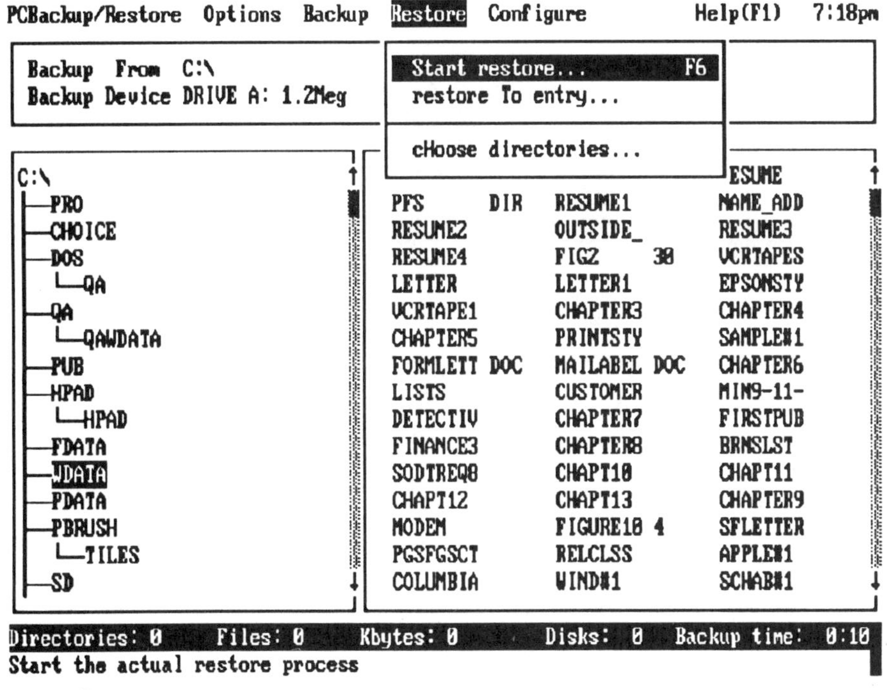

14-20 Restore menu

The Start restore command lets you begin restoration by asking you to insert the first disk from your backup set. You will be prompted for new disks as the backup takes place. If you are overwriting an existing file, you will be warned that a file already exists and asked if you want to overwrite it. When the disk is restored, it will give you a message to that effect in a message box.

The Restore to entry command lets you restore certain directories, as shown in Fig. 14-21. Type in the appropriate drive and directories.

The Choose directories command lets you restore selected files and directories. The last backup disk contains a list of all the files and directories that have

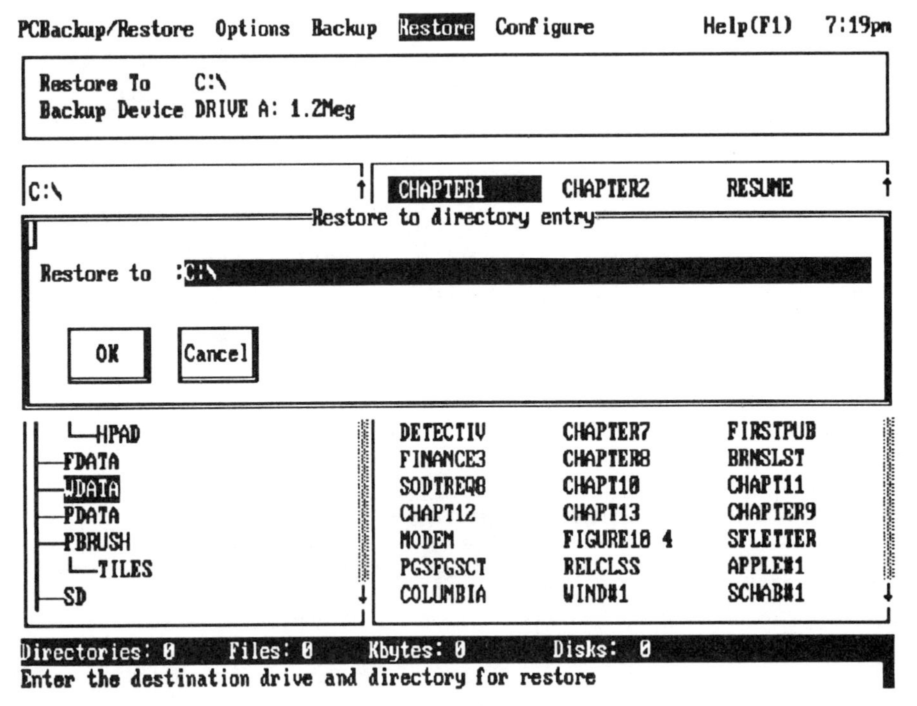

14-21 Restore to entry command

been backed up. You will be asked to insert that disk first so that this list can be displayed on the screen in the Tree list. Select the files and directories that you want to restore, and press the F6 key or select Start restore.

The dialog box in Fig. 14-22 might appear when you try to restore a file that already exists. Use the Overwrite option when you want to overwrite the existing file with the one being restored. Use the Overwrite with newer file only option when the file on the backup disk has a later date than the file on the hard drive. Use the Skip this file option when you do not want to override the file on the hard disk. Use the Repeat for all later files option in conjunction with any of the first three options.

When the files are restored, a message to that effect will appear in a dialog box, as shown in Fig. 14-23.

If the directory disk is lost or damaged, PC Backup can recreate the directory from the other disks. Select Start restore from the Restore menu. If the last disk is missing, a message will appear in a dialog box asking you to rebuild the lost directory. Select Rebuild from the dialog box. If you insert the wrong disk, select the Retry option. You will be prompted for each disk, and, when you are finished, you will be asked to save the new directory. Use a new disk to save the directory, and select Start restore again.

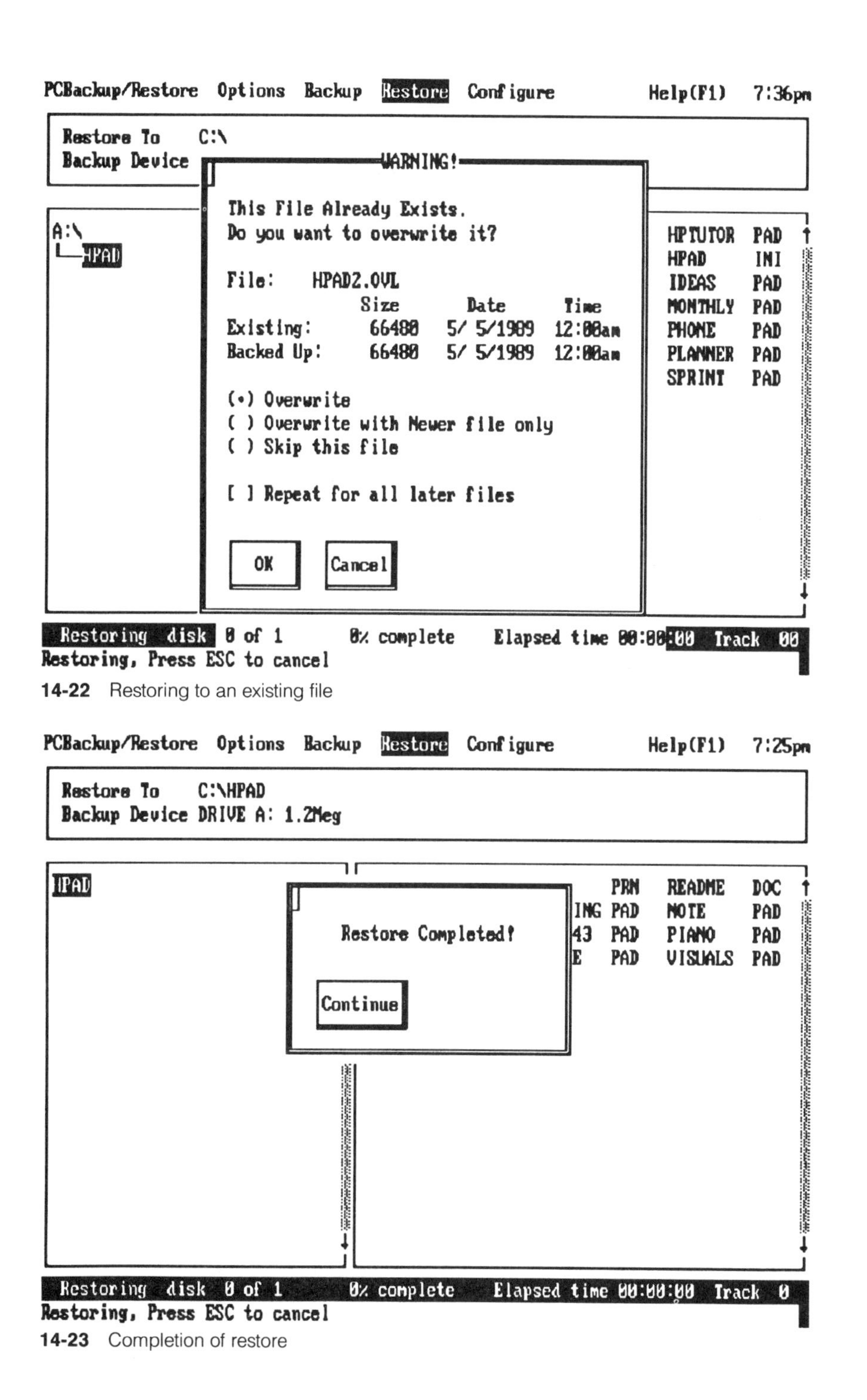

14-22 Restoring to an existing file

14-23 Completion of restore

Configure menu

The Configure menu, shown in Fig. 14-24, allows you to change your backup settings. If you select the Choose drive and media command, Fig. 14-25 appears, letting you choose the drive and define the disk media for a backup and/or a restore.

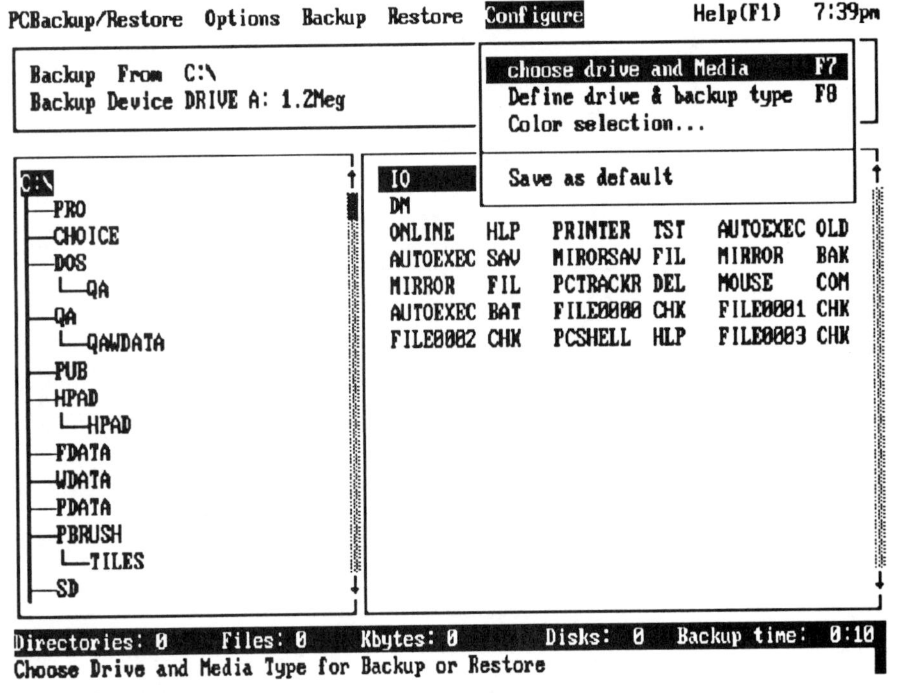

14-24 Configure menu

The Define drive & backup type command, shown in Fig. 14-26, lets you change the drive and media type for backups, and also lets you change the method of backup.

The Color selection command, shown in Fig. 14-27, lets you change the colors of the foreground and background for the menu bar, main screen, dialog box, and error message box.

PCBDIR

PCBDIR is a function of PC Backup that determines the number of disks required for a backup, and generates a report to the screen, disk file, or printer. Run it from the DOS prompt by typing PCBDIR A:, where A: is the location for the backup disk. It can be any floppy drive you have. A PCBDIR report to the screen

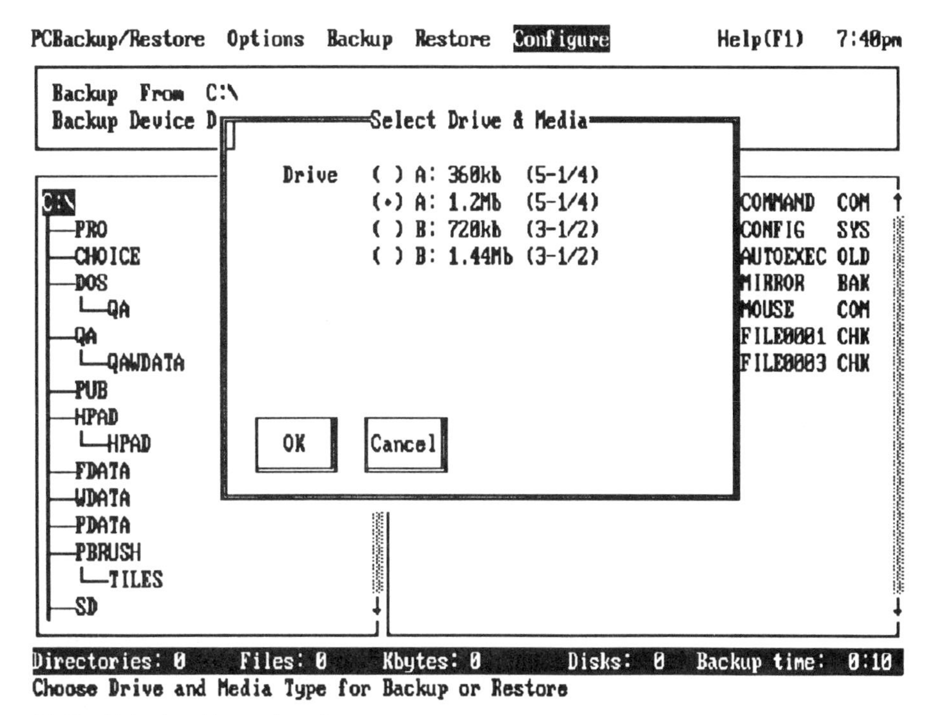

14-25 Selecting drive and media

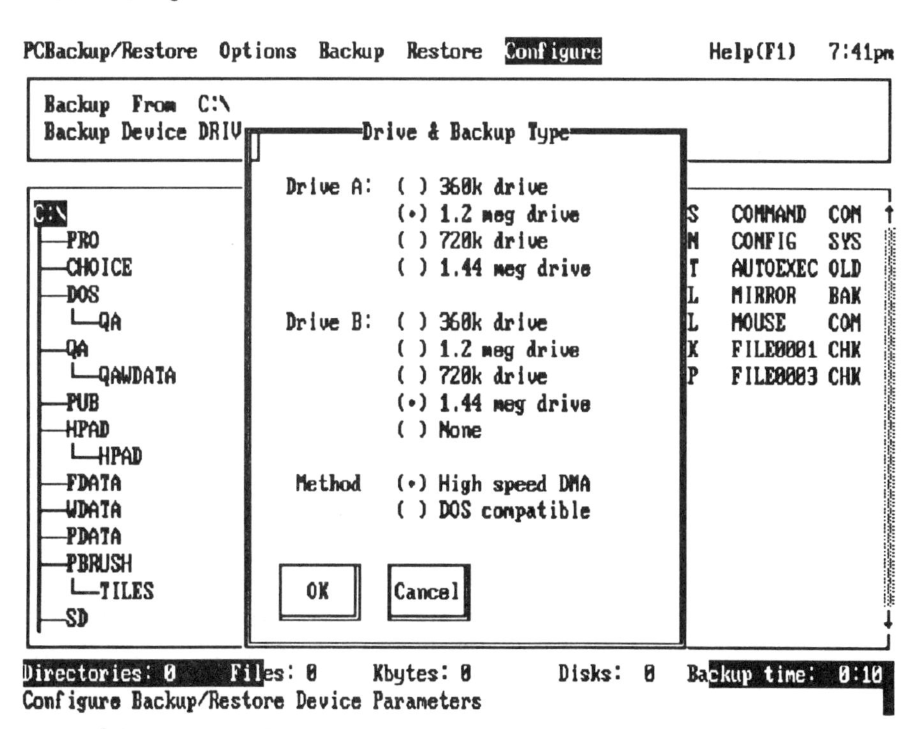

14-26 Selecting drive and backup tape

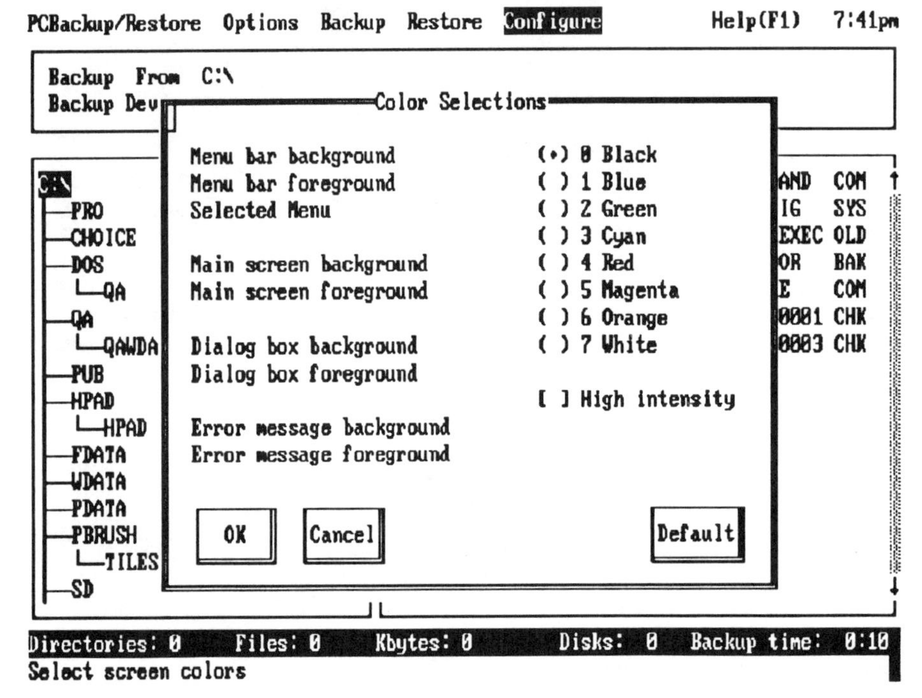

14-27 Color Selection menu

is shown in Fig. 14-28. You can generate a report that contains complete information if the disk is the last disk in a set of backups. The printed report shown in Fig. 14-29 shows the directory specification, total number of files, filenames, size of files, date and time created, and compression status.

```
C:\PRO>pcbdir a:

PC Backup directory report 5.5
Copyright 1989 Central Point Software, Inc.  All rights reserved.

Diskette is number 1 of a backup made  1-01-90  7:15pm
Total directories: 1.  Total files: 17.
Files begin with directory  WDATA\
Format of backup: 80 tracks, 16 sectors.

A PC Backup 'directory' exists on that disk.  A report can be generated.

The destinations allowed for the report are:
S - screen, F - a disk file, P - printer, N - no report.
Which report do you want? N

The current selection is A:
What drive:path contains the backup [A: - E:, Quit]? Q
```

14-28 PCBDIR command

```
PC Backup directory report 5.5
Copyright 1989 Central Point Software, Inc.  All rights reserved.
Backup performed on 12-27-89   2:38pm

Backup method: High-speed DMA
Drive type: 1.2M          Media type: 1.2M

Total directories:     1
Total files:          12
Disks used:            1

      Name          Size      Date      Time     Attr   Disk#

Directory  C:\

Directory  C:\WDATA
  PCTCH1            10690  12-15-89   8:24am            1
  PCTCH2            28188  11-28-89  10:14pm            1
  PCTCH3             7021  11-29-89   5:47pm            1
  PCTCH4            24241  12-01-89   3:06pm            1
  PCTCH5            16811  12-06-89   3:28pm            1
  PCTCH6            26064  12-10-89  10:57am            1
  PCTCH7            16426  12-12-89  10:05am            1
  PCTCH8            17275  12-17-89   4:57pm            1    Compressed
  PCTCH9            24770  12-18-89   7:30pm            1
  PCTCH11            9337  12-20-89   8:03pm            1    Compressed
  PCTCH12            8911  12-26-89   2:45pm            1
  PCTCH10           13987  12-27-89   1:16pm            1

Total bytes:    199K
```

14-29 PC Backup report

15

CHAPTER

DiskFix, user levels, and other enhancements

This chapter highlights some of the new commands included with PC Tools, version 6.0. This includes DiskFix, changing user levels, fax support, using a laptop, and other enhancements. Changes in the backup facility include a faster backup, a higher rate of data compression during backup, and backup to tape drive capability. The PC Cache module has faster disk efficiency and can use expanded and extended memory.

DiskFix

If you get an error message when reading or booting a hard or floppy disk, you can run DiskFix to correct the problem. DiskFix is a powerful utility that can detect and correct many disk problems. It provides you with a description of the problem, the symptoms of that problem, and an opportunity to correct it.

DiskFix can be run from any DOS prompt simply by typing DISKFIX. The DiskFix Options menu shown in Fig. 15-1 appears. You are given the opportunity

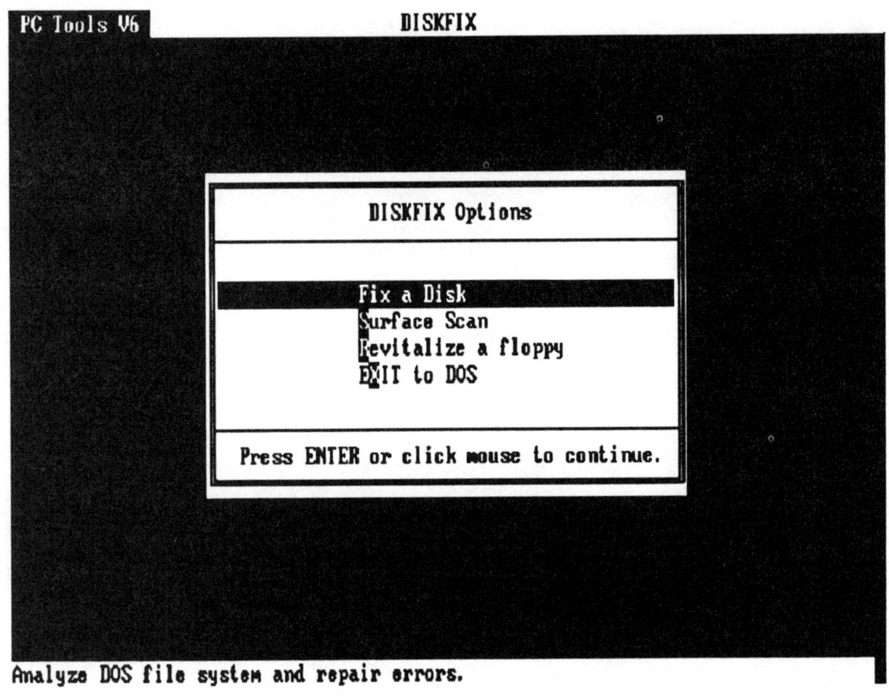

15-1 DiskFix options

to examine a disk for a problem, perform a surface scan to check the media surface, revitalize a floppy that has read errors, or exit to DOS. If you select Fix a disk, Fig. 15-2 appears asking you which disk drive to analyze.

After you select a drive, the program performs a drive analysis and gives you the status of the disk, as shown in Fig. 15-3. It determines if the information in the boot sector is not damaged, that it is the correct type of disk (media descriptor), that the file allocation table (FAT) is readable, that the directories are readable, that no two files occupy the same space (are cross linked), and that there are no lost clusters. It also gives you the opportunity to check the surface media. You can also print out a report.

User levels

PC Tools lets you customize the menus in PC Shell and PC Backup by selecting the appropriate user level. You will also be asked to select a user level during the installation of PC Tools. You can use the Configuration command to change your user levels in PC Shell, as shown in Fig. 15-4.

The Beginner user level lets you perform routine DOS tasks such as copying, comparing, and renaming files. It lets you copy, compare, and format disks, change drives, and create directories. The Beginner's Backup menu lets you start the backup and choose the directories to back up.

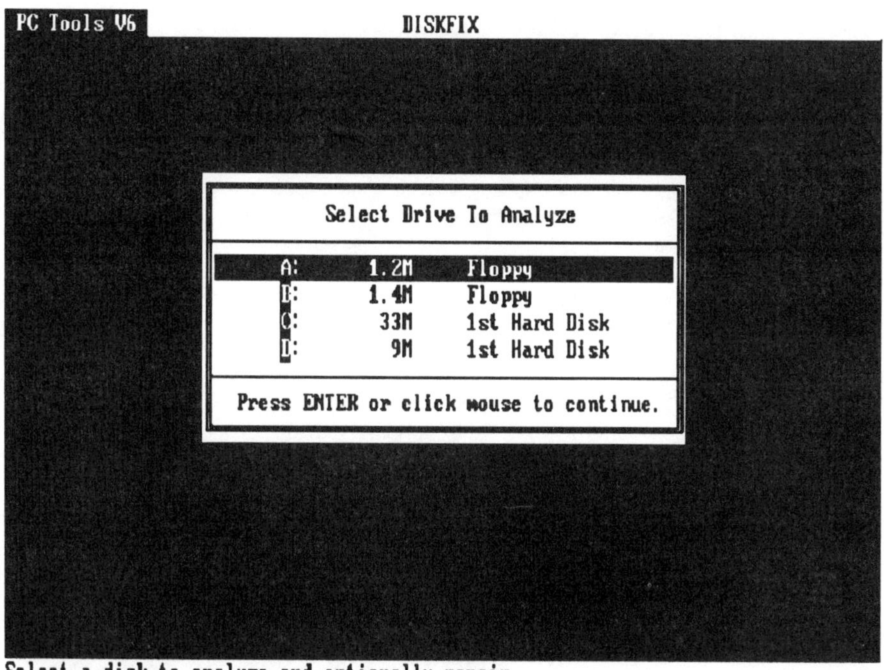

15-2 Drive options

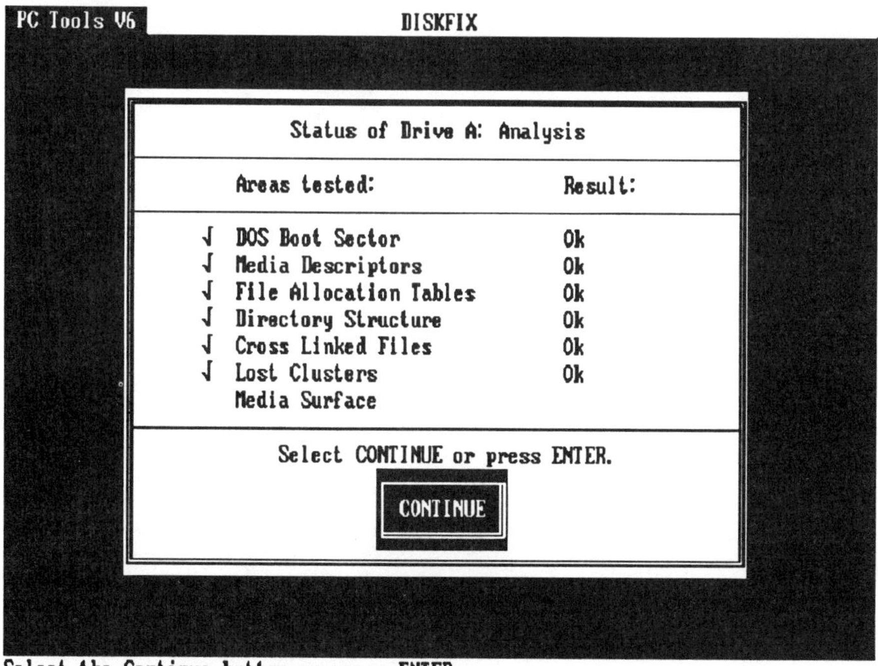

15-3 Status of disk drive analysis

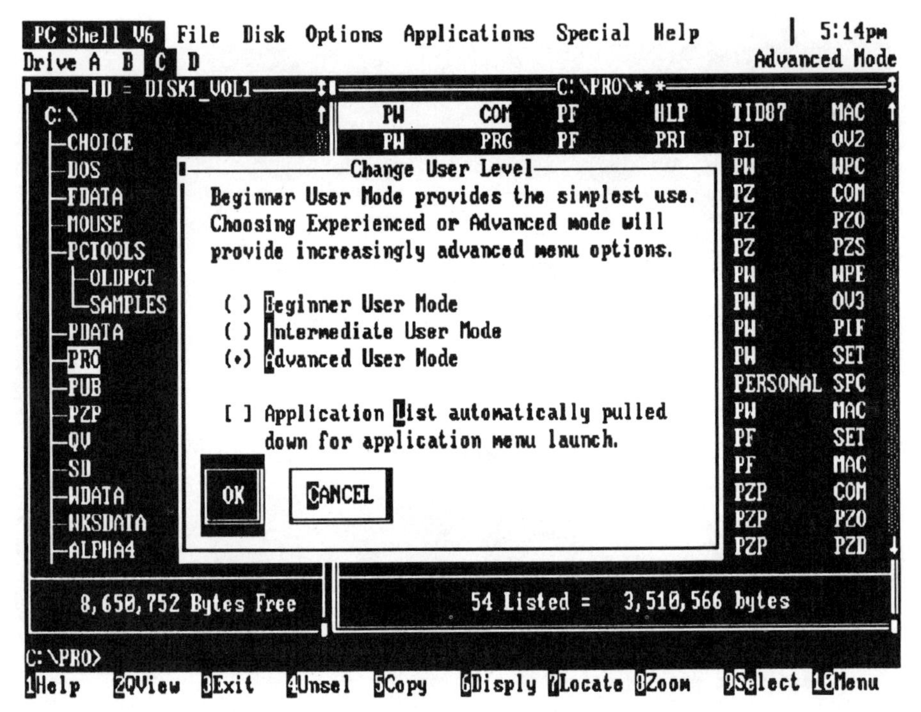

15-4 Changing user level

The Intermediate user level gives you all the commands of the Beginner level, plus moving, deleting, editing, and printing files. It also provides you with more information about your system. The Intermediate backup level lets you include reporting; include/exclude files, overwrite warnings, and date range selections; include subdirectories; and exclude attributes.

The Advanced user level lets you access all the commands of PC Tools. It also lets you compress and verify the disk.

Using a laptop

PC Shell can now be used to transfer selected files from a laptop to a desktop machine, or vice versa. The two machines must be connected with a serial cable, and you must be using LapLink from Traveling Software to transfer data between machines. Once the two machines are connected with a cable, run PC Shell, select the LapLink/QC command from the Special menu, select the files you are copying, and select the directory where the files are being copied.

The DOS command line

Each menu in version 6.0 of PC Shell contains a DOS command line where DOS commands can be entered and executed.

Desktop manager

The desktop manger, version 6.0, contains several new functions, as specified below.

Using a fax board or machine

The Desktop menu, shown in Fig. 15-5, now includes a menu item that allows you to send a fax message to any remote facsimile machine or any computer with a fax board. A fax board, which is an add-on board inserted into the computer, lets your computer operate as a facsimile machine.

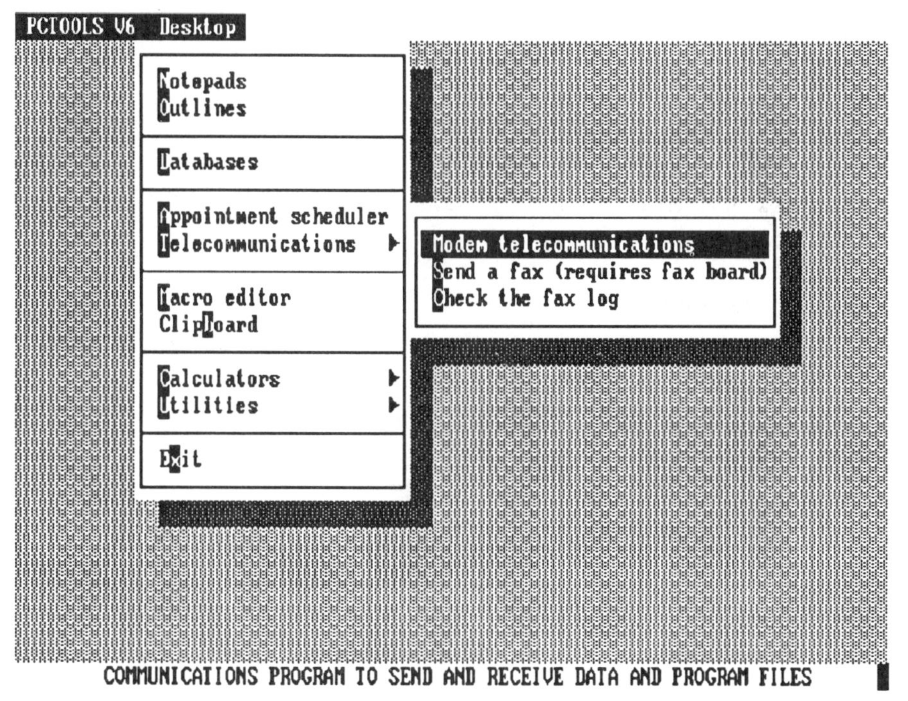

15-5 Fax menu

Before sending a fax message, you should create a directory on your hard disk to hold the fax before it is sent. A directory can hold up to 99 fax documents. Using your fax software, create the following statement in your ITLFAX.EXE file:

SHARE = directory (name of directory you just created)

The fax board software and the ITLFAX.EXE file should also be specified in the AUTOEXEC.BAT file's path statement.

The Fax Drive dialog box appears when you select the Fax drive command from the Configure menu. Enter the fax directory name, the page length (the default is 11 inches), whether you are sending a cover letter or not, the time format you want to use, and your name.

To send a fax, select the Send a fax command from the Telecommunications option. Choose Add a new entry from the Actions menu. Enter the date, time, your name, the name of who is receiving the fax, the fax number, comments (which identify the fax entry), and the type of resolution. Use Normal resolution to send text and Fine resolution to send graphics. If you are sending a fax to another machine that has a fax board, select Fax board to fax board.

If you have several documents saved in the fax directory and want to select one of them to send, choose the Select files and send command, which allows you to choose the correct document to send. Documents can be deleted from the fax directory by selecting Delete the current entry from the Actions menu.

The Check the fax log command of the Telecommunications option lets you check the status of faxes that have been sent and received. The word *Sent* will be placed in the status column for each fax that has been sent. The word *Received* will be placed in the status column to indicate that the fax has been received. The word *Aborted* or other error messages in the status column indicates that the fax encountered phone line problems and was canceled during transmission.

Video size command

The Video size command, shown in Fig. 15-6, has been added to the Window menu. Selecting this command displays the dialog box shown in Fig. 15-7, which allows you to change the number of lines displayed on the screen if you are using an EGA or VGA display.

The File menu

The File menu, shown in Fig. 15-8, has several new commands that work with databases. The Transfer command lets you transfer records from one database to another. This allows you to combine information from several smaller databases into one large one, or to break a large database into several smaller ones. This can be useful if you want to separate clients by town, county, ZIP, etc.

To transfer records, open the database containing the records you want to transfer, select the records with the Select records command, select Transfer from the File menu, choose the destination database where you want the records transferred, and press the Enter key. If the database being sent contains a field or several fields not in the receiving database, the information for those fields will not be sent. Be sure to check the destination database after the transfer to confirm that all the selected records have been transferred.

The Append command lets you add all records in one database to the end of the currently opened database. The Browse command lets you display records in horizontal format, as shown in Fig. 15-9. Each record with all its fields occupies one line. Use the mouse or the PgUp or PgDn keys to scroll through all the records in the database.

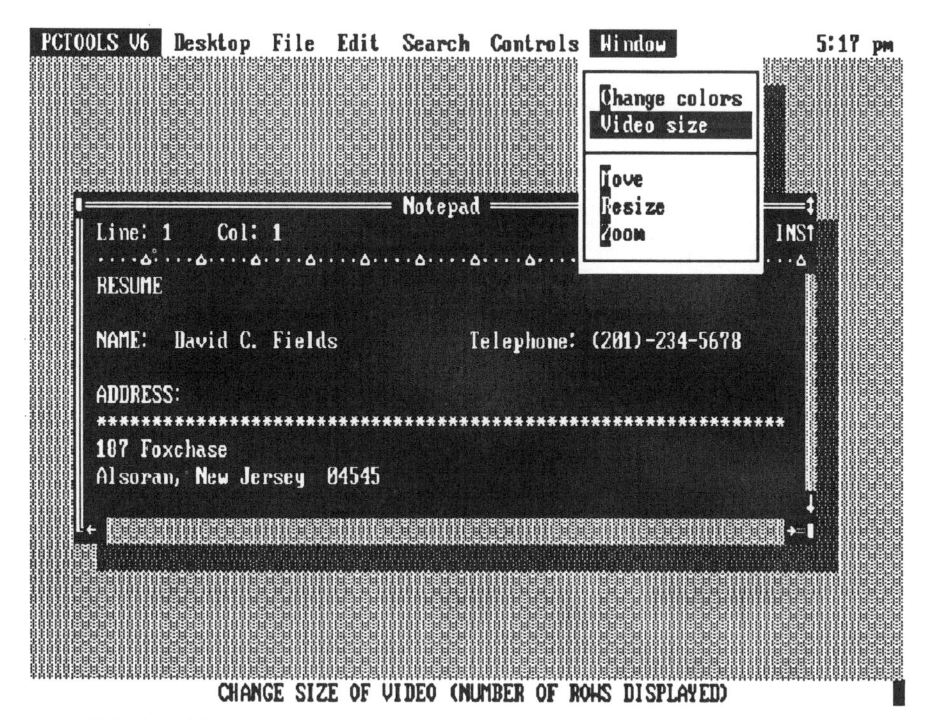

15-6 Selecting video size

15-7 Video size window

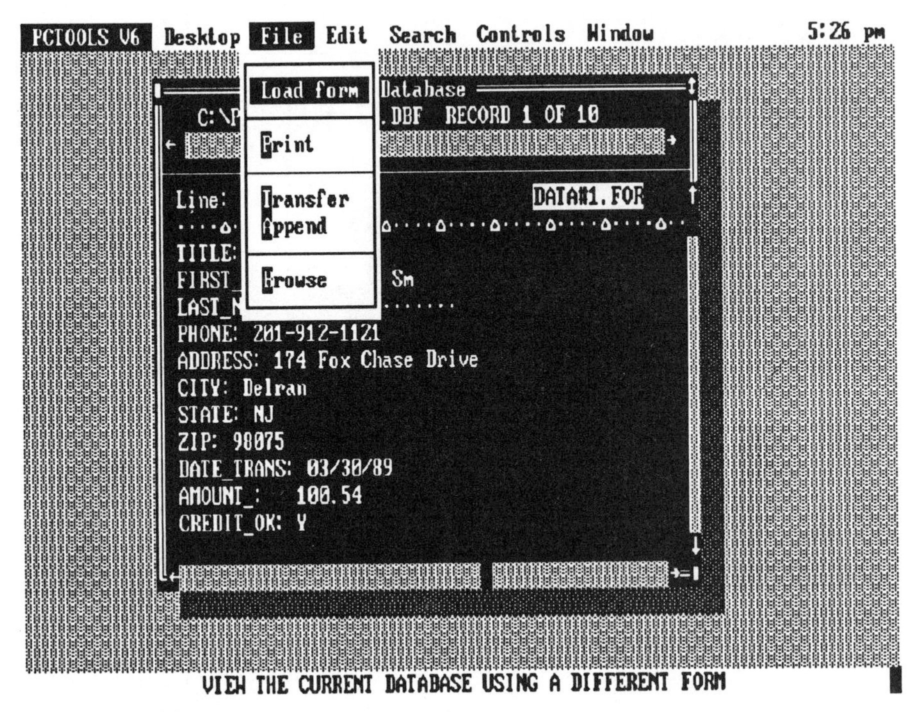

15-8 Desktop File menu

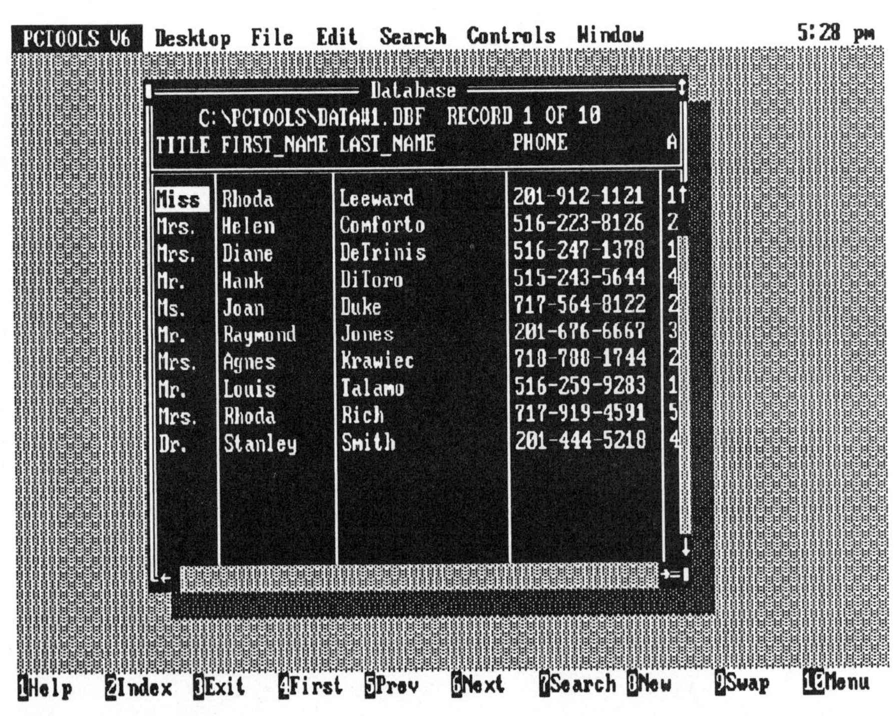

15-9 Browse command

You can add new records in the Browse mode by pressing the F8 key to create an empty record, typing in the new information, and pressing the Enter key to add the information into the record. You can delete records when you are in the Browse mode by moving the cursor to the record you want to delete and selecting Delete record from the Edit menu. You can also edit records by placing the cursor over the field you want to edit and retyping the data you want to change.

PC Shell

The PC Shell facility of PC Tools also has some version 6.0 improvements.

Locate command

The File menu, shown in Fig. 15-10, lets you locate, undelete, and clean files, and gives you a quick file view. The Locate file command generates the File Locate dialog box shown in Fig. 15-11. This provides the names of the files on your disk. Suppose you want to find all the word processing documents that contain the name of the town Massapequa. Select the word processing program that contains those files. After you select the appropriate file, a Search For dialog box will appear, prompting you for the word to search for. Type Massapequa or any other search word, as shown in Fig. 15-12. Selecting Search will cause PC Tools to display a list of all word processing files that contain that word.

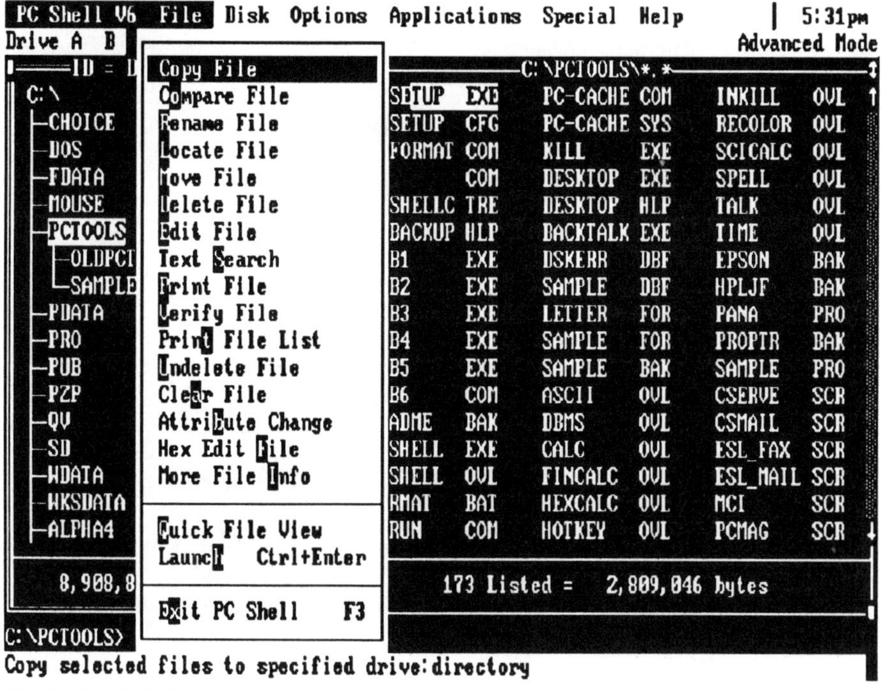

15-10 PC Shell File menu

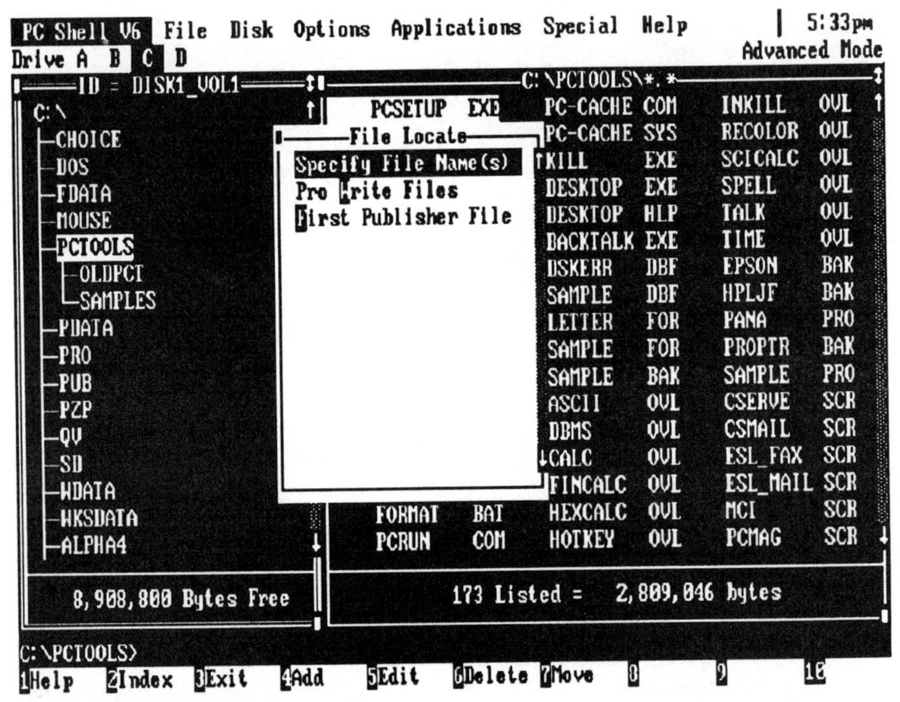

15-11 Locate command

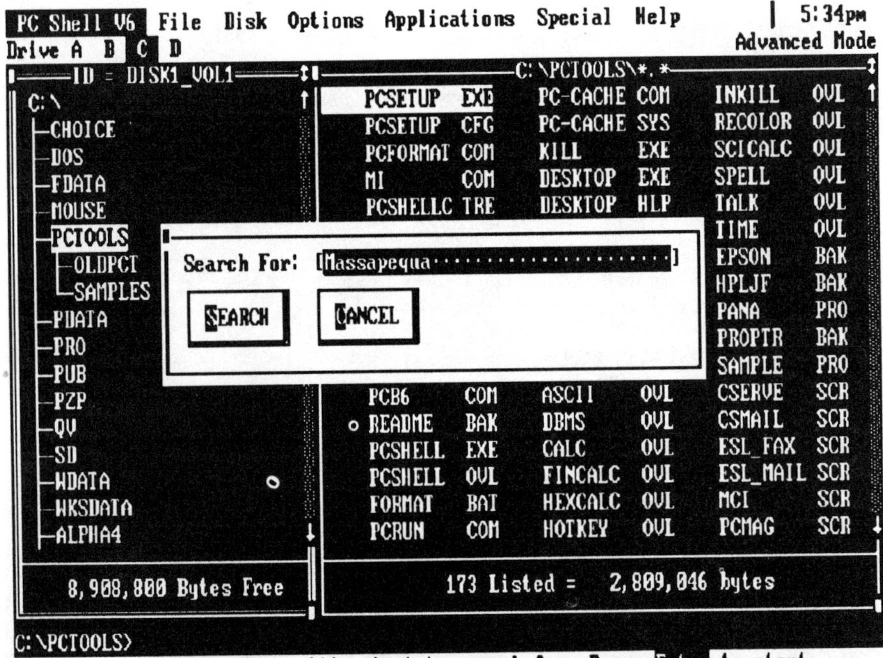

15-12 Search For command

Clean file command

The Clean file command physically erases information in files that you have deleted from the disk. When a file is deleted from a disk, the filename is removed, but the data in the file remains until rewritten by another file. The Clean file command erases this data even before it is overwritten by data from another file. Once a file is cleaned, the information in that file can never be recovered.

Quick file view command

The Quick file view command lets you see selected files in a zoomed view window. Most files can be viewed in their native format, including Lotus, dBASE, R:Base, Paradox, Symphony, MS Works, WordStar, WordPerfect, and straight ASCII text.

The Options menu

The Options menu, shown in Fig. 15-13, lets you change the user level and define function keys in the setup configuration. The Define Function Keys menu, shown in Fig. 15-14, lets you change the commands of the function keys in the message bar, except for the F1, F3, and F10 keys, which cannot be changed. Simply highlight the appropriate function key along with the available function and it will be replaced on the message bar. The message bar can be completely customized.

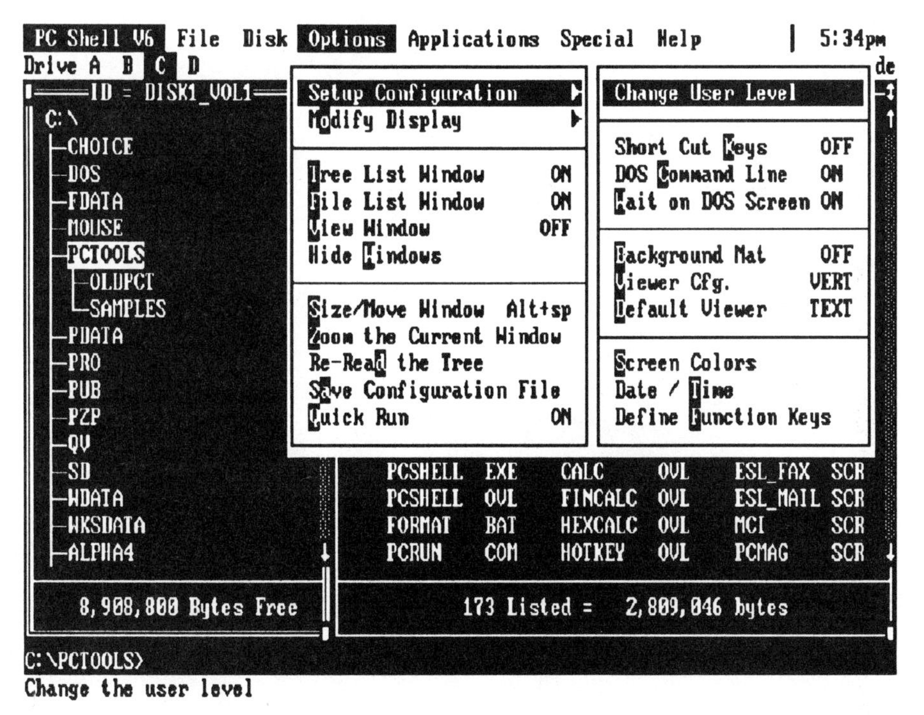

15-13 Options menu

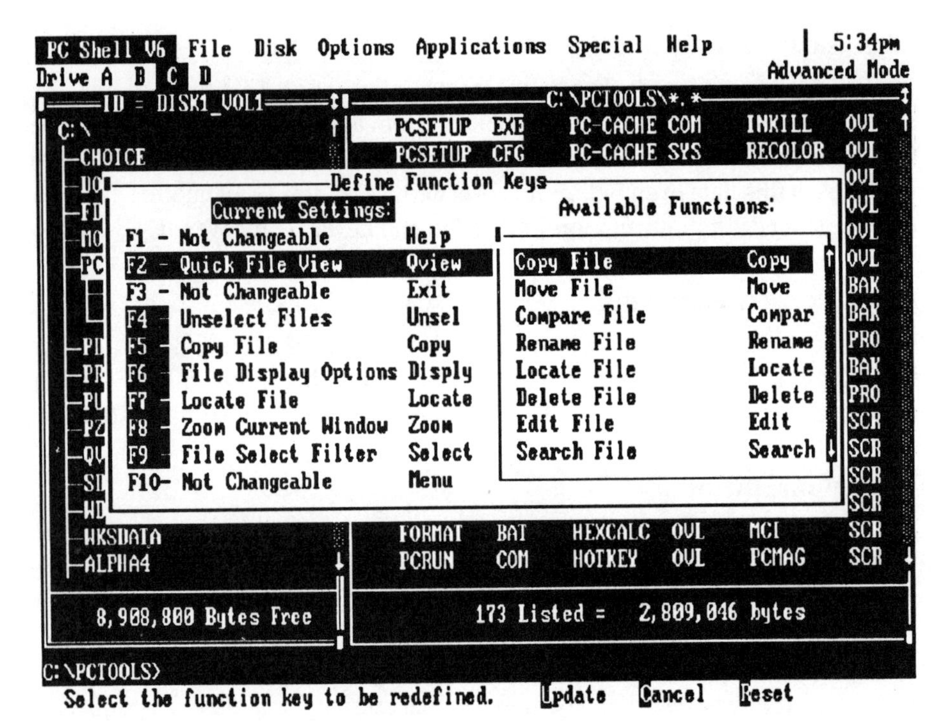

15-14 Define Function Keys menu

Index

D

E

N

O

P

T

U

V

W

X